Technology Manual

Augustin Vukov | Singh Kelly | Pamela Lancaster

University of Toronto *Pensacola Junior College* *University of Kentucky*

Statistics ELEVENTH EDITION

McClave | Sincich

PEARSON
Prentice
Hall

Upper Saddle River, NJ 07458

Vice President and Editorial Director, Mathematics: Christine Hoag
Editor-in-Chief, Mathematics & Statistics: Deirdre Lynch
Print Supplement Editor: Joanne Wendelken
Associate Managing Editor: Bayani Mendoza de Leon
Senior Managing Editor: Linda Mihatov Behrens
Project Manager, Production: Raegan Keida Heerema
Supplement Cover Manager: Paul Gourhan
Supplement Cover Designer: Victoria Colotta
Operations Specialist: Ilene Kahn
Senior Operations Supervisor: Diane Peirano

© 2008 Pearson Education, Inc.

Pearson Prentice Hall

Pearson Education, Inc.

Upper Saddle River, NJ 07458

Pearson Prentice Hall™ is a trademark of Pearson Education, Inc.

The author and publisher of this book have used their best efforts in preparing this book. These efforts include the development, research, and testing of the theories and programs to determine their effectiveness. The author and publisher make no warranty of any kind, expressed or implied, with regard to these programs or the documentation contained in this book. The author and publisher shall not be liable in any event for incidental or consequential damages in connection with, or arising out of, the furnishing, performance, or use of these programs.

Printed in the United States of America

10 9 8 7 6 5 4 3 2 1

ISBN-13: 978-0-13-206956-4

ISBN-10:   0-13-206956-3

Pearson Education Ltd., *London*
Pearson Education Australia Pty. Ltd., *Sydney*
Pearson Education Singapore, Pte. Ltd.
Pearson Education North Asia Ltd., *Hong Kong*
Pearson Education Canada, Inc., *Toronto*
Pearson Educación de Mexico, S.A. de C.V.
Pearson Education—Japan, *Tokyo*
Pearson Education Malaysia, Pte. Ltd.

THE MINITAB MANUAL

Augustin Vukov
University of Toronto

Statistics ELEVENTH EDITION

McClave | Sincich

PEARSON
Prentice
Hall

Upper Saddle River, NJ 07458

Preface

This Minitab Manual is designed to carefully integrate the statistical computing package MINITAB™ into an introductory Statistics course, using Statistics by McClave and Sincich, 10th edition. It provides worked out illustrative examples using data from examples and exercises in the text (data available at www.prenhall.com/mcclave and on the text CD-ROM), which are suitable for solution using Minitab. I have presented a gradual one step at a time approach, teaching just what you need to know for the current topic, rather than bombarding you with too much information too quickly. More ambitious students can easily use Minitab's Help menu to delve more deeply.

There are many statistical software packages available, but Minitab is ideal for an introductory course, due to its ease of use. With Minitab, you are freed from the tedium of long calculations, and instead can focus on what is important – learning the concepts, determining which procedures are relevant, checking their applicability, interpreting the results of the analysis

With Minitab, you issue commands that tell Minitab to do certain calculations or to produce graphs. But, instead of writing down commands, and the consequent need to learn syntax, you will use the *menu driven* approach, i.e. you select the appropriate procedure from the menu selections, and then fill in some required information in dialog boxes. If you make an error in your entries, Minitab will give you direct feedback, with an error message, pointing out the problem or omission.

Minitab is proprietary software, and is not free. You have to purchase it, or use it on computers at your school where it has been purchased or leased. Minitab software is available for various operating systems, but I will only deal with Minitab for Windows. There is a full *Professional version* and a less expensive abbreviated *Student version*. The latter is more than adequate for this course, and will also enable you to perform powerful statistical analyses of data for other courses. If you wish to work at home, you have to purchase (or rent) and install Minitab on your PC or Mac (for use on a Mac, you need Windows emulation software). I recommend that you purchase the Student version, which is available bundled together with the textbook at a *greatly discounted* price. For further information about Minitab, go to www.minitab.com, where you can download a 30-day free demo, or find information about renting Minitab for a semester.

Minitab is continually being updated and improved. Hence, there are different release numbers. This manual is based on Minitab (Student) Release 14. Release 15 is out now, but not in a Student version. Generally, the differences between release 14 and release 15 are slight, but in case you may have access to release 15, I have added some helpful comments or instructions, here and there, where improvements have been made.

Augustin Vukov
Toronto, Ontario, Canada
October. 30, 2007

Table of Contents
[M/S = Statistics, 11th ed., by McClave and Sincich]

Chapter 1 - Getting Started with Minitab

Before doing any of the examples or exercises in the text, let's start with a brief overview and simple first session.

Installing the software on your PC
If you have purchased the student version of Minitab, you have to install it on your computer. Close down all programs on your computer. Insert the CD into your computer's CD-ROM drive. An installation wizard should appear to guide you through the installation process. Keep hitting 'next' until the installation is completed. Choose the default 'Complete' setup rather than the 'Custom' setup (for advanced users), and if you wish, choose to install the Minitab icon to your desktop.

If you have installed the student version of Minitab on your computer, the first step is to open it up, as follows:

Start>Programs>MINITAB 14 Student>MINITAB 14 STUDENT

where the symbol '>' above leads you through the appropriate sequence of menu selections. Or if you installed the Minitab icon on your desktop, just double-click it.

If you are working at a computer at your school, follow log-on instructions given to you at your school for accessing Minitab.

The first view that you will have of Minitab will look like:

There are three key sections of the screen above:

(1) The main menu bar, at the top:

File Edit Data Calc Stat Graph Editor Tools Window Help

If you click on any of these, a submenu will drop down, displaying the available options. Below, I clicked on **Stat** from the main menu bar first, and from the **Stat** pull-down menu, I slid the cursor to **Nonparametrics,** which has it own sub-menu immediately appearing to the side, and from the **Nonparametrics** menu, I slid the cursor over to **Kruskal-Wallace**. Try this yourself.

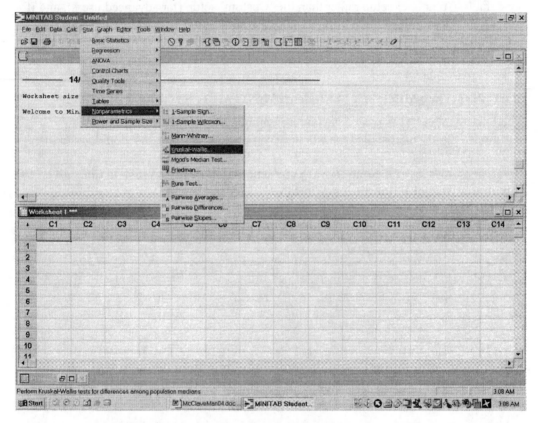

Click on **Kruskal-Wallace**, and up pops the **Kruskal-Wallace** dialog box, shown below.

If you wanted to apply the Kruskal-Wallace statistical procedure to some data, you would now fill in the required information in this dialog box (**Response**, **Factor**), and hit **OK**. Since we have not entered any data yet, just click on **Cancel**.

You will learn gradually about the procedures available from this menu, but in brief, the key functions of each main menu item are:

File: Open and save various types of files. Print windows

Edit: Copy, cut, paste data and text

Data: Manipulate the data in the data window

Calc: Perform basic calculations and transformations on the data. Calculations and random sampling for common probability distributions

Stat: Execute common statistical procedures

Graph: Construct various plots of the data

Editor: Graph, data, and session window editing tools. Enable use of the command language.

Tools: Supplemental tools, e.g. calculator, notepad, toolbar display options

Window: Manage the display of the various windows

Help: Get help on Minitab procedures

(2) The Data (Worksheet) Window, occupying the lower half of the screen
This is where the data sit. At the start, you have a new blank worksheet, named **Worksheet 1**:

Data obtained from subjects in a study are entered in a very specific way. The columns represent the variables, and the rows represent the subjects.

Suppose that you have weight and height measurements on three subjects. Start by typing in an appropriate name for each column (Weight, Height), directly underneath the column number, but above row **1** as shown. Stick to 16 letters or less, since in some output, longer column names may be truncated to the first 16 letters.

Next, type in the data shown below. If you hit *Enter* on your keyboard, after entering 120, you will be thrown into the cell directly below - because of that downward arrow in the upper left corner of the worksheet. Hit *Ctrl + Enter* (i.e. both together) on your keyboard, to move to the top of the next column. You may also just point and click on any cell, to get you there. If you click on the arrow, it will point to the right, and *Enter* will now move you over one column instead of down one row.

↓	C1	C2	C3	C4
	Weight	Height		
1	120	62		
2	140	65		
3	166	71		
4				
5				

(3) The Session Window, occupying the upper half of the screen
This is where the output of your statistical procedures will appear, except for graphs, which pop up in separate new graphical windows. If you select (click, drag, click) from the menu as follows: **Stat>Basic Statistics>Display Descriptive Statistics**, you will see the following dialog box:

We want to display some descriptive measures for the subjects' weights and heights, so we need to type in the column names under **Variables**. We can type either *C1* or *Weight* to indicate column 1, and then *C2* or *Height* to indicate the second column or variable. But instead of typing, try double-clicking (left mouse button) on 'C1 Weight' at the left. It will be copied over to where the active cursor is blinking, which should be under **Variables**. Repeat for C2. Alternatively, delete your previous entries, and instead try the following: left-click on C1, hold down and drag to C2, so that both are highlighted. Hit **Select**, and both should appear under **Variables**, if your cursor was blinking there. You now should see:

We will indicate the above series of selections, in this manual, by writing:

Stat>Basic Statistics>Display Descriptive Statistics: Variables: *Weight Height*

where bold entries indicate fixed menu and dialog items on display within Minitab, and italicized entries indicate items that you have to fill in, depending on the particular study.

Now, left-click on **OK**, and you get the following output in the Session Window occupying the upper half of your screen:

Session

14/08/2004 3:46:43 PM ---

Worksheet size: 10000 cells.

Welcome to Minitab, press F1 for help.

Descriptive Statistics: Weight, Height

Variable	N	N*	Mean	SE Mean	StDev	Minimum	Q1	Median	Q3	Maximum
Weight	3	0	142.0	13.3	23.1	120.0	120.0	140.0	166.0	166.0
Height	3	0	66.00	2.65	4.58	62.00	62.00	65.00	71.00	71.00

To print this window, click on **File>Print Session Window**. If you do not see this selection, it is because the session window is not currently the *active window*. Clicking on a window makes it *active*. Click on the data window. You will see **File>Print Worksheet**. Click on the session window, and instead you will see **File>Print Session Window**.

Now would also be a good time to check out the printer settings. Click on **File>Print Setup** and check that the selections are as you desire.

Graphical windows: Click on **Graph>Plot** and you will see:

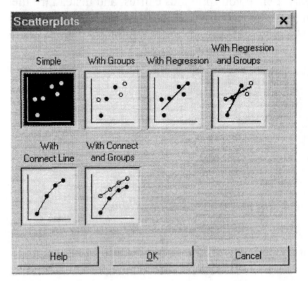

Click **OK** since all we want here is the simple scatterplot. You get another dialog box:

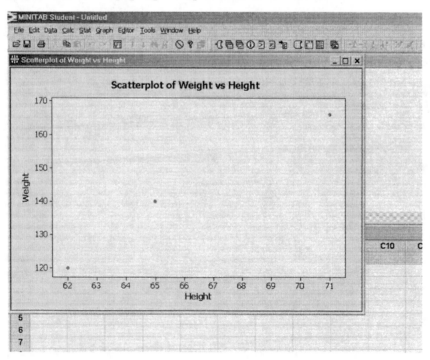

Above, we entered *Weight* and *Height* under **Y** and **X** respectively, then hit **OK**. A new *graphical* window pops up on top of the other windows:

To print out a hard copy of the graph, use **File>Print Graph**.

You may choose to add labels, such as titles, to a graph if you wish. In the dialog box for the graph, look for **Labels...** and click it:

Fill in your desired **Title**, e.g. *Exercise 3a* and perhaps a **Subtitle** *Plot of Weight vs. Height for College Students.* Add additional labels if you wish. You can also label data in the plot, if you click on **Data Labels**. **OK, OK**, to get the plot again, with your labels.

Click anywhere on the data window, and it comes to the top (becomes the active window). Click now on the session window, and the graph appears to be completely gone. It is just hiding! Click on **Window** and you see:

Select **4** **Scatterplot of Weight vs Height**, and the window with this graph will be back on top.

Project Manager Window: This is minimized at startup. Look to the bottom left, and you will see it lurking below the data window:

Click on the left button, to restore the window, and you see:

There are various folders displayed to the left:

Session folder which lists the procedures you have executed in the session window as well as the graphs produced; useful for managing your output, e.g. *right-click* on either entry above to see your options: e.g. **Bring to Front**, **Delete**, **Print**, **Append to Report** (see ReportPad below)

History folder which lists commands you have used in this session;

Graphs folder for managing graphs;

ReportPad folder, a basic word processing tool for editing reports of your work;

Worksheets folder containing one folder for each open worksheet, where each worksheet folder provides basic info about that worksheet; e.g. click on **Columns** above and you will see the name, count, and type (numeric/text) of each column in worksheet 1.

You may enter data into several different worksheets, but only one worksheet can be the *current* or *active* one (indicated by 3 asterisks). If you wish to switch to another worksheet, you may use **Window** from the main menu bar, to select it. Or you can also proceed using window operations described below.

Window operations: In the upper right corner of the graphical, session, and worksheet windows, you will likely see three square buttons:

These three buttons perform certain operations on the window. To hide (minimize) the window, click on the left one, to maximize its size (or to restore its size), click on the middle one, and to be completely rid of the window, click on the right one. You may also use the icon in the upper *left corner* of a window, to execute any of these operations. Left-click on this icon and a menu of operations will appear.

If you wish to move a window, left-click on the upper (title) bar of that window, and drag the window.

For windows that are hidden, you can find and restore them using **Window** from the main menu. Or you can look for the hidden window's title bar, which will be lurking somewhere at the bottom of your screen, possibly hidden behind some windows. After hiding all the windows, I found the following four shortened title bars at the bottom of my screen:

The icon at the left tells you what type of window it is. To restore any one of these, click on the left button, or click anywhere on the bar, for a pop-up menu.

Stored Constants: In the course of doing certain calculations, you may wish to calculate and store for later use, *one single number*, rather than an entire column of data. Do this by writing k1 (k2, etc.) rather than a column designator, C1, C2, etc. for the place of storage requested in the dialog box being used. The **Project Manager** window will list all such stored constants.

Missing Data: Occasionally, some data are missing or known to be erroneous. Suppose that the height of the second listed subject was incorrectly recorded, but the weight measurement is fine. Click on the second row in C2, and hit *Backspace* on your keyboard. An '*' will appear to indicate the missing value (or type '*' from your keyboard).

↓	C1 Weight	C2 Height
1	120	62
2	140	*
3	166	71

Note that if you hit *Delete* instead, you mess up the data set, since the third height measurement of 71 moves up a cell to become the second subject's height.

Saved Data Files

In the example above, we typed the data directly into a blank worksheet. While you can always fall back on this approach, if necessary, the data sets used in the text have *already been entered and saved for you*. However, there are different formats in which data may be saved. The format used is indicated by attaching an appropriate suffix to the file name. Some common formats, with corresponding suffixes (shown in parentheses) are:

(i) **Minitab Worksheet file (*.mtw)**

This is the standard and most efficient Minitab format for saving data. However, it is not transportable to other operating systems running Minitab, nor is it usable in other statistical software packages. And there may be problems going back and forth between newer and much older versions of Minitab. To open up such a file, just use **File>Open Worksheet: Files of Type: Minitab (*.mtw)**. Select this format when you wish to save data for later analysis using Minitab on your computer (via **File>Save Worksheet**). If copying it for a friend, it may be safer to use one of the other formats listed below.

(ii) **Minitab Portable Worksheet file (*.mtp)**   [*the format used for our saved files*]
This 'portable' format may be read correctly by any version of Minitab, regardless of the operating system running Minitab. To open up these data files, use **File>Open Worksheet: Files of Type: Minitab Portable (*.mtp)**.

(iii) **Text file (*.dat or *.txt)**
This is a very general format that allows you to use the data file for analysis by any statistical software, but you will have to give Minitab some help in interpreting the information in the file. When you open up such a file, you have to make sure that the **Options** are set properly in the **File>Open Worksheet: Options** dialog box. If not, you will get a confusing jumble of data, perhaps with multiple entries appearing in each cell, or with data in the row where the column names should be. Play around with the **Options...** and the **Preview...** buttons, until it looks right for opening up.

(iv) **Excel spreadsheet (*.xls)**
You should have no problem opening up Excel spreadsheets containing data structured as in a Minitab worksheet. Specify **Files of Type: Excel (*.xls)**

Data Files Used in this Manual
The examples and sample exercises in this manual use data sets that have already been saved on the text CD-ROM, in the *Minitab Portable format*. The name of the data file appears in your textbook next to the corresponding example/exercise. Retrieve the data file from your CD using **File>Open Worksheet**. In the dialog box, select **Files of Type: Minitab Portable (*.mtp)**. At the top, next to **Look in**, select the folder on your CD with the data sets, and select the *Minitab* folder. Then choose *Exercises* or *Examples* folder:

Double-click on the desired data set. Hit **OK** in response to Minitab's query about adding this to the current project. The data will appear in a new worksheet (but in the same *project file*– discussed later).

> *Note that if the files are of one type, and you specify a different type in the dialog box above, no files will be visible!!*

When opening files saved in the *Minitab Worksheet format*, proceed similarly, but switch to **Files of type: Minitab (*.mtw, .mpj)** in the box. If not sure which format was used for saving the data, try **Files of type: All** to view the files, but then switch to the correct file type, or certain file open options may not be available.

Exiting from Minitab
  When finished using Minitab, click on **File>Exit** and you get

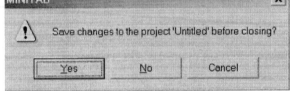

Generally, you will select **No**, which results in none of your work being saved. If you want to save some or all of your work, click on **Yes** and you will see:

The default name of 'MINITAB' should be replaced by a more useful name, and decide where to save the file (**Save in**). Hit **Save**, and all your work (session, worksheets, graphs) will be saved in a *Minitab Project file*, designated by the suffix **.mpj*. To open it up again, use **File>Open Project**. However, at any time during your session, you can save the content of any of your windows, by making it the active window (click on it), then selecting **File>Save...**, or you can save all your work up to that point by using

File>Save Project. If you want to delete from your project some of your graphs, just close (not hide) each graph window directly or via the **Project Manager Window**.

Are you ready to write your report? If so, before exiting, go to the **Project Manager Window**, and use the **ReportPad**. E.g. click on the **Session folder** or the **Graphs folder**. Right-click an item, and choose **Append to Report**. Right-click on the **ReportPad** folder, and from the selections, you may choose to print your report, or to save it (rich text format or web page), or to edit it in your word processor.

Chapter 2 – Methods for Describing Sets of Data

In this chapter, we will construct various data plots using the commands:
Graph>Pie Chart
Graph>Bar Chart
Graph>Histogram
Graph>Stem-and-Leaf
Graph>Dotplot
Graph>Boxplot
Graph>Scatterplot

Do some simple calculations on the data, using
Calc>Column Statistics
Calc>Calculator
Calc>Standardize

Open saved worksheets, and new worksheets, using
File>Open Worksheet
File>New>Minitab Worksheet

Organize and summarize data, using
Stat>Tables>Tally Individual Variables
Stat>Basic Statistics>Display Descriptive Statistics

Manipulate data in the worksheets, using
Data>Sort
Data>Code

Qualitative data

Qualitative data can be summarized in a table using **Stat>Tables>Tally Individual Variables** or with a bar chart or pie chart using **Graph>Bar Chart** or **Graph>Pie Chart**

Figures 2.1-2.4 - M/S pages 29-30

We want to summarize and display the data on 22 adult aphasiacs shown in Table 2.1. The data in this example have been saved in a Minitab worksheet file named APHASIA on your text CD-ROM. Insert the CD into the appropriate drive on your computer (or if working at computer facilities at your school, find out where the data sets have been saved, and in what format). Open up Minitab. Click on **File>Open Worksheet**.

  The dialog box below will appear. Next to **Files of type:**, choose **Minitab Portable (*.mtp)**. At the top, next to **Look in**, your current folder is displayed. You need to find the folder with this data file. Click the little pointer at the right to see a list of available folders/drives and select the drive with your CD (or use the yellow folder icon at the top, with an up arrow, to go up one folder level at a time). Once you find the drive with the text CD, select the folder with data files. Choose the MINITAB folder, and then the EXAMPLES folder. You should now see a file named APHASIA.

Double-click on APHASIA or select it and click on **Open**. Hit **OK** when asked about adding it to your current project. You will then see the saved worksheet in the bottom half of your screen:

*Why do we see C1-T, rather than just C1 at the top of the first column? The '-T' indicates that this is a text column. A column must be either text or numeric, and cannot be a combination of the two. Your first entry determines the format, unless you choose to format the column before data entry, which is done by right-clicking the column, then from the pop-up menu, choosing **Format Column>Text**. If you use numbers in a text column, they will be interpreted as text labels or names, without numerical meaning.*

Let's summarize the data. First, produce the frequency distribution table, via **Stat>Tables>Tally Individual Variables**. Fill in as follows:

Tally Individual Variables   ✕

C1   Type   Variables:
                   Type|

                   Display
                   ☑ Counts
                   ☑ Percents
                   ☐ Cumulative counts
                   ☐ Cumulative percents

   Select

   Help   OK   Cancel

I only selected **Counts** and **Percents**, but you might wish to make additional selections like the **Cumulative percents**. Hit **OK**. You will then see in the Session Window in the upper half of your screen:

Tally for Discrete Variables: Type

Type	Count	Percent
Anomic	10	45.45
Brocas	5	22.73
Conduction	7	31.82
N=	22	

Counts are the same as frequencies, and Percents are the same as relative frequencies except for a factor of 100. This neatly tells us how this variable is distributed, so e.g. we see that nearly half have the Anomic type of aphasia.

Now for some pictures: We can produce a Pie Chart with **Graph>Pie Chart**

Pie Chart   ✕

                ◉ Chart raw data
                ○ Chart values from a table
                   Categorical variables:

      Pie Chart Options...   Labels...

      Select   Multiple Graphs...   Data Options...

   Help   OK   Cancel

Everything is blank! But just click in the box under **Categorical variables**, and '**C1 Type**' will appear in the box on the left side (where Minitab produces a list of candidate variables). Double click on '**C1 TYPE**' now (or click it, then **Select**) to copy this column over to the box on the right side, i.e. to where your cursor is now blinking:

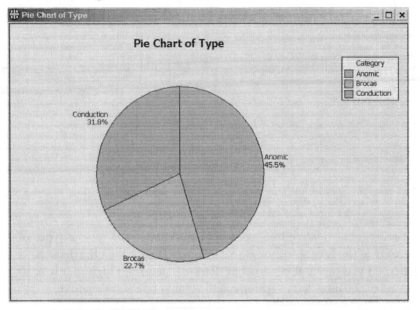

Hit **OK** now to get

You will be seeing this in some nice bright colors! In order to display the names and percents alongside the pie, I also selected in the main dialog box: **Labels>Slice Labels** and ticked off **Category name** and **Percent**. Comparing areas, we readily see that the Anomic type dominates at nearly 50% of cases, followed by Conduction, then Broca's.

To produce the same information in a Bar Chart, click on **Graph>Bar Chart**. In the first box you see, click **OK** for the default **Simple** bar chart choice. Likewise, don't mess with what the **Bars represent:** *Counts of unique values*. Hit **OK,** and a second box appears:

In the second box, we entered *Type* under **Categorical variables**. Hit **OK**, to get:

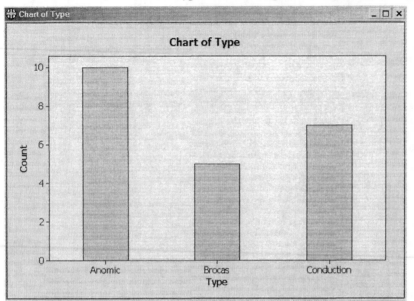

In this display, we easily see the relative frequency of occurrence of the various types of aphasia, by a comparison of the heights of the respective bars (rather than the distinct areas of the Pie Chart). If you prefer them ordered, *Pareto* style, from high to low or from low to high, just use the **Bar Chart Options** in the dialog box shown above.

To print the session window, or one of the graphical windows, just click on it to make it the active window, the use **File>Print Session Window** or **File> Print Graph**, as the case may be (after making sure that **File>Print Setup** has the right specs).

Example 2.1 - M/S page 31

The data in this example have been saved in a Minitab worksheet file named BLOODLOSS on your text CD. Click on **File>Open Worksheet**. Specify **Files of Type: Minitab Portable (*mtp).** Find the drive with your text CD, and select the folder with text examples, in Minitab format. Double-click on the file BLOODLOSS, to open it. You will now see the saved worksheet. The first 6 rows are:

	C1-T PATIENT	C2-T DRUG	C3-T COMP
1	1A	NO	REDO
2	2A	NO	NONE
3	3A	NO	NONE
4	4A	NO	NONE
5	5A	NO	NONE
6	6A	NO	NONE

Ignore C1 (patient designation). In C2, we see whether or not the patient in that row used the drug, and in C3 we have the type of complication, if any.

We want to compare the distribution of complications for those taking the drug (drug = yes) and those not taking the drug (drug = no). For side-by-side bar charts of the two distributions, there are two possible ways to proceed, either by using *multiple graphs in separate panels,* or by asking for a *cluster* style bar chart:

(i) **Graph>Bar Chart-Simple>Categorical Variables:** *Comp*

   Multiple Graphs: By Variables (with groups in separate panels): *Drug*
   Bar Chart Options: Show Y as Percent

                              OR

(ii) **Graph>Bar Chart-Cluster>Categorical Variables:** *Drug Comp*

   Bar Chart Options: Show Y as Percent and **Take Percent Within categories at level 1 (outermost)**

(i)   **(ii)**

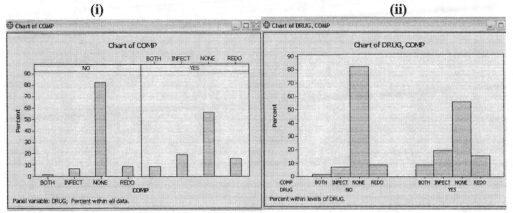

We see that complications are more frequent when using the drug. Note the importance of using *percents, not counts* - which would make for a difficult comparison, if there were more subjects in one group than in the other.

For the exact counts/percents, use **Stat>Tables>Cross Tabulation and Chi-Square**

which produces:

Tabulated statistics: DRUG, COMP

Rows: DRUG   Columns: COMP

	NONE	BOTH	INFECT	REDO	All
NO	47	1	4	5	57
	82.46	1.75	7.02	8.77	100.00
YES	32	5	11	9	57
	56.14	8.77	19.30	15.79	100.00
All	79	6	15	14	114
	69.30	5.26	13.16	12.28	100.00

Cell Contents: Count
                      % of Row

You can also produce the above summary via: **Stat>Tables>Descriptive Statistics**

where you have to click **Categorical Variables** in the main box.

Ordering the categories:

Note the *default alphabetical ordering* of categories in the summary above and in the earlier displays. If you wish to change the ordering, click anywhere on the column, and select **Editor>Column>Value Order**, or right-click the column and from the pop-up menu, choose **Column>Value Order**. Then choose **User-specified order** or **Order of occurrence in worksheet**.

Exercise 2.182 - M/S page 106

Use **File>Open Worksheet** to open up the worksheet HEARLOSS. Here we have *raw data* on types of hearing loss in C1. To summarize, graphically, use
Graph>Pie Chart: **Chart raw data** or
Graph>Bar Chart-Simple: **Categorical Variables**: *C1*, with **Options: Show Y as Percent**.
Hit **OK**, and you get

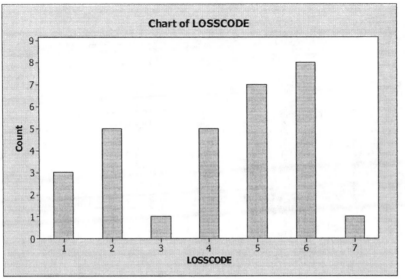

Chart of LOSSCODE

You might wish to use more informative labels for the categories. To do this, you can use **Data>Code>Numeric to Text**, and enter meaningful names or abbreviations.

Exercise 2.177 - M/S page 104

Click on **File>Open Worksheet**, find the data file DOLPHIN on your text CD, and **Open** it up. We want to graphically summarize the types (categories) of dolphin whistles. Note that 'o' does not indicate a type of whistle but rather 'other types'.

NOTE that this is not *raw data*, which would consist of a list of 185 whistle classifications, but rather a *summary table* of the raw data, showing whistle categories in C1 and counts in C2. For a pie chart, use **Graph>Pie Chart**, and tick off **Chart values from a table** rather than **Chart raw data**. Fill in as below, and be sure to include some sort of **Labels** for the **Slice**s in your pie.

Click on **OK, OK,** and we get

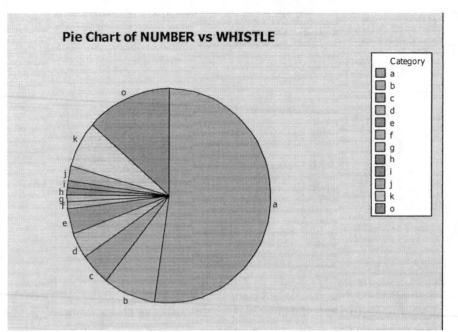

You can also click on **Percent** in the sub-dialog box for **Labels**, but I chose to omit this information here because the resulting display is just too messy in some parts. We easily see e.g. that more than half of whistles are type a.

For the bar chart, there are two possible ways to proceed with summarized data. Select **Graph>Bar Chart**. We have to change from the default setting at the top, indicating what the bars in your display will represent:

Bar Charts

Bars represent:

Counts of unique values ▼

Click on the arrow to the right, and you will see 3 choices. Switch to the choice below:

Bar Charts

Bars represent:

Values from a table ▼

We still want a **Simple** bar chart. Hit **OK**, and then fill in:

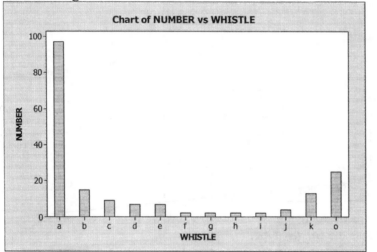

OK and we get:

Again, we see the prevalence of type a whistles, though the percentage is not as clear as with the pie chart. To switch from counts to percents above, use the **Bar Chart Options: Show Y as Percent**.

Alternatively, you could stick with

Bars represent: *Counts of unique values*, **Simple** chart,
Categorical variables: *WHISTLE*,
Data Options: **Frequency**: **Frequency variable(s)**: *NUMBER*

Exercise 2.19 - M/S page 37

Click **File>Open Worksheet**. Locate and **Open** the MTBE data file, the first part of which looks like:

↓	C1	C2	C3	C4-T	C5-T	C6	C7	C8-T	C9
	pH	DissOxy	Industry	WellClass	Aquifer	Depth	Distance	Detection	MTBE
1	7.87	0.58	0.00	Private	Bedrock	60.960	2386.29	Below Limit	0.20
2	8.63	0.84	0.00	Private	Bedrock	36.576	3667.69	Below Limit	0.20
3	7.11	8.37	0.00	Private	Bedrock	152.400	2324.15	Below Limit	0.20
4	7.98	0.41	0.00	Private	Bedrock	*	788.88	Below Limit	0.20
5	7.88	1.44	0.00	Private	Bedrock	91.440	1337.88	Below Limit	0.20
6	8.36	0.18	0.00	Private	Bedrock	115.824	2396.74	Below Limit	0.20
7	6.92	1.85	0.00	Private	Bedrock	92.964	1425.59	Detect	0.51

These are *raw* data, not *summarized data*. For (a), use **Graph>Bar Chart-Simple, Bars represent:** *Counts of unique values*, with **Categorical Variables:** *WellClass, Aquifer, Detection.* Hit **OK** to get the three bar charts in separate graphs. Here is one of them:

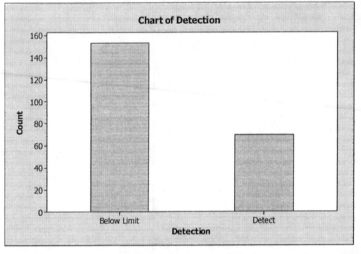

For part (c) we want to compare contamination rates by aquifer. So, instead of the *simple* bar chart, let's choose the *cluster* type:

Graph>Bar Chart-Cluster, Bars represent: *Counts of unique values*, with **Categorical Variables:** *Aquifer Detection,* and then select: **Bar Chart Options: Show Y as Percent** and **Take Percent Within categories at level 1 (outermost). OK, OK,** and we get:

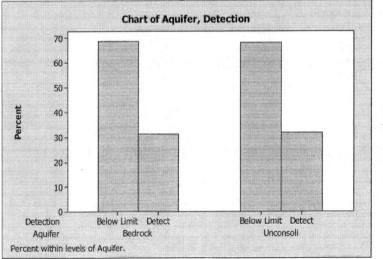

If you want *gaps* between the bars, just click on the x-axis, and use **Edit Scale> Scale**

If instead, you were interested in comparing pie charts, side-by-side, you would use:
Graph>Pie Chart: **Chart raw data**, **Categorical Variable:** *Detection*, and select
Multiple Graphs>By Variables: By variables with groups on same page: *Aqufier*

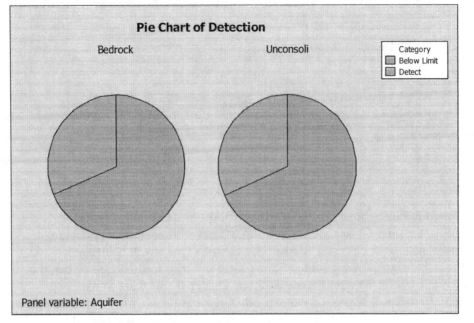

Quantitative Data: Graphical Displays

Dotplots, histograms, stem & leaf plots, and boxplots (discussed later) are the customary methods for displaying quantitative data. You will find these listed under **Graph**.

Example 2.2 - M/S page 42

The 50 default rates in Table 2.4 are saved in a file named LOANDEFAULT. Click **File>Open Worksheet.** Locate the file and open it. You will see in C1 the 50 default rates:

↓	C1	C2
	Default	
1	12.0	
2	19.7	
3	12.1	
4	12.9	

For a histogram, click on **Graph>Histogram-Simple** and plug in '*Default*' under **Graph Variables**, as below.

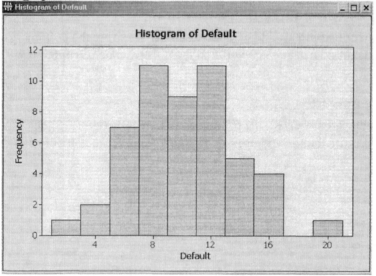

There are many sub-dialog boxes displayed, which allow you to change various characteristics of the graph – you will not need to use any of these unless you want to compare histograms, in which case, the **Multiple Graphs** option will be needed.

Keep in mind that if you change some of the *default settings* in any dialog boxes, the changes will persist through your session, until you change them again, or exit.

Hit **OK** and Minitab will do a good job of producing a reasonable histogram, with the classes and labeling decided by Minitab. *You can change the class intervals or other aspects of the graph later, if you wish, simply by double-clicking on various portions of the graph.*

If you want to produce a histogram exactly like the one in Figure 2.12 in the text, with midpoints of 1.5, 4.5, 7.5, ... , 19.5, then you need to tell Minitab to make some changes to the display:

Double-click on any of the bars in the histogram, and you will see the **Edit Bars** dialog box. Or click once on the bars, and then right-click or go to the main menu and select **Editor>Edit Bars** (be sure that all the bars are selected, *not just one bar*, in which case you will only be able to *edit the characteristics of that single bar*). Classes are also called 'bins', so choose **Binning** from the four choices shown, select **Interval Type:** *Midpoint*, and fill in as below:

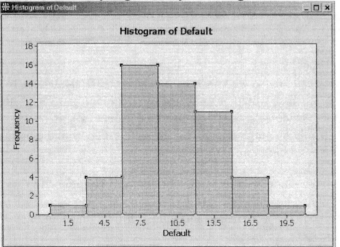

Note that you can specify either the midpoints of the classes or the limits (cutpoints) of the classes, or you can just specify the number of classes that you would like. To specify the midpoints above, you may also write *1.5:19.5/3*, which is a short way to say: run from 1.5 to 19.5 in steps of 3.

Hit **OK**, and now you get exactly the histogram in the text display:

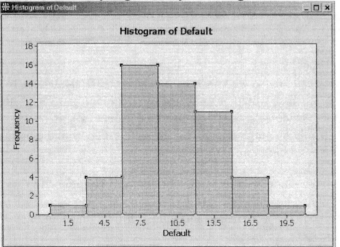

If you prefer to display percents on the y-axis, double-click on any numbers on the y-axis, to produce the **Edit Scale** dialog box, and select **Type: Scale Type:** *Percent*

For a stem and leaf plot, like in Fig. 2.13, select **Graph>Stem-and-Leaf**

```
Stem-and-Leaf                                          ×

C1    Default         Graph variables:
                      Default                              ▲


                                                          ▼

                      By variable:


                      ☐ Trim outliers


                      Increment:

        Select

    Help                          OK              Cancel
```

Fill in the relevant **Graph variable**, hit **OK**, and you will see the plot in the Session Window (a separate graphical window is not needed for this type of display):

Stem-and-Leaf Display: Default

```
Stem-and-leaf of Default   N  = 51
Leaf Unit = 1.0


   1    0  2
   5    0  4455
  14    0  666667777
 (12)   0  888888899999
  25    1  000011111
  16    1  2222223
   9    1  4444555
   2    1  6
   1    1  9
```

The Leaf Unit is shown at the top as '1.0', so we know that the data are 02, 04, 04, … 16 (and not say 0.2, 0.4, 0.5, … , 1,6 , which would be indicated by 'Leaf Unit = 0.1') . The last figure (tenths of percent) for each default % gets chopped off (truncated), as computer programs prefer to use only one-digit leaves. So the smallest entry in the plot, the '2', could actually have been anything from 2.0% to 2.9%.

What are those numbers in the left column? To find out, try Minitab's **Help** function. Go back to the **Stem and Leaf** dialog box and select **Help** (or from the main menu, select **Help>Help**, then **Search**).   You will see an example and some information about this plot, including the following remark:

- *Counts (left) – If the median value for the sample is included in a row, the count for that row is enclosed in parentheses. The values for rows above and below the median are cumulative. The count for a row above the median represents the total count for that row and the rows above it. The value for a row below the median represents the total count for that row and the rows below it.*

Increment: You can alter the display by your choice of **Increment**, in the dialog box. The increment is the difference between the lowest possible value in one row and the lowest possible value in either adjacent row (or the stem unit divided by the number of splits). In the display above, this equals $(4 - 2)$ or $(16 - 14) = 2$. Choose '1' for the increment, and the number of rows will double. Choose '5' and there will be fewer rows. The only possible values of **Increment** are .01, .02, .05, .1, .2, .5, 1, 2, 5, 10, 20, 50, etc. since you can only repeat the same stem once, twice, or five times, and still maintain equal width classes (rows).

How about a dotplot? Use **Graph>Dotplot - One Y, Simple**, then enter the variable 'Default' under **Graph Variables**:

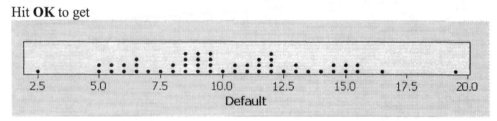

Hit **OK** to get

Figure SIA2.5 - M/S page 45

To produce these side by side histograms, open up the EYECUE data set, then select **Graph>Histogram-Simple: Graph Variables:** *Deviation,* with **Multiple Graphs: By Variables (with groups in separate panels):** *Gender,* **Scale: Y-Scale Type:** *Percent*

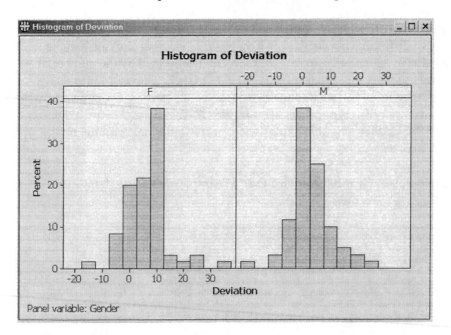

Exercise 2.27 - M/S page 46

The actual data are not given here, just a grouped distribution with classes and relative frequencies. So, start with a new worksheet via **File>New: Minitab Worksheet** unless you have just opened up Minitab, in which case a blank worksheet is already sitting there. Enter the measurements' midpoints in C1, relative frequencies in C2, and name your columns as below:

↓	C1	C2
	Measurement	RelFreq
1	1.5	0.10
2	3.5	0.15
3	5.5	0.25
4	7.5	0.20
5	9.5	0.05
6	11.5	0.10
7	13.5	0.10
8	15.5	0.05

Since we want the counts, multiply C2 by 500, and place these counts into C3 by using **Calc>Calculator**. Next to **Store result in variable:** type *Counts,* and under **Expression**: enter '*500 * C2*', exactly as below. Hit **OK**. No column is yet named '*Counts*', so Minitab assigns this name to the first free column in the worksheet.

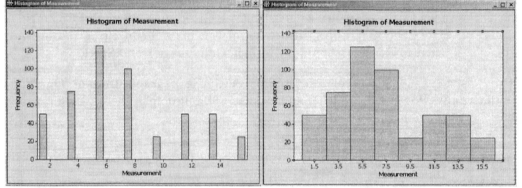

Now, in your worksheet, you will see the counts or frequencies in C3. Let's try **Graph>Histogram-Simple: Graph Variables:** *Measurement*, **Data Options: Frequency: Frequency Variables:** *Count*. You see the display on the left below.

Whoops, it seems Minitab assumed that the frequencies (counts) were attached exactly to each midpoint value, and not spread out over the entire class. Double-click on the x-axis, then use **Edit Scale: Binning: Interval Type:** *Midpoints*, **Midpoint positions:** *1.5:15.5/2* (which means from 1.5 to 15.5 in steps of size 2). Now, you get the proper histogram as shown on the right above.

   Alternatively, you could try **Graph>Bar Chart-Simple, Bars represent:** *Values from a table*: **Graph Variables:** *Count*, **Categorical Variable**: *Measurement*. To fatten the bars, double click on the x-axis values, then **Edit Scale: Scale: Gap between clusters:** *0*.

Exercise 2.37 - M/S page 48

Click **File>Open Worksheet**. Find and open BRAINPMI on your text CD. The data are in C1. Use **Graph>Dotplot-One Y, Simple,** with **Graph Variables:** *C1*. You get:

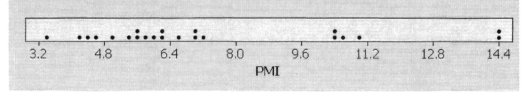

Note the distinct groups.

Exercise 2.41 - M/S page 49

(b) Use **File>Open Worksheet**, find and open LISTEN. Note the structure of the data in the worksheet, with three columns:

↓	C1-T	C2	C3
	Group	Response Rate	Percent Correct
1	Control	250	23.6
2	Control	230	26.0
3	Control	197	26.0
4	Control	238	26.7

This is a very convenient structure, referred to as **stacked** data, where *each subject corresponds to one row,* and in that row we record all the data on that subject, i.e. the subject's Group Membership, Response Rate and Percent Correct. Hence, all the Response Rates are *stacked* in column C2, Correct Percentages *stacked* in C3, and Group memberships *stacked* in C1. The data would have been **unstacked** if they appeared like the display in your textbook, with 6 columns in total: C1: Response Rates for Controls, C2: %Correct for Controls, C3: Response Rates for Treatment Group, etc. [To convert from one structure to the other, use **Data>Unstack Columns** or **Data>Stack**]

If you wish to use dotplots to compare several groups, select: **Graph>Dotplot: One Y, With Groups.** You have to switch from the default **Simple** dotplot choice. Note how the picture helps you choose the correct type of dotplot:

Fill in as follows:

Note the *apostrophe marks* around the variable: 'Percent Correct'. These are inserted, at times, for clarity, when the *variable name has a space, or certain other special symbols, in it.*

Minitab now knows it should separate the data in C3 based on the distinct group settings it finds in the column C1 = Group. We get:

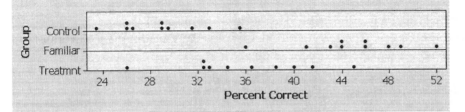

[If the data had been *unstacked* like in the textbook display, you would select **Graph>Dotplot-Simple, One Y,** list all three variables under **Graph Variables,** and select **Multiple Graphs: Multiple Variables: In separate panels…, Same Scales…]**

Likewise, we can produce *separate stem-and-leaf plots by Group*, except that ***Minitab's Stem-and-Leaf procedure requires Group to be a numerical variable.*** So recode the three group names in C1 into the numbers 1,2,3 in a new column (C4) named 'GroupNumber' using **Data>Code>Text to Numeric** as follows:

Code - Text to Numeric

C1	Group
C2	Response Rat
C3	Percent Corr

Code data from columns:

`c1`

Into columns:

`GroupNumber`

Original values (eg, red "light blue"):   New:

`Control`	`1`
`Treatmnt`	`2`
`Familiar`	`3`

Now, select **Graph>Stem and Leaf**, and then fill in:
Graph Variables: *C3*, and **By Variable:** *GroupNumber*,
You get, in the session window:

Stem-and-Leaf Display: Percent Correct

```
Stem-and-leaf of Percent Correct  GroupNumber = 1    N = 10
Leaf Unit = 1.0

   1    2  3
  (6)   2  666999
   3    3  12
   1    3  5

Stem-and-leaf of Percent Correct  GroupNumber = 2    N = 10
Leaf Unit = 1.0

   1    2  6
   5    3  2234
   5    3  68
   3    4  014

Stem-and-leaf of Percent Correct  GroupNumber = 3    N = 10
Leaf Unit = 1.0

   1    3  6
   5    4  1344
   5    4  6689
   1    5  2
```

Minitab uses the same *increment*, in each plot, in order to maintain comparability. [For *unstacked data*, you would list all three variables under **Graph Variables** and specify one common increment]

For comparative histograms, use **Graph>Histogram-Simple** and fill in:
Graph Variables: *Percent Correct*, **Multiple Graphs>By Variables: By Variables with groups in separate panels:** *Group*. You get

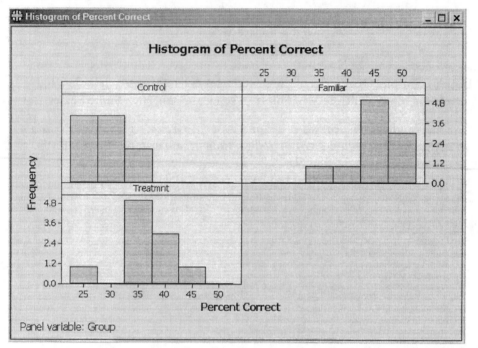

We readily see the shift up in accuracy as we go from the Control group to the Treatment group to the Familiarization group. [For *unstacked data*, you would use **Multiple Graphs: Multiple Variables,** rather than **Multiple Graphs: By Variables**]

Quantitative Data: Measures of Center and Spread

We can use Minitab to calculate individual measures such as mean or standard deviation with **Calc>Column Statistics** or get a batch of useful measures with **Stat>Basic Statistics>Display Descriptive Statistics**.

Example 2.4 - M/S page 52

Open up the file EPAGAS, where the 100 EPA mileage ratings of Table 2.2 have been stored in C1. To find the mean rating, use **Calc>Column Statistics**, tick the statistic needed, and input the variable:

Column Statistics ✕

C1	MPG	Statistic

Statistic

- ○ Sum
- ● Mean
- ○ Standard deviation
- ○ Minimum
- ○ Maximum
- ○ Range

- ○ Median
- ○ Sum of squares
- ○ N total
- ○ N nonmissing
- ○ N missing

Input variable MPG

Store result in [   ] (Optional)

Select

| Help | | OK | Cancel |

Note that there are many other statistics you can choose to calculate for a column of numbers. The output is:

Mean of MPG

```
Mean of MPG = 36.994
```

You also have the option, in the dialog box, to store the result as a saved constant (k1,k2,…). Storing it as a constant may come in handy if you are going to use this single number as part of some future calculation.

Example 2.6 - M/S page 54

To get the median mileage rating, we can proceed as in Example 2.4, choosing **Median** in the dialog box, or we can just ask Minitab for a good selection of useful summary statistics, via **Stat>Basic Statistics>Display Descriptive Statistics**

Display Descriptive Statistics ✕

C1	MPG	Variables:

Variables:
MPG

By variables (optional):

Select

	Statistics...	Graphs...
Help	OK	Cancel

Note the '**By variables**' option, which we would use if e.g. there were a categorical variable, like gender in C2, and we wished separate stats for males and for females. You can also produce a histogram or boxplot (studied later) for the data via the **Graphs...** selection in the dialog box.

In the session window, we see:

Descriptive Statistics: MPG

Variable	N	N*	Mean	SE Mean	StDev	Minimum	Q1	Median	Q3
MPG	100	0	36.994	0.242	2.418	30.000	35.625	37.000	38.375

Variable	Maximum
MPG	44.900

Under Median, we see 37.0, telling us that half of the mileages are in excess of 37.0. You can choose additional statistics or delete some of those displayed above, by clicking on **Statistics** in the dialog box.

In the output, **StDev** stands for 'standard deviation', which will be discussed in section 2.5. **Q1** and **Q3** stand for first (or lower) and third (or upper) quartile, respectively (see sec. 2.8). These are also called the 25th and 75th *percentiles* (see sec. 2.7). **Minimum** and **Maximum** are self-explanatory. **N** is the number of EPA ratings. **N*** indicates the number of missing observations in the column.

Example 2.7 - M/S page 55

Open up a fresh worksheet, via **File>New>Minitab Worksheet** and enter the 10 numbers in column 1. To find the mode, use **Stat>Tables>Tally Individual Values: Variables:** *C1*,   and you get

Tally for Discrete Variables: C1

C1	Count
5	1
6	1
7	2
8	2
9	3
10	1
N=	10

Since this command lists all distinct outcomes and their frequencies, you can readily see which outcome is the most common (i.e. the mode).

Exercise 2.65 - M/S page 60

(c)-(e) Open up the data file SLI. The DIQ scores are all stacked into one column, with a group indicator in another column, so we can use **Stat>Basic Statistics>Display Descriptive Statistics** with **Variables:** *DIQ* and **By variables:** *Group*, and we get:

Descriptive Statistics: DIQ

Variable	Group	N	N*	Mean	SE Mean	StDev	Minimum	Q1	Median	Q3
DIQ	OND	10	0	101.90	3.06	9.69	87.00	93.50	103.00	110.75
	SLI	10	0	93.60	2.87	9.08	84.00	86.00	91.50	100.25
	YND	10	0	95.30	2.38	7.51	86.00	90.00	92.00	101.25

Variable	Group	Maximum
DIQ	OND	113.00
	SLI	110.00
	YND	110.00

You can easily pick out the means or medians for each of the three groups from the above.

Example 2.11 - M/S page 69

Open up the RATMAZE file, containing the times for the 30 rats in C1. Use **Stat>Basic Statistics>Display Descriptive Statistics** with **Variables:** *Time*, and we get:

Descriptive Statistics: RUNTIME

Variable	N	N*	Mean	SE Mean	StDev	Minimum	Q1	Median	Q3	Maximum
RUNTIME	30	0	3.744	0.401	2.198	0.600	1.960	3.510	5.165	9.700

As in the text, the 2 standard deviation interval about the mean works out to be (-.66, 8.14). If you don't have a calculator handy, use Minitab's: **Calc>Calculator** to work out these numbers.

We want to count how many observations fall in this interval. We could use **Data>Sort**: **Sort column:** *C1*, **By column:** *C1*, **Store in:** *Original Column*, to sort the numbers in C1 in ascending (or descending) order, making a direct count easy.

Alternatively, we can use **Calc>Calculator** with the following entries:

If you put a 'true or false' statement under **Expression**, Minitab checks it out for each row, and enters a '1' if true, '0' if false, into the storage column that you specify at the top. Such a true/false expression is called a *Logical Operation*. My **Expression** above says that the entry in C1 is greater than -.66 and also is less than 8.14, i.e. within the 2 standard deviation interval. You will see now 0's and 1's in C2, named 'Within2StDev'. Now, tally the 1's and 0's with **Stat>Tables>Tally Individual Variables:** *C2*

Tally for Discrete Variables: Within2StDev

```
Within2StDev   Count
          0       2
          1      28
         N=      30
```

We see that 28 of 30, or about 93% lie within the interval, quite close to the 95% predicted by the empirical rule, even though the distribution is not very bell-shaped.

Exercise 2.178 - M/S page 105

Open up the PAF file. For a stem and leaf plot of the PAF Inhibition percentages shown in C1, use **Graph>Stem and Leaf: Graph variables:** *C1*, to get

Stem-and-Leaf Display: PAF Inhibition

```
Stem-and-leaf of PAF Inhibition  N  = 17
Leaf Unit = 1.0

   6   0   000009
   8   1   25
  (2)  2   45
   7   3   13
   5   4   0
   4   5
   4   6   2
   3   7   057
```

From the above, we can identify the mode as 0. For the other descriptive statistics, use **Stat>Basic Statistics>Display Descriptive Statistics**, with **Variables:** *C1*, and you see:

Descriptive Statistics: PAF Inhibition

Variable	N	N*	Mean	SE Mean	StDev	Minimum	Q1	Median
PAF Inhibition	17	0	27.82	6.76	27.88	0.000000000	0.000000000	24.00

Variable	Q3	Maximum
PAF Inhibition	51.00	77.00

The *Mean*, *Median*, and *StDev* (standard deviation) are apparent. For the range, subtract the *Minimum* from the *Maximum*, and square the *StDev* to get the variance, or if you prefer, click on **Statistics** in the dialog box for this procedure, and tick off *Range* and *Variance* as additional choices.

Quantitative Data: Boxplots

This useful display for quantitative data provides measures of center, spread, and shape, helps detect outliers, and is very useful for comparing distributions. Look for it under **Graph**.

Example 2.16 - M/S page 82

Open up the LOANDEFAULT file. To produce a boxplot, use **Graph>Boxplot-One Y, Simple,** and fill in the obvious entries:

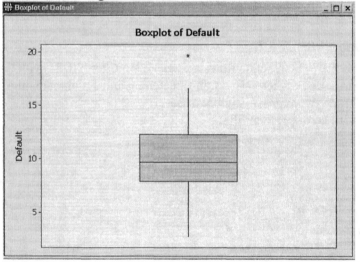

Hit **OK**, and we get

In the above, the asterisk points out one unusual observation, the nearly 20% default rate for Alaska.

In these Minitab boxplots, all unusual observations are denoted with an '*', *regardless* of how far beyond the inner fences they may lie, i.e. no 'outer fences' are computed by Minitab, as mentioned in the textbook.

To read values more precisely from the plot, you might wish to show more ticks on the Y-axis. Double-click on any of the numbers shown on the y-axis, to display the **Edit Scale** dialog box:

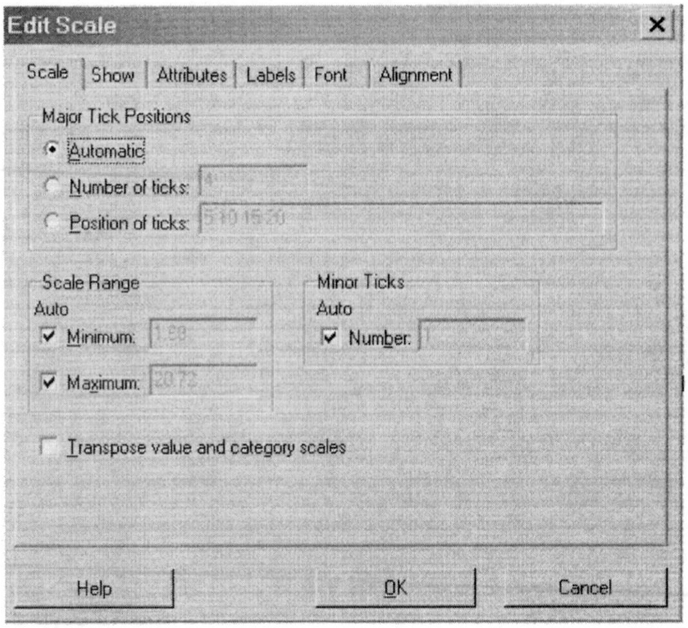

Edit Scale [×]

Scale | Show | Attributes | Labels | Font | Alignment

Major Tick Positions

● Automatic

○ Number of ticks: `4`

○ Position of ticks: `5 10 15 20`

Scale Range
Auto
☑ Minimum: `1.88`
☑ Maximum: `20.72`

Minor Ticks
Auto
☑ Number: `1`

☐ Transpose value and category scales

Help   OK   Cancel

Using this, you can change the **Major Tick Positions** or their **Number**, as well as the **Number** of **Minor Ticks** (unclick box under **Auto**, then enter desired **Number** of ticks), or reverse (**Transpose**) the orientation of the plot, so that the Default axis stretches from left to right instead of vertically.

Example 2.17 - M/S page 82

Open up the REACTION data file. The reaction times are stacked in C2 = TIME, with the stimulus type in C1 = STIMULUS, where NT = non-threatening and T = threatening. To compare reaction times for the two types of stimulus, use
Graph>Boxplot - One Y, With Groups, then fill in:

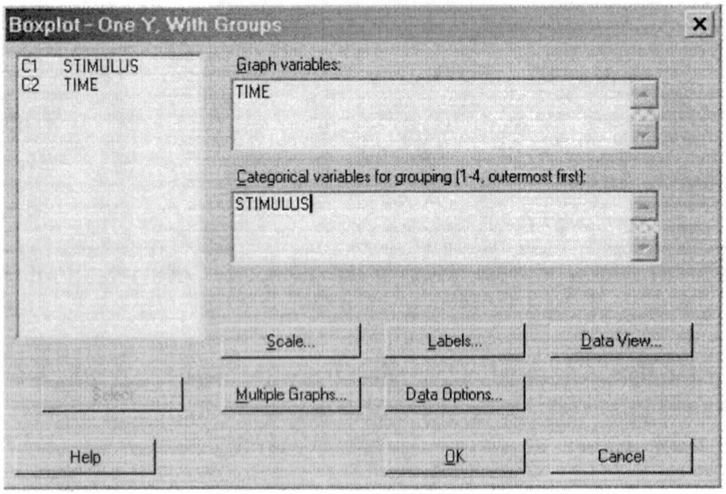

Boxplot - One Y, With Groups [×]

C1 STIMULUS
C2 TIME

Graph variables:

`TIME`

Categorical variables for grouping (1-4, outermost first):

`STIMULUS`

Select   Scale...   Labels...   Data View...

Multiple Graphs...   Data Options...

Help   OK   Cancel

We get

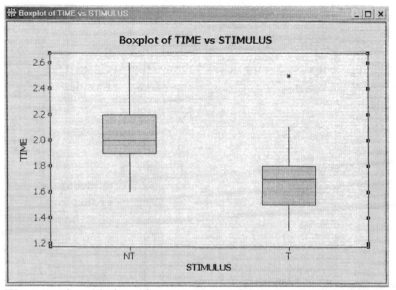

We get a nice side-by-side comparison of the two samples, showing generally higher, i.e. slower, reaction times for the non-threatening stimulus, similar variability in the two groups, and one possible reaction time outlier in the threatening stimulus group.

Exercise 2.126 - M/S page 87

Open up the worksheet LM2_126, showing the Sample (A or B) in C1 and the Data in C2. Since the data are all stacked in C2, we readily get side-by-side boxplots, via **Graph>Boxplot: One Y, With Groups: Graph Variables:** *C2*, **Categorical Variables**: *C1*.

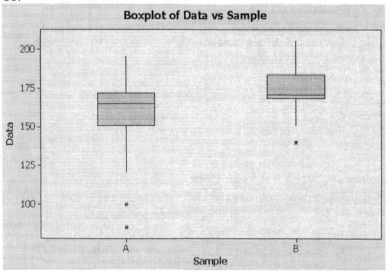

We see a lower center and greater variability for sample A, which exhibits some left (negative) skewness, while right (positive) skewness is exhibited by sample B. Using the scale on the left side, you can read off approximate numerical values for the two suspect outliers in sample A, and the single suspect outlier in sample B.

Exercise 2.133 - M/S page 88

Open up the SHIPSANIT file. The vessel name is in C1, cruise line in C2, and the sanitation score is in C3. For a boxplot of the downtimes, use **Graph>Boxplot-One Y, Simple: Graph Variables:** *SCORE*

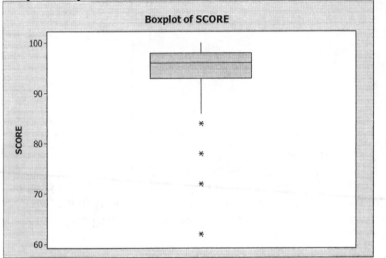

(a) The distribution is negatively skewed, with four suspect outliers. You can read off the scores approximately from the graph, or *point your mouse at any outlier*, say the most extreme one above, and a Minitab pop-up will ID it for you:

> *
> Outlier symbol , Row 149: SCORE = 62

Go to row 149, and you can see that this is the Stad Amsterdam vessel. Or, to make things easier, sort the data first via **Data>Sort: Sort Columns:** *C1-C3*, **By column:** *C3*, **Store in:** *Original columns*. You will then see the worst offenders at the top of your worksheet:

Stad Amsterdam	*Stad Amsterdam*	*62*
Legacy	*Windjammer Barefoot Cruises /International Marine*	*72*
Nautilus Explorer	*Lever Diving*	*78*
Sea Bird	*Lindblad Expeditions*	*84*

Shall we brush?

We can also learn about the outliers by *brushing*. Click on the graph to make it the active window. Click **Editor>Brush**, which produces a special *brushing cursor* in the graphical window. Encircle the four outliers. In a little floating window off to the side, called the *brushing palette*, the row numbers 62, 72, 78, 84 will appear. Look through the worksheet, and you will see markers next to these three rows. Select **Editor>Set ID Variables: Variables:** *C1 C2 C3*, to display complete information about the highlighted observations right in the brushing palette:

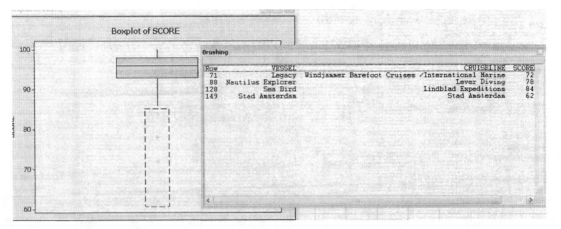

Brushing allows you to highlight extreme points on a graph in order to learn more about them.

(c) To standardize all the downtimes, use **Calc>Standardize,** with entries as below

which puts the z-scores in C4. Use **Data>Sort: Sort:** *C1-C4,* **By:** *C4 (or C3),* as described earlier, and at the top of the worksheet, you will see:

VESSEL	CRUISELINE	SCORE	zscores
Stad Amsterdam	Stad Amsterdam	62	-6.82053
Legacy Windjammer	Barefoot Cruises /International Marine	72	-4.74813
Nautilus Explorer	Lever Diving	78	-3.50469
Sea Bird	Lindblad Expeditions	84	-2.26125

The top three vessels seem to clearly stand apart from the rest in sanitation.

Bivariate Relationships – The Scattergram

With two quantitative variables, we may use **Graph> Scatterplot** to plot one variable versus the other in a scattergram (also known as a scatterplot), in order to assess the nature of the relationship between the variables.

Example 2.19 - M/S page 89

Open up the MEDFACTORS worksheet, which shows Length of Stay in C2 and #Factors in C1, for 50 coronary patients. To plot Factors on the y-axis versus LOS on the x-axis, use **Graph>Scatterplot-Simple: Y variables:** *Factors*, **X variables:** *LOS*.

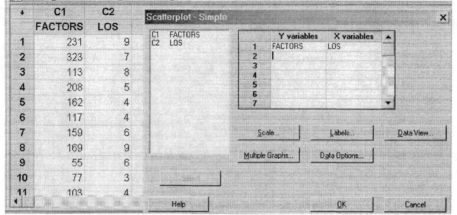

Hit **OK** to produce the scattergram:

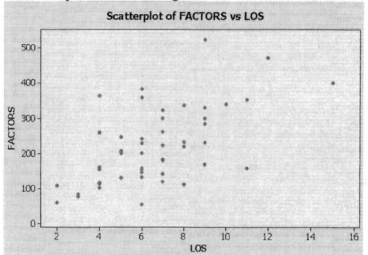

from which we see a roughly linear positive relationship.

Exercise 2.180 - M/S page 105

Open up the worksheet SLI worksheet. We can investigate the relationship between DIQ and Pronoun Errors for all children and also just for the SLI children, with a very useful scattergram variant. Select **Graph>Scatterplot-With Groups** and fill in as below, to produce a plot using *different symbols* for each GROUP.

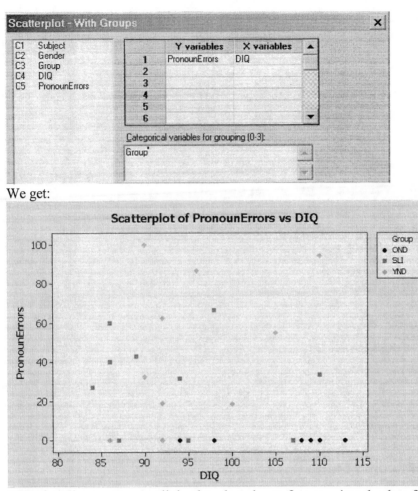

We get:

This plot allows us to see all the data, but also to focus on just the data from the SLI group, indicated by red square symbols (with the bright colors that you will be seeing, the differences between groups are much clearer). We see little relationship between the two variables here, overall, as well as for the SLI group.

If you want to plot only the 10 SLI observations, select and delete (using the ***Delete*** key on your keyboard) the other 20 rows in the worksheet.

   Or use

Data>Delete Rows: *1:10 21:30* **From:** *c1-c5*,

   Or you can isolate data for each group in a single new worksheet with:

Data>Split Worksheet: By Variables: *GROUP*

Then **Graph>Scatterplot-Simple: Y:** *PronounErrors*, **X:** *DIQ*, produces:

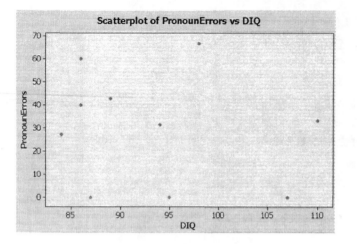

Chapter 3 – Probability

In this chapter, we will produce random data and draw random samples, using
Calc>Random Data>Bernoulli
Calc>Random Data>Integer
Calc>Set Base
Calc>Random Data>Sample From Columns

Summarize qualitative data using
Stat>Tables>Tally Individual Variables

Sort the data from low to high, with
Data>Sort

Enter a simple set of patterned numbers into a column, with
Calc>Make Patterned Data>Simple Set of Numbers

Display data in the session window, using
Data>Display Data

Do some calculations on the data in the worksheet using:
Calc>Calculator

Random Events

We can simulate various types of random experiments, like coin tossing, using
Calc>Random Data.

Figure 3.2 & the Law of Large Numbers - M/S page 117

If probability is just long run relative frequency, then the proportion of heads obtained when flipping a balanced coin repeatedly, in a truly random fashion, should get quite close to .5 after a large number of flips. Instead of actually flipping a real coin, we can simulate flipping a coin by using **Calc>Random Data>Bernoulli** as indicated below

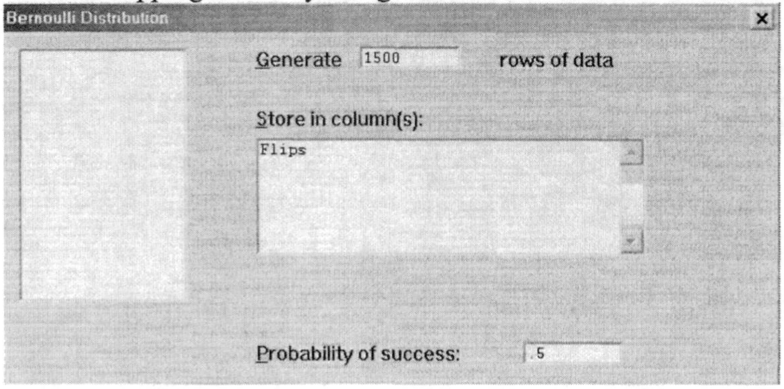

This will generate 1500 rows of random data, where in each row there is a .5 probability of '1' (termed a *success*) and .5 probability of '0'. Consider '1' to be a head, and '0' to be a tail. The individual flips are stored in *Flips* = C1.

Let's put the running or cumulative total of heads in C2, using a function called *Partial Sums (PARS)*, from the **Calc>Calculator**

Calculator	☒

C1   Flips
C2   Number of Heads

Store result in variable: Number of Heads

Expression:

PARS('Flips')

Enter the *Number of Flips*: 1:1500 in C3, using **Calc>Make Patterned Data> Simple Set of Numbers**

Simple Set of Numbers	☒

C1   Flips
C2   Number of F
C3   Number of F

Store patterned data c3

From first   1

To last   1500

In steps of: 1

List each value   1   times

List the whole   1   time

Divide c2 by c3, giving the current proportion of heads in c4 = *Proportion of Heads*:

Calculator	☒

C1   Flips
C2   Number of Heads
C3   Number of Flips

Store result in variable: c4

Expression:

'Number of Heads' / 'Number of Flips'

Here is some of the worksheet:

↓	C1	C2	C3	C4
	Flips	Number of Heads	Number of Flips	Proportion of Heads
1	0	0	1	0.000000
2	1	1	2	0.500000
3	0	1	3	0.333333
4	0	1	4	0.250000
5	1	2	5	0.400000
6	1	3	6	0.500000
7	1	4	7	0.571429
8	0	4	8	0.500000
9	1	5	9	0.555556
10	1	6	10	0.600000

Now, plot the proportion of heads versus flip number, with:

Graph>Scatterplot-Simple with
Data View: Data Display: Connect Line;
Data Options: Subset: Include: Row Numbers: *6:1500/30*
(i.e. showing only every 30th point).
Two different simulations of 1500 flips are shown below:

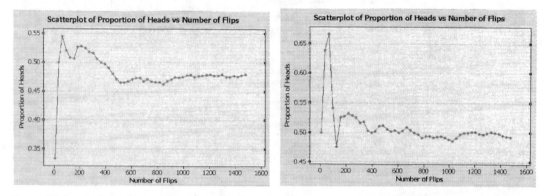

In each simulation above, the proportion of heads can be seen to be gradually converging toward 0.5. If you continue flipping, the proportion of heads must surely converge to the true probability of obtaining a head, in the long run. [This coin-flipping situation may be classified and studied further under a fascinating part of Probability theory called *Random Walk* theory]

Exercise 3.156 - M/S page 172

Open up the EVOS data set. Column C4 indicates oil contamination (yes/no) for each of the 96 transects. The probability that a randomly selected transect is contaminated with oil may be estimated by the proportion in the sample that are contaminated, which you can find using **Stat>Tables>Tally: Individual Variables:** *C4*, **Display:** *Percents*

Tally for Discrete Variables: Oil

Oil	Count	Percent
no	36	37.50
yes	60	62.50
N=	96	

Random Samples

To make inferences about a population from a sample, we must use some type of *random* (*probability*) sample. To select a *simple random sample* (analogous to drawing names from a hat), you can use a random number table like the one in the back of the textbook, or you can use random number generators present in all statistical packages.

Example 3.22 - M/S page 154

The households have been numbered 00000-99999. To select say fifty households, use **Calc>Random Data>Integer** as indicated below. The randomly generated data then appears in C1 in the worksheet, as shown to the left.

You might wish to print out the 50 selected households in order, so use **Data>Sort: Sort Columns:** *C1*, **By column:** *C1*, **Store in:** *Original column*, to sort the values in C1 into ascending order, and then use **Data>Display Data: Columns to display:** *C1*, to print out the values in your session window:

Data Display

```
HouseID
    144    2794    4136    6566    7163    7646    8925   13356   13744
  14362   14956   15671   17175   18147   18692   22548   25879   29100
  35129   36088   39839   40256   40261   40432   42345   42595   42772
  43236   45043   46323   51465   52914   53369   54453   56696   56723
  56850   57117   60458   63202   64185   68299   76966   81114   82184
  83230   87275   87545   89214   95971
```

The numbers that you get though will *not be the same* as these, as there is a random 'starting point' mechanism at work with Minitab. If you want your random selection to be repeatable, then before using **Calc>Random Data**, select **Calc>Set Base** and enter some number - this tells Minitab's random number generator where to start, in its sequence of pseudo-random numbers. Set the same base, before generating random data again, and you will get the same results.

If there are any repeated households in the random sample, just make additional random selections (or ask for a bit more than actually needed at the start).

Example 3.23 - M/S page 156

If you have a population listed explicitly in say C1, you can select a portion of them randomly, using **Calc>Random Data>Sample From Columns**. If this is an experiment, the ones selected will get the active treatment, and the rest will get the placebo. The dialog box and results are shown below:

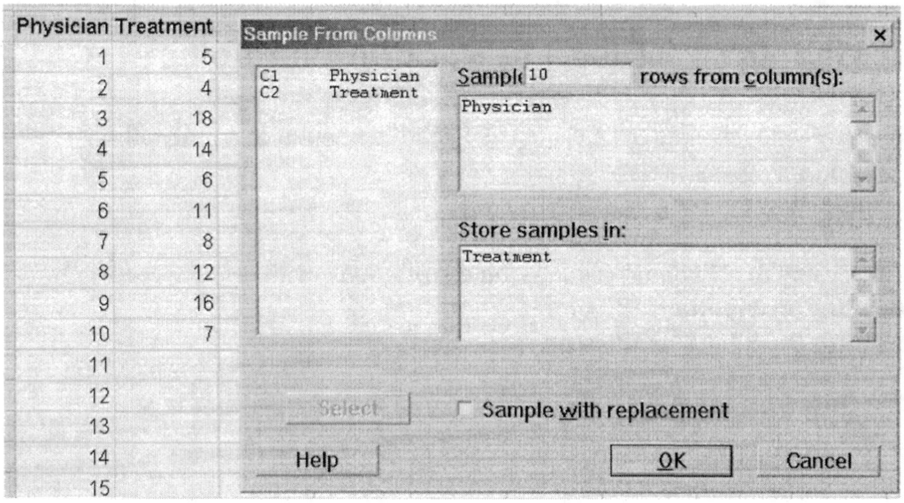

Physician	Treatment
1	5
2	4
3	18
4	14
5	6
6	11
7	8
8	12
9	16
10	7
11	
12	
13	
14	
15	

Sample From Columns   ✕

C1   Physician
C2   Treatment

Sample 10   rows from column(s):

Physician

Store samples in:

Treatment

Select   ☐ Sample with replacement

Help   OK   Cancel

Exercise 3.101 - M/S page 157

(b) We want to generate 10 random numbers from the population of all possible phone numbers: 0000000 – 9999999, or if no phone numbers start with 0 or 1, as is often the case, from 2000000-9999999. Use **Calc>Random Data>Integer: Generate:** *10* **rows, Store in:** *Phone*, **Minimum:** *2000000*, **Maximum:** *9999999*. Display them with **Data>Display Data: Columns to display:** *Phone*

Data Display

```
Phone
    6270999    9868830    9243011    4784577    7187920    6837347
    2308035    5489204    8153160    2482625
```

(c) To generate 5 numbers starting with 373, use **Calc>Random Data>Integer: Generate:** *5* **rows, Minimum:** *3730000*, **Maximum:** *3739999*

Chapter 4 – Discrete Random Variables

In this chapter, we will calculate the mean and standard deviation of a probability distribution, using
Calc>Calculator

Do calculations using the Binomial and Poisson distributions, with
Calc>Probability Distributions>Binomial
Calc>Probability Distributions>Poisson

Plot a probability distribution, with
Graph>Bar Chart

Enter a simple set of patterned numbers into a column, with
Calc>Make Patterned Data>Simple Set of Numbers

Examine the contents of the worksheet, using
Window>Project Manager

Mean and Standard Deviation

To calculate the mean and standard deviation of a probability distribution, we need to enter the outcomes in one column, the corresponding probabilities in a second column, and then use **Calc>Calculator** to construct these measures.

Example 4.7 - M/S page 192

Let's calculate the mean and standard deviation for the specified distribution using Minitab. First, we enter the distribution itself into two columns, as follows:

	C1	C2	C
↓	Outcome	Prob	
1	0	0.002	
2	1	0.029	
3	2	0.132	
4	3	0.309	
5	4	0.360	
6	5	0.168	

Now, use **Calc>Calculator**. Find the function **Sum**, in the **Functions** list. Double-click it to make it appear under **Expression,** where you will now see: SUM(*number*), with the argument *number* highlighted. For the mean, we want to sum products of the two columns, i.e. replace *number* with *C1*C2*. Type it in yourself, or you can double-click 'C1', from the left side, then click the multiplication symbol '*', then double-click 'C2', to copy things over. This calculation is displayed below:

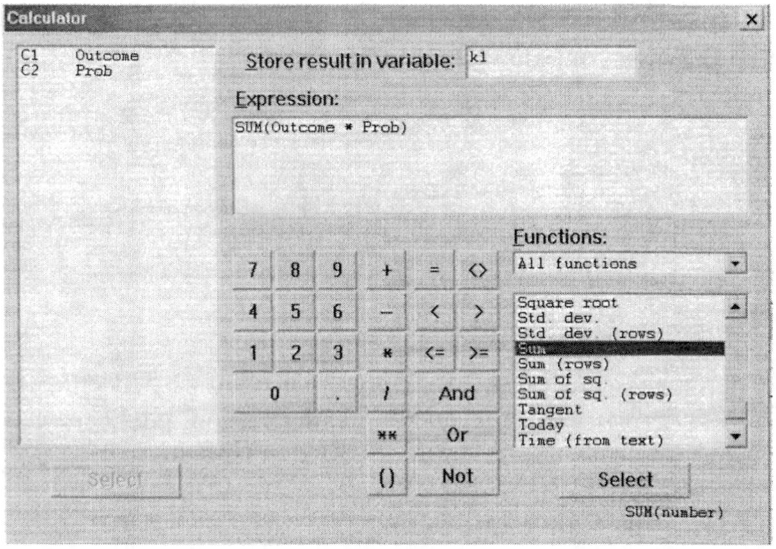

Note that I have stored the result in the form of a *stored constant*, denoted 'k1'. Use **Window>Project Manager** (or **Data>Display**) to see the value of k1:

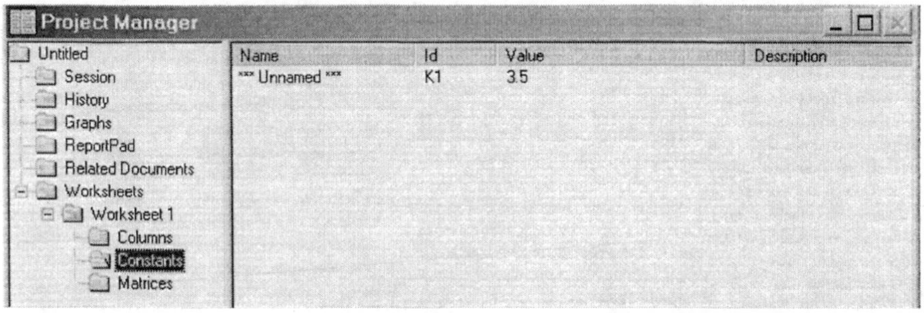

So, the mean is 3.5. To calculate the standard deviation, type in the **Expression** below (or select *Square root*, then *Sum*, etc.), being very careful with all the parentheses:

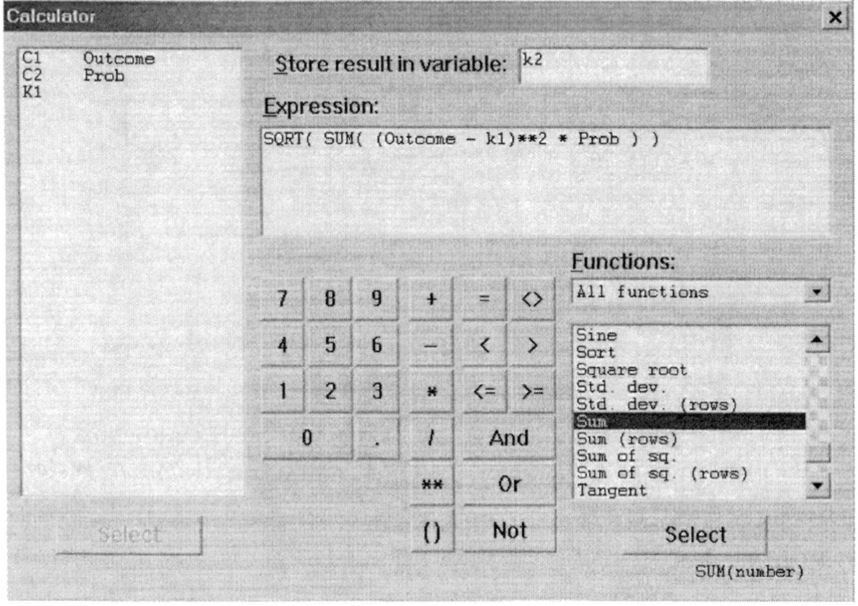

I chose to **Store result in:** k2.   Then, **Window>Project Manager**

Untitled	Name	Id	Value	Description
Session	*** Unnamed ***	K1	3.5	
History	*** Unnamed ***	K2	1.02372	
Graphs				
ReportPad				
Related Documents				
Worksheets				
Worksheet 1				
Columns				
Constants				
Matrices				

tells us that the standard deviation equals 1.024.   To remind yourself of what k1 or k2 actually represent, you can *right-click* on ***Unnamed*** in the window, producing a pop-up menu letting you either **Rename** it, e.g. 'mean', or **Set Description…**

You can graph p(x) via **Graph>Bar Chart-Simple, Bars represent:** *Values from a table:*

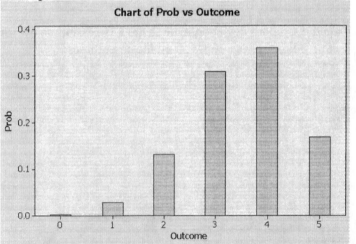

which produces

If you prefer a histogram look, you can reduce the space between bars to zero, by double-clicking on any of the x-values, producing the **Edit Scale: Scale** dialog box, shown below, in which you would unclick the check mark next to **Gap between clusters**, and then type '*0*'.

Edit Scale ✕

Scale | Show | Attributes | Labels | Font | Alignment

Ticks for Innermost Displayed Category Labels
Auto
☑ Tick start: [   ]
☑ Tick increment: [   ]

Space Between Scale Categories
Auto
☑ Gap within clusters: [0]   (-1,1)
☐ Gap between clusters: [0]   (0-5)

☐ Transpose value and category scales

| Help | | OK | | Cancel |

Exercise 4.41 - M/S page 195

Enter the number of homes in C1 = *Homes* and the probabilities in C2 = *Prob* as below

↓	C1 Homes	C2 Prob
1	0	0.09
2	1	0.30
3	2	0.37
4	3	0.20
5	4	0.04

To calculate the mean, **Calc>Calculator** and enter the expression

Expression:
```
SUM( Homes * Prob )
```

Store the mean in 'k1'.
Then, for the standard deviation, use **Calc>Calculator** with expression

Expression:
```
SQRT( SUM( (Homes-k1)**2 * Prob ) )
```

Store the result in 'k2'. To see the values of the stored constants k1 and k2, use **Data>Display Data** or check your worksheet's contents via **Window>Project Manager**. You can then calculate $\mu \pm 2\sigma$ and determine the probability within this interval.

The Binomial Probability Distribution

Many real world situations give rise to random variables, whose behavior can be well modeled by the *Binomial distribution*. Probabilities can be easily calculated using statistical software.

Example 4.12 - M/S page 204

The number who favor the candidate is Binomially distributed with n = 20 and p = .6. To find probabilities of single outcomes, like P(x = 11), we may use
Calc>Probability Distributions>Binomial

Binomial Distribution	×
⊙ <u>P</u>robability	
○ <u>C</u>umulative probability	
○ <u>I</u>nverse cumulative probability	
Nu<u>m</u>ber of trials:	20
Pro<u>b</u>ability of success:	.6
○ Input co<u>l</u>umn:	
Optional s<u>t</u>orage:	
⊙ Input co<u>n</u>stant:	11
Optional stora<u>g</u>e:	
Se<u>l</u>ect	

We have plugged in for n and p, and clicked on **Probability** among the three choices at the top. We **input** the **constant** *11*. If we wanted to find the probabilities for 11, 12 and 13 successes, we would type all of these numbers into say C1, and then use **Input column** with entry *C1*.

The output is:

Probability Density Function

```
Binomial with n = 20 and p = 0.6

 x   P( X = x )
11     0.159738
```

Minitab uses the term 'Probability Density Function' in the output, though strictly speaking this term applies only to the case of continuous random variables, to be discussed later on.

To find the probability that x falls in some range, it is easier if we use the selection **Cumulative Probability** rather than finding a bunch of individual probabilities and summing them. So, for P(x≤10), we use **Calc>Probability Distributions>Binomial** with

Binomial Distribution ☒

○ **P**robability
● **C**umulative probability
○ **I**nverse cumulative probability

N**u**mber of trials: `20`
Pro**b**ability of success: `.6`

○ Input c**o**lumn:
    Optional st**o**rage:

● Input co**n**stant: `10`
    Optional sto**r**age:

which produces:

Cumulative Distribution Function

```
Binomial with n = 20 and p = 0.6

  x   P( X <= x )
 10      0.244663
```

The output gives the probability up to and including the number entered. If you wanted P(x<10), you would input '9', since < 10 is equivalent to ≤ 9.

To find P(x>12), note that this equals 1 – P(x≤11), and then find the **Cumulative Probability** for the input constant '11'.

If you needed the probability for a range of values like P(8≤ x ≤16), find the **Cumulative Probability** for both '7' (not 8) and '16' and subtract.

To graph this function, enter values for x in say C1 (name it 'x') and then let Minitab compute the probabilities, which you would store in say C2 (name it 'prob'). Graph C2 versus C1. See the steps below.

Calc>Make Patterned Data>Simple Set of Numbers

Simple Set of Numbers ☒

Store patterned data in: `x`

From first value: `0`
To last value: `20`
In steps of: `1`

List each **v**alue `1` times
List the **w**hole sequence `1` times

Calc>Probability Distribution>Binomial

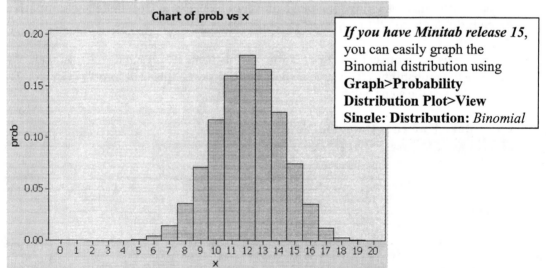

Binomial Distribution

C1   x

- Probability
- Cumulative probability
- Inverse cumulative probability

Number of trials: `20`
Probability of success: `.6`

- Input column: `x`
  Optional storage: `prob`
- Input constant: ``
  Optional storage: ``

Select

Help   OK   Cancel

This gives you a worksheet where the first 5 rows are:

	C1	C2
	x	prob
1	0	0.000000
2	1	0.000000
3	2	0.000005
4	3	0.000042
5	4	0.000270

Now, use **Graph>Bar Chart-Simple, Bars represent:** *Values from a table*: **Graph variables:** *prob*, **Categorical variable:** *x*

Chart of prob vs x

> *If you have Minitab release 15*, you can easily graph the Binomial distribution using **Graph>Probability Distribution Plot>View Single: Distribution:** *Binomial*

To reduce the space between bars to zero, I double-clicked on the x-values, producing the **Edit Scale** dialog box, then **Scale: Space Between …: Gap between clusters**: *0*

You could also use **Graph>Scatterplot** with **Data View... Data Display: Project lines**, to graph 'prob' vs 'x', with skinny bars projecting down.

$\mu \pm 2\sigma$ works out to be (7.6, 16.4) here (see text). Since the distribution is fairly bell-shaped, we would then expect about .95 in probability to lie between 8 and 16 (inclusive). Verify this by adding up nine probabilities from the graph or worksheet above; or you can evaluate $P(8 \le x \le 16)$ using **Calc>Probability Distributions>Binomial: Cumulative probability** (input '7', input '16', and subtract).

Exercise 4.61 - M/S page 208

(c) Here we have 20 simulated tracks, with 80% chance of detection. Enter these parameters and calculate the requested probability $P(x = 15)$: **Calc>Probability Distributions> Binomial: Probability, Number of Trials:** *20*, **Probability of Success:** *.8*, **Input Constant:** *15*, and we get:

Probability Density Function

```
Binomial with n = 20 and p = 0.8

  x   P( X = x )
 15    0.174560
```

(d) **Calc>Probability Distributions>Binomial: Cumulative probability, Number of Trials:** *20*, **Probability of Success:** *.8*, **Input Constant:** *14*   gives

Cumulative Distribution Function

```
Binomial with n = 20 and p = 0.8

  x   P( X <= x )
 14     0.195792
```

Hence, by the complementary rule, 1 - .196 gives the requested probability of at least 15 detections.

Exercise 4.70 - M/S page 209

Instead of finding the cumulative probability up to a specified x_0, here we need to find the value of x_0, which possesses a certain cumulative probability. Hence, rather than the Cumulative Probability function, we use the *Inverse* Cumulative Probability function. The desired cumulative probability is .95 (5% passing means 95% failing), so input this constant using **Calc>Probability Distributions>Binomial: Inverse cumulative probability, Number of Trials:** *20*, **Probability of Success:** *.5*, **Input Constant:** *.95*

Inverse Cumulative Distribution Function

```
Binomial with n = 20 and p = 0.5

  x   P( X <= x )       x   P( X <= x )
 13     0.942341       14     0.979305
```

Minitab shows the outcomes that come close to our requirement. With 15 as the passing grade, 98% will fail (obtain 14 or less) and 2% will pass. But if 14 is the passing grade, 6% (1 - .9423) will pass - too many!

Poisson Distribution

This distribution tends to arise, when we are counting the number of events of a certain type occurring over a region of time or space, where events occur independently of each other, and at a constant rate or density.

Example 4.13 - M/S page 211

To answer questions (b) – (d), using just cumulative probabilities, we will need the cumulative probabilities for x = 1, 5, 4, so enter these three numbers into C1, and then: **Calc>Probability Distributions>Poisson** with the following entries:

Hit **OK**, and we get

Cumulative Distribution Function

```
Poisson with mean = 2.6

x   P( X <= x )
1      0.267385
5      0.950963
4      0.877423
```

So, now you can easily answer (b) P(x<2) = .2674, (c) P(x>5)= 1 – P(x≤5) = 1 - .9510, (d) P(x=5) = (x≤5) – P(x≤4) = .9510 - .8774

Alternatively, for part (d), we could just select **Probability** rather than **Cumulative probability** in the **Poisson Distribution** dialog box, and input the constant '5'.

Example 4.137 - M/S page 224

Let x = number of admitted patients. We need $P(x \geq 5) = 1 - P(x \leq 4)$. Use **Calc>Probability Distributions>Poisson: Cumulative probability, Mean:** *2.5*, **Input Constant:** *4*

```
Cumulative Distribution Function

Poisson with mean = 2.5

x   P( X <= x )
4      0.891178
```

So, 1 - .8912 is the probability of at least 5 arrivals, hence too few beds.

Hypergeometric distribution

This distribution arises when we count the number of objects of type 'S' (Success) in a random sample of size n from a collection of N objects containing r elements of type S.

Example 4.14 - M/S page 217

(b) There are 10 applicants, 6 male, 4 female. We select at random 3 of the applicants. Find the probability that no females are selected. Select **Calc>Probability Distributions>Hypergeometric**:

Note: M replaces the textbook symbol r

We get:

```
Hypergeometric with N = 10, M = 4, and n = 3

x   P( X = x )
0      0.166667
```

Chapter 5 – Continuous Random Variables

In this chapter, we will do calculations using the normal and exponential distributions, using
Calc>Probability Distributions>Normal
Calc>Probability Distributions>Exponential

Plot and compare normal curves, with
Graph>Scatterplot: Multiple Graphs

Construct a normal probability plot, using
Graph>Probability Plot

Draw a histogram with a normal curve superimposed, using either
Stat>Basic Statistics>Display Descriptive Statistics: Graphs: Histogram with normal curve or **Graph>Histogram - With Fit**

The Normal Distribution

Use **Calc>Probability Distributions>Normal** to find normal curve areas and percentiles, or even the height of the normal curve at any point.

Figure 5.6: Drawing normal curves - M/S page 234

Let's draw two of the three normal curves shown, i.e. Normal(0,1.5) and Normal(3,1) Enter –5 to +6 in steps of 0.1 in C1, via **Calc>Make Patterned Data: Simple Set of Numbers**. Then use **Calc>Probability Distributions>Normal**, as indicated below.

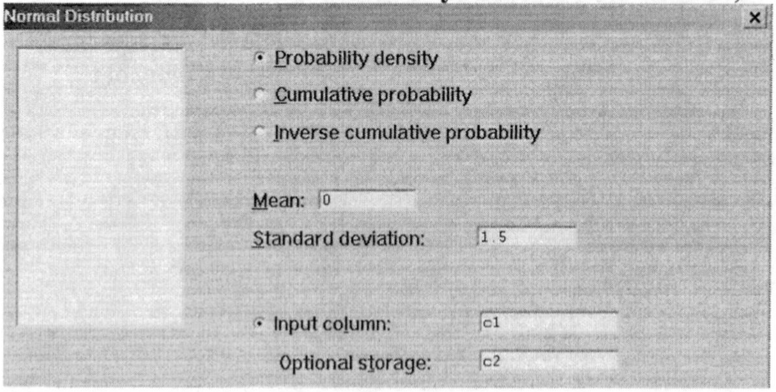

Repeat, changing the mean and standard deviation to 3 and 1, and the storage column to c3.

Then, **Graph>Scatterplot-Simple,** with **Multiple Graphs** as below:

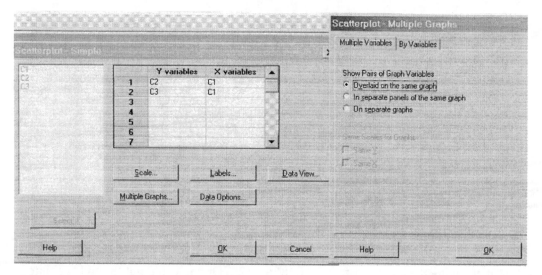

We get:

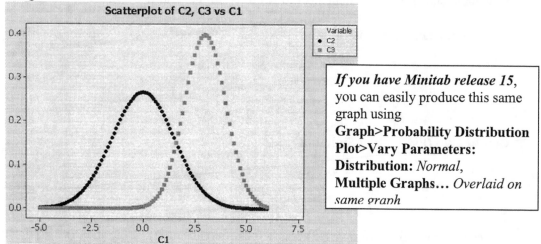

If you have Minitab release 15, you can easily produce this same graph using **Graph>Probability Distribution Plot>Vary Parameters: Distribution:** *Normal,* **Multiple Graphs...** *Overlaid on same graph*

Note the difference in mean and spread, and also how the height is greater for the less dispersed curve, in order to ensure the same total area under each curve.

Example 5.6 - M/S page 238

The normal curve table in the text gives areas under a *standard* normal curve only, so when using that table, you need to standardize your values first. But when using Minitab for normal curve calculations, there is *no need to standardize*. Just specify the correct mean and standard deviation in the dialog box. Here, $\mu = 10$ and $\sigma = 1.5$. We want to find $P(8 \le x \le 12)$. To find the area under the normal curve to the left of 8, we must select **Cumulative Probability** from the choices listed at the top of the dialog box for **Calc>Probability Distributions>Normal**. First input the constant '8':

Repeat changing the input constant from '8' to '12'. Or you could put '8' and '12' into one column and enter that column next to **Input column.** You will see:

Cumulative Distribution Function

```
Normal with mean = 10 and standard deviation = 1.5

x   P( X <= x )
8     0.0912112
```

Cumulative Distribution Function

```
Normal with mean = 10 and standard deviation = 1.5

 x   P( X <= x )
12      0.908789
```

The desired probability $P(8 \leq x \leq 12)$ equals the difference of $P(x \leq 12)$ and $P(x \leq 8)$ or .9088 - .0912 = .8176. In general, $P(a \leq x \leq b) = CumProb(b) - CumProb(a)$, for a continuous distribution.

Example 5.10 - M/S page 244

Here, the problem is reversed from the example above. Given an area, we want to find the corresponding value of x_0. We want to find the 90th percentile, i.e. a value x_0 such that $P(x \leq x_0) = .90$. Instead of using the Cumulative Probability function, we use the **Inverse cumulative probability** selection from **Calc>Probability Distributions > Normal** which gives us

Inverse Cumulative Distribution Function

```
Normal with mean = 550 and standard deviation = 100

P( X <= x )        x
      0.9    678.155
```

The answer is 678. This was simpler than the calculation in the text, because we did not have to start with a standard normal and then convert to the actual test score distribution.

Sometimes, the information given is not a left-side area. Then, you have to determine what is the area to the left of the unknown x_0 since this is the input constant required by Minitab. See exercise 5.47 below, e.g.

Figure SIA5.2 - M/S page 245

In this example, various cumulative probabilities have to be found for a $N(\mu=5, \sigma)$ distribution with varying choices for σ. For say $\sigma = 2$, enter the desired x-values in C1 (named 'x'), then use **Calc>Probability Distributions>Normal** as below, including a **storage** column. **OK**, and the probabilities in C2 below will appear.

Then e.g. $P(9 \leq x \leq 11) = .9987 - .9773$.   Repeat for the other two values of σ.

Exercise 5.117 - M/S page 270

To find the three probabilities requested, we will need areas corresponding to 45, 55, 51, 52, so enter these numbers into C1. Then, **Calc>Probability Distributions>Normal: Cumulative Probability, Mean:** *50*, **Standard Deviation:** *3.2*, **Input Column:** *C1*

Cumulative Distribution Function

```
Normal with mean = 50 and standard deviation = 3.2

  x   P( X <= x )
 45      0.059085
 55      0.940915
 51      0.622670
 52      0.734014
```

(a) $P(x > 45) = 1 - .0591$
(b) $P(x < 55) = .9409$
(c) $P(51 < x < 52) = .7340 - .6227$

Exercise 5.47 - M/S page 249

We want to find the number d such that 30% of the tree diameters exceed d. This implies that 70% have diameters less than d, so we input *0.70* into **Calc>Probability Distributions>Normal: Inverse Cumulative Probability, Mean:** *50*, **Standard Deviation:** *12*, **Input Constant:** *.0.70*

Inverse Cumulative Distribution Function

```
Normal with mean = 50 and standard deviation = 12

P( X <= x )          x
        0.7    56.2928
```

So only 30% of trees will have diameters in excess of 56.3 cm.

Assessing Normality: The Normal Probability Plot

A common diagnostic tool for assessing normality of data is the *normal probability plot*. We plot the data versus the average values that would arise if sampling from a standard normal. The latter are called the 'normal scores' by Minitab and may be calculated using **Calc>Calculator** and choosing **Normal scores** from the **Functions** list. However, **Graph>Probability Plot** provides a quicker one-step approach.

Example 5.11 - M/S page 250

Open up the EPAGAS file, containing the EPA mileages first encountered in table 2.3 in chapter 2. We want to assess whether the EPA mileages may be roughly normal in distribution. To construct a normal probability plot, select: **Graph>Probability Plot-**

Single. Under **Graph variables,** enter *MPG*, then select **Distribution** and choose **Distribution:** *Normal (default choice),* as shown below:

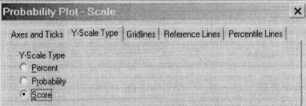

Also click on **Scale** in the dialog box above, then **Y-Scale Type**, and choose **Score** (the expected normal score), instead of the default **Percent** selection, if you want to match the graph in your textbook (this difference in labeling is completely inconsequential):

OK, OK, and you get:

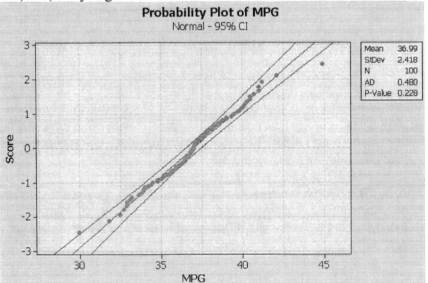

Since the points in our normal probability plot above fall reasonably close to a straight line, we may say that the distribution of the mileages appears to be close to normal.

Note some extra features in the plot above. A straight line, and two (95%) confidence interval bands have been drawn in for some extra guidance re linearity of the plot. If the points lie within the bands, the normal distribution is a close fit to the data. While the mild S-shape here suggests a small departure from normality, the plot as a whole is quite close to being straight, with only one point outside the confidence bands. The distribution is nearly normal.

Using the **Probability Plot: Distribution: Data Display** sub-dialog box (default settings shown below), you can remove from the plot, if you wish, the confidence interval bands, by un-clicking in the box: **Show confidence interval**. To remove the straight line (i.e. the **Distribution fit**), change to: **Symbols only,** in the box.

Alternatively, once you have produced the graph, you can make the type of changes discussed above by right-clicking or double-clicking on relevant parts of the graph, and using the pop-up dialog boxes.

For some alternative assessments of normality, you can use **Stat>Basic Statistics> Display Descriptive Statistics**: **Variables:** MPG, which produces:

Descriptive Statistics: MPG

Variable	N	N*	Mean	SE Mean	StDev	Minimum	Q1	Median	Q3
MPG	100	0	36.994	0.242	2.418	30.000	35.625	37.000	38.375

Variable	Maximum
MPG	44.900

Now, you can compute the IQR/s ratio, and compare with 1.3. Or you can calculate the 1, 2, 3 standard deviation intervals around the mean, for comparison with 68%, 95%, 100%, respectively. [**Data>Sort** or **Stat>Tables>Tally: Display: Cumulative percents** will put the data into ascending order for easier counting]

When asking to **Display Descriptive Statistics** for *MPH*, above, you can ask at the same time for a histogram with superimposed normal curve (like Fig 5.18a) by clicking on **Graphs…** in the dialog box and then choosing **Histogram with normal curve**. We get the following graph (along with the descriptive stats above):

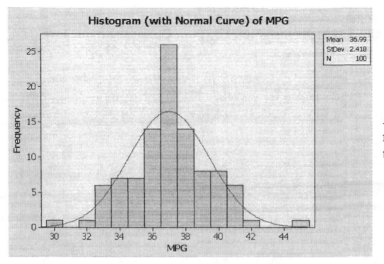

A little peaked in the middle, but quite close to a normal shape.

You can also produce this plot using **Graph>Histogram - With Fit: Graph variables:** *MPG* (as illustrated in the next exercise).

Exercise 5.62 - M/S page 256

Open up the DARTS file on your text CD. We have ***grouped data*** here, with (rounded) distance from the target in c1, and corresponding frequencies in c2, so in any dialog box, you will have to tell Minitab not just which column the distances are in, but also where to find these ***frequencies*** – If you omit this information, Minitab will assume that each listed distance is observed only once!

We want to assess whether the distances from the target may be normally distributed. For a normal probability plot, use:

Graph>Probability Plot-Single: Variables: *C1*,
   Distribution: Distribution: *Normal,*
   Data Options: Frequency: Frequency variable: *C2,*
   Scale: Y-Scale Type: *Score*

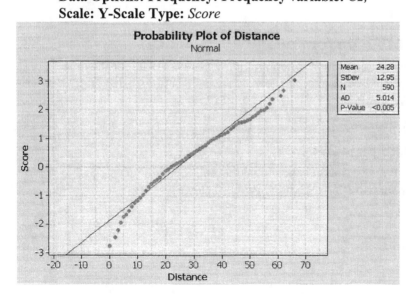

There is some clear curvature - how do you interpret this?

Let's also examine the histogram via:
Graph>Histogram - With Fit: Graph variables: *C1,* **Data Options: Frequency: Frequency variable:** *C2.*

With **Histogram - With Fit**, there is a default setting of: **Data View: Distribution: Fit Distribution:** *Normal.* If instead you had selected the **Histogram - Simple**, you would then have to select and click your way through **Data View**, to have the *Normal* fit added to the histogram.

We get:

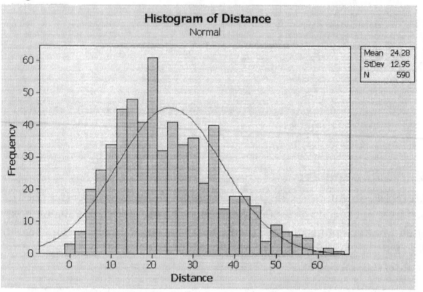

Above, we could not use **Stat>Basic Statistics>Display Descriptive Statistics: Graphs: Histogram with normal curve**, since it makes no allowance for a **frequency** column.

Exponential Distribution (optional)

This is another continuous distribution, often providing a good model for the time between occurrences of random events. It is one of the distributions listed under **Calc>Probability Distributions**.

Example 5.13 - M/S page 264

To find $P(x > 5) = 1 - P(x < 5)$, for an exponential distribution with mean equal to 2, we use **Calc>Probability Distributions>Exponential** as below:

Note that you should plug in the desired mean of *2* next to **Scale**, and keep the **Threshold** at its default value of *0.0* . Hit **OK**, and we get

Cumulative Distribution Function

```
Exponential with mean = 2

x  P( X <= x )
5     0.917915
```

So, 1 - .918 is our answer.

Exercise 5.100 - M/S page 266

The time to goal, x, is exponentially distributed with a mean of 9.15. We want P(x < 3) in (a) and P(x > 20) in (b), so enter '3' and '20' in C1, and then **Calc>Probability Distributions>Exponential: Cumulative Probability, Scale (= Mean):** *9.15*, **Input Column:** *C1*

Cumulative Distribution Function

```
Exponential with mean = 9.15

 x      P( X <= x )
 3       0.279543
20       0.887611
```

Hence, P(x < 3) = .2795, and P(x > 20) = 1 − P(x < 20) = 1 - .8876.

Chapter 6 – Sampling Distributions

In this chapter, we will simulate random samples, using
Calc>Random Data>Uniform
Calc>Random Data>Integer

Calculate statistics for random samples, stored in rows, using
Calc>Row Statistics

And calculate probabilities for a sample mean, using
Calc>Probability Distributions>Normal

[Also, you can learn much about sampling distributions by playing with the Java applet at
http://www.ruf.rice.edu/~lane/stat_sim/sampling_dist/index.html]

Example 6.2 - M/S page 282

We want to approximate the sampling distribution of the sample mean and sample
median for samples of size 11 from a uniform population. We will generate 700 samples
of 11 observations each (rather than 1000 samples, due to Student Minitab worksheet size
limitations*). For ease of computation, we put one sample of 11 observations in each
row, rather than using the columns for the samples. Generate the random data using
Calc>Random Data>Uniform

Uniform Distribution	
Generate 700 rows of data	
Store in column[s]:	
c1–c11	
Lower endpoint: 150	
Upper endpoint: 200	
Help	OK Cancel

*Note: The maximal
worksheet capacity with
the student version of
Minitab is only 10,000
cells, so if we generated
1000 rows × 11 columns,
we would run over the
capacity, and be forced to
delete some cells in the
worksheet. And we will
also need additional
worksheet space for some
upcoming calculations.
But if you are using the
full professional version of Minitab, you will have much greater storage capacity.

You will see entries ranging from 150 to 200, in each of c1-c11. Plot the data in any of
the columns (**Graph>Histogram**), to confirm that the data is indeed uniformly
distributed between 150 and 200, like in Fig 6.4 in the text. For each row, calculate the
sample mean, and store it in a column named 'Mean' (c12), using **Calc>Row Statistics**

Row Statistics

| C1 |
| C2 |
| C3 |
| C4 |
| C5 |
| C6 |
| C7 |
| C8 |
| C9 |
| C10 |
| C11 |

Statistic

- ◦ Sum
- ● Mean
- ◦ Standard deviation
- ◦ Minimum
- ◦ Maximum
- ◦ Range

- ◦ Median
- ◦ Sum of squares
- ◦ N total
- ◦ N nonmissing
- ◦ N missing

Input variables:

C1–C11

Select

Store result in: Mean

Help   OK   Cancel

Repeat, changing the **Statistic** from **Mean** to **Median**, and store the results in a column named 'Median', which will be c13. Now, the worksheet looks like:

↓	C1	C2	C3	C4	C5	C6	C7	C8	C9	C10	C11	C12 Mean	C13 Median
1	175.180	154.637	173.427	169.181	151.506	197.434	190.747	186.361	162.034	198.649	178.748	176.173	175.180
2	171.837	188.318	153.214	177.713	181.342	171.862	166.958	155.728	167.868	190.775	155.993	171.055	171.837
3	195.832	194.057	151.200	184.486	161.999	159.282	193.147	180.937	160.723	156.329	193.082	175.552	180.937
4	151.249	174.310	176.326	177.713	161.462	198.128	181.274	161.588	155.862	168.667	161.356	169.812	168.667
5	152.919	194.543	179.285	157.862	157.330	189.174	171.665	189.100	186.927	192.123	150.065	174.636	179.285

Check that the mean and median of the 11 numbers in row 1 are indeed 176.173 and 175.180, as displayed above.

Since we have selected a sample of size 11 many times, and recorded the mean of each sample in c12, a histogram for c12 will show us, approximately, the sampling distribution of the sample mean (for a sample of size 11 from a uniform population). Similarly, a histogram of c13 approximates the sampling distribution of the sample median. Let's examine these two distributions, side by side, with **Graph>Histogram - Simple: Graph variables:** *Mean Median,* **Multiple Graphs: Multiple Variables:** *In separate panels, Same Y, Same X*

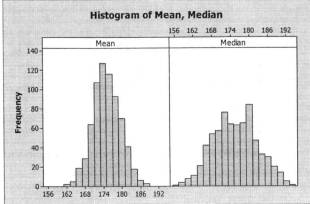

If you prefer to display percents on the y-axis (good idea if unequal sample sizes), just double-click on the y-axis, and use the pop-up **Edit Scale>Type**

Both sampling distributions are mound-shaped (unlike the population), appear to center at 175 (the population center), and have less dispersion than the population. The sample means are also less dispersed than the sample medians, suggesting that the mean would be the superior estimator of the population center (if unknown, as is usually the case) for this type of population.

Exercise 6.9 - M/S page 284

We want to approximate the sampling distribution of the sample mean and also the sample variance for samples of size 25 from a discrete uniform population consisting of the integers 00 - 99. To accommodate space constraints in the student version of Minitab, generate 300 samples (rather than the requested 500).

　Calc>Random Data>Integers: *300* **rows of data, store in** *C1-C25*, **Min:** *0*, **Max:** *99*, gives you 300 rows of random data with 25 numbers per row.

　For each row (= sample), compute the mean and store in c26 ('mean'), via **Calc>Row Statistics: Statistic: Mean, Input Variables:** *C1-C25*, **Store result in:** *mean*. In c26, the row means will appear, as below.

　Repeat your **Calc>Row Statistics**, but now click on **Standard deviation,** and store the results in c27. To convert the numbers in c27 into variances, you need to square them, so use: **Calc>Calculator: Store result in:** *C27*, **Expression:** *C27 * C27*. Give c27 a name.

↓	C14	C15	C16	C17	C18	C19	C20	C21	C22	C23	C24	C25	C26	C27
													mean	variance
1	15	74	30	86	92	30	25	7	12	83	6	29	53.76	912.84
2	22	25	99	0	40	99	0	18	42	52	7	41	45.92	1048.44
3	47	42	67	45	62	52	79	79	65	69	91	86	54.60	782.42

Produce the histograms with **Graph>Histogram - Simple: Graph Variables:** *C26, C27*

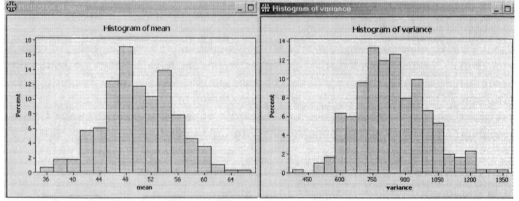

Both look roughly bell-shaped, though the variances exhibit slight positive skewness. If you use the professional version of Minitab with 500 or more samples (rows), the shapes will be smoother and closer to the 'true' limiting shape.

Example 6.7 - M/S page 291

(b) We want the probability that the sample mean is greater than 82, if we draw a random sample of size 36 from a population with mean and standard deviation of 80 and 6 respectively. By the Central Limit Theorem, we know that the distribution of the sample mean can be approximated by a normal curve, where μ for the curve is 80, and σ for the curve is 6/√36 or 1. There is no need to standardize, just feed in these parameters to Minitab:

Calc>Probability Distributions>Normal

We get:

Cumulative Distribution Function

```
Normal with mean = 80 and standard deviation = 1

  x      P( X <= x )
 82      0.977250
```

By the complementary rule, the answer is 1 - .9772 .

Exercise 6.60 - M/S page 299

The population mean and standard deviation are 5.1 and 6.1 resp. We will draw a random sample of 150 victims. By the Central Limit Theorem, the distribution of the sample mean will be normal, with mean = 5.1, and standard deviation = 6.1/√150 = 0.50.

For (c), we want the probability that the sample mean exceeds 5.5, and for (d), the probability that it lies between 4 and 5. So, we need areas corresponding to 5.5, 4, and 5. Enter these three numbers into C1. Select **Calc>Probability Distributions>Normal: Cumulative Probability, Mean:** *5.1*, **Standard Deviation:** *.5*, **Input Column:** *C1*

Cumulative Distribution Function

```
Normal with mean = 5.1 and standard deviation = 0.5

   x    P( X <= x )
 5.5     0.788145
 4.0     0.013903
 5.0     0.420740
```

Hence, the answers are: (c) 1 - .788, (d) .421 - .014

Chapter 7 - Confidence Intervals from Single Samples

In this chapter, we will draw a *t*-curve, using
Calc>Probability Distributions> t

Find a confidence interval for a population mean, using
Stat>Basic Statistics>1-Sample t
Stat>Basic Statistics>1-Sample Z   [if σ known]

Find a confidence interval for a population proportion, using
Stat>Basic Statistics>1 Proportion

Confidence Interval for a Population Mean

If σ is known (very unusual), use **Stat>Basic Statistics>1-Sample Z**, to calculate a
confidence interval for the mean (you will be prompted for the value for σ). Otherwise,
you should use **Stat>Basic Statistics>1-Sample t**, for which *s* will be calculated from
the sample and used in the C.I.

For a large sample, the text uses a 'z' value in the C.I., whereas Minitab uses the more
accurate 't' value, which differs little from the 'z' value in this case. [If you *really* want
to precisely imitate the text, for large samples, you would first find *s* for the sample, then
plug in this number for σ in **1-Sample Z**, but this would truly be a waste of time!]

Example 7.1 - M/S page 310

Open up the AIRNOSHOWS data file. To find a 90% confidence interval for the
population mean, select **Stat>Basic Statistics>1-Sample t,** and tell Minitab where to find
the **Sample**, then click on **Options**, and change the default **Confidence level** from *95* to
90 percent, as below:

OK, OK, and we get some results:

One-Sample T: NOSHOWS

```
Variable    N      Mean    StDev   SE Mean         90% CI
NOSHOWS   225   11.5956   4.1026    0.2735   (11.1438, 12.0473)
```

Note that if we had the sample ***summary stats***, rather than raw data, we would just click on **Summarized data** above, and enter the sample size, mean, and standard deviation.

Exercise 7.22 - M/S page 315

Open up the data file ATTIMES, where the attention times may be found in c1. To find a 90% confidence interval for the mean time, select **Stat>Basic Statistics>1-Sample t,** and fill in: **Sample in column:** *C1* and set **Options: Confidence level:** *90* . **OK, OK,** and we get:

One-Sample T: Time

```
Variable    N     Mean    StDev   SE Mean         90% CI
Time       50   20.8480  13.4138   1.8970   (17.6676, 24.0284)
```

The 90% C.I. appears above. If you calculate the C.I. from the mean and stdev above, using a *z* value of 1.645, your result will differ slightly from the interval above, because Minitab uses the more accurate *t* value of 1.677 (see M/S section 7.3).

Figure 7.7: Compare *t* and *z* curves - M/S page 317

Let's draw this comparison of the standard normal curve and *t* curve (with 4 df), using Minitab. Enter -3 to +3 in steps of .1 in C1, via **Calc>Make Patterned Data>Simple Set of Numbers**. Then **Calc>Probability Distributions> t**

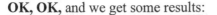

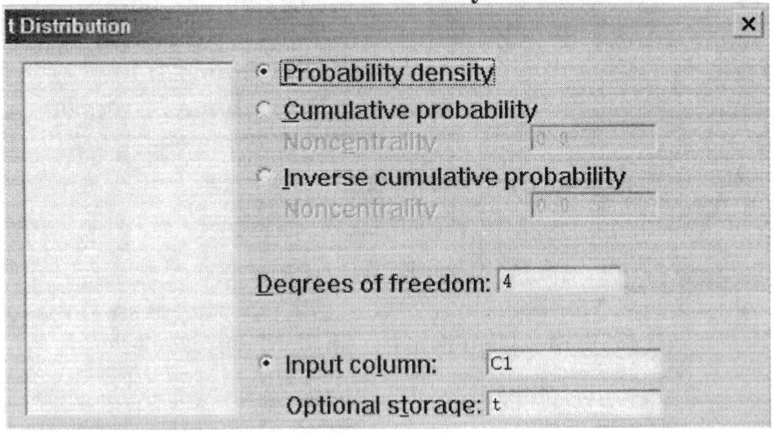

```
t Distribution                                    [x]

            (•) Probability density
            ( ) Cumulative probability
                Noncentrality          0 0

            ( ) Inverse cumulative probability
                Noncentrality          0 0


            Degrees of freedom: 4


            (•) Input column:       C1
                Optional storage: t
```

Above, we have calculated the probability density curve values for the numbers in C1, and stored them in a column named 't'.

   Repeat this, but switch to **Normal** distribution, with **Mean:** *0*, **Standard Deviation:** *1*, and **Input column:** *C1*, **Optional storage:** *z*

Now plot 't' and 'z' vs. the C1 values, using **Graph>Scatterplot – With Groups,** as below; and in the dialog box, also select **Multiple Graphs: Multiple Variables:** *Overlaid on the same graph*

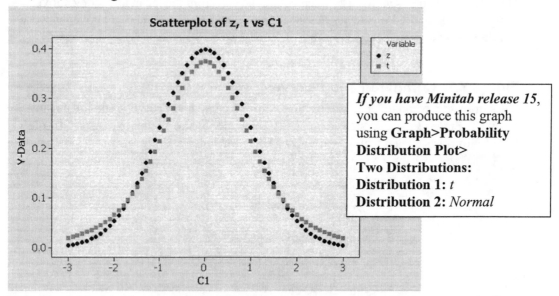

OK, OK, and we get

> *If you have Minitab release 15,* you can produce this graph using **Graph>Probability Distribution Plot> Two Distributions: Distribution 1:** *t* **Distribution 2:** *Normal*

Different symbols and colors are used for each curve. We see that the shapes are similar, but the *t* curve is more dispersed (with fatter tails) than the standard normal curve.

Example 7.2 - M/S page 320

Open up the PRINTHEAD data file, showing the number of characters in c1. To produce a 99% confidence interval for the mean number of characters printed before failure, use **Stat>Basic Statistics>1-Sample t** as shown below, and don't forget to set the desired

Confidence level with the **Options**. To help check the distributional assumptions, also select **Graphs...** in the dialog box, and tick off some of the choices.

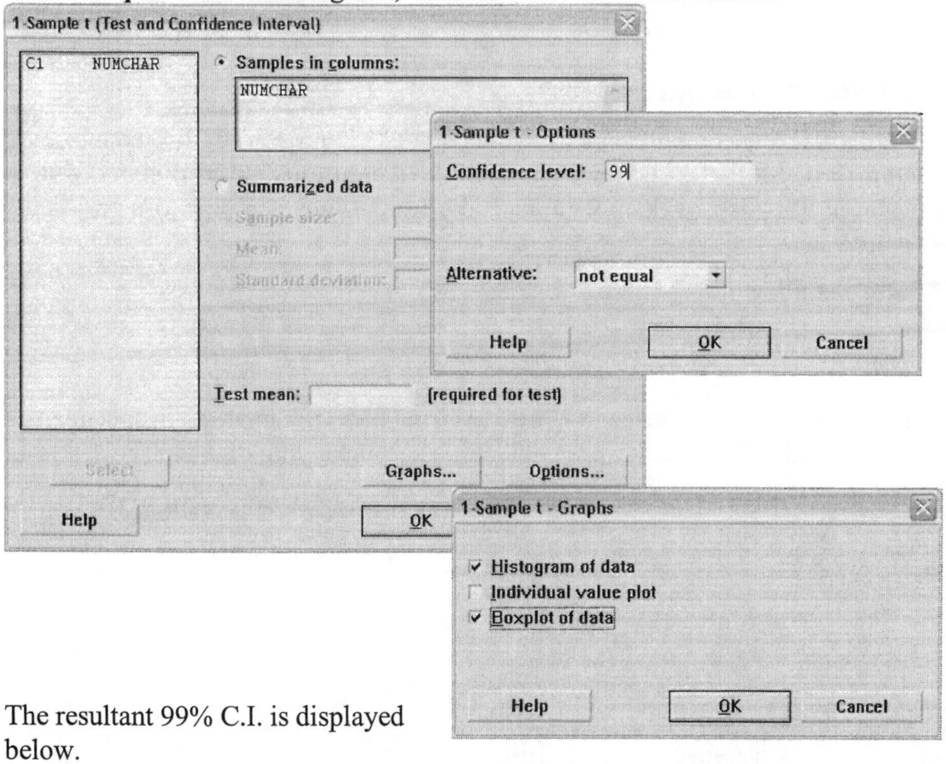

The resultant 99% C.I. is displayed below.

One-Sample T: NUMCHAR

```
Variable    N     Mean    StDev   SE Mean        99% CI
NUMCHAR    15  1.23867  0.19316  0.04987  (1.09020, 1.38714)
```

Note that 'SE Mean' in the output stands for *standard error of the mean*, and equals $s/\sqrt{n}$, which is $0.1932/\sqrt{15}$ here.

To assess our underlying assumption of normality, have a look at the graphs displayed:

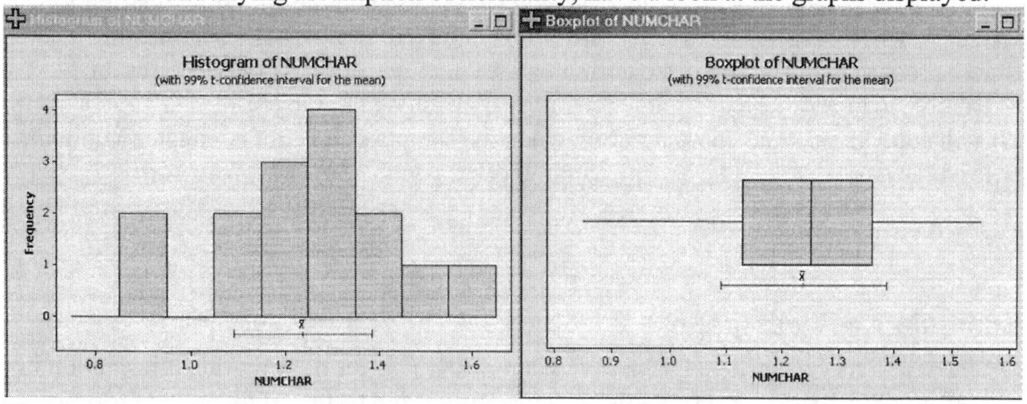

The data appear nearly normal, so our 99% level of confidence may be trusted. The C.I. is also displayed in each of these graphs, directly below the plot. You can also produce a stem and leaf plot, like in the text, using **Graph>Stem-and-Leaf**, or a normal probability plot, using **Graph>Probability Plot**, as discussed in earlier chapters.

Exercise 7.41 - M/S page 325

Open up BRAINPMI, containing the measurements in c1. For a 95% CI for the mean PMI, use **Stat>Basic Statistics>1-Sample t: Sample in column:** *C1*, **Options: Confidence level:** *95*. Also, click on **Graphs** and tick off whichever ones you prefer. **OK**, **OK**, gives you

One-Sample T: PMI

```
Variable    N      Mean    StDev   SE Mean        95% CI
PMI        22   7.30000  3.18493  0.67903   (5.88788, 8.71212)
```

So, we can say, with 95% confidence that the true mean PMI of human brain specimens lies between 5.89 and 8.71. Or can we? Look at the accompanying data plots:

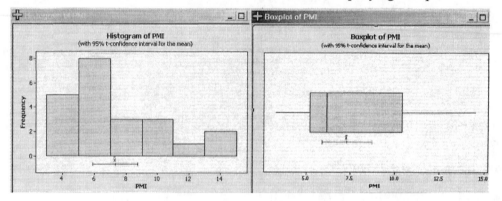

Confidence Interval for a Proportion

The text emphasizes use of the *large sample C.I.* for a proportion, which uses the normal approximation. Minitab will produce an *exact* C.I., based on the Binomial distribution *unless* you select **Options: Use test and interval based on normal distribution**. The exact approach may be used if the normal approximation is not applicable or as an alternative to the 'Adjusted Confidence Interval' discussed in the text.

 If you apply the normal approximation when the sample size is not adequate (Minitab's criterion is more lax than the text's), Minitab gives you the *following warning*:

```
* NOTE * The normal approximation may be inaccurate for small samples.
```

Example 7.4 - M/S page 328

We have 257 successes in 484 trials. We want a 90% C.I. for p. Enter this information in **Stat>Basic Statistics>1 Proportion,** under **Summarized data**. Click on the **Options...** and enter **Confidence Level:** *90*. Tick **Use test and interval based on normal**

distribution, as shown below. The items **Test Proportion** and **Alternative** under **Options** relate to *hypothesis testing (chapter 8)* - just ignore these items for now.

OK, OK, and you get:

Test and CI for One Proportion

```
Test of p = 0.5 vs p not = 0.5

Sample    X    N   Sample p        90% CI         Z-Value  P-Value
1        257  484  0.530992  (0.493681, 0.568303)   1.36    0.173
```

So, with 90% confidence, the proportion of all Florida consumers who are confident about the economy lies between .494 and .568. [Ignore the 'Z-value' and 'P-value' items, which are useful only for hypothesis testing]

Exercise 7.102 - M/S page 344

Open up the OILSPILL file again (see exercise 2.186) showing causes of 50 oil spills in c2. Use **Stat>Tables>Tally Individual Variables: Variables:** *C2,* to find the count for hull failures. We see that 12 of 50 were from hull failure. Plug these numbers into **Stat>Basic Statistics>1 Proportion: Summarized Data:** *50* **trials,** *12* **successes,** with **Options: Confidence level:** *95,* **Use test and interval based on normal distribution**

Test and Confidence Interval for One Proportion

```
Test of p = 0.5 vs p not = 0.5

Sample    X    N   Sample p        95.0 % CI       Z-Value  P-Value
1        12   50  0.240000  (0.121621, 0.358379)   -3.68    0.000
```

We find that the proportion of oil spills caused by hull failure is between .12 and .36, with 95% confidence.

Exercise 7.105 - M/S page 344

(a) Open up SUICIDE, showing data on 37 suicides committed by inmates of correctional facilities. We want to estimate the proportion of suicides that were committed by inmates charged with murder/manslaughter. Y/N in C5 indicates whether or not the inmate was charged with murder/manslaughter. This is *raw categorical data*. While we could do a tally for C5, and then use the summary counts, let's proceed instead by selecting **Stat>Basic Statistics>1 Proportion,** and plug into **Samples in Columns**: *C5*. Make sure **Options** show a *95%* confidence level, and use of the **normal distribution**. We get:

```
Test and CI for One Proportion: Murder?

Test of p = 0.5 vs p not = 0.5

Event = Y

Variable    X    N   Sample p           95% CI          Z-Value   P-Value
Murder?    14   37   0.378378   (0.222109, 0.534648)    -1.48     0.139
```

Minitab decided that the **Event** of interest, or *Success* is **'Y'**, and counted these up, reporting a total of X = 14 successes. Why was 'Y' and not 'N' deemed a success? By default, the entry which is last in alphabetical order, is deemed the 'success' (like S follows F, or 1 follows 0).

It is easy to change the way Minitab orders the categories in a column. The default is alphabetical, i.e. 'N, Y'. Suppose we wanted to change the ordering to 'Y, N'. Right-click anywhere on C5 in the worksheet, then choose **Column>Value Order.** Or at the top, in the main menu, choose: **Editor>Column>Value Order.** Click on **User-specified order**. Type *Y*, press *Enter*, then type *N*. Hit **OK.** Now, if you run the same procedure, the C.I. will be for the proportion of N's rather than Y's, since N now follows Y in the ordering.

(b) We want a C.I. for the proportion of suicides at Night. The raw data on *time* of suicide are in C6, showing three different times of day: D (day), A (afternoon), N (night). But this procedure only works for *dichotomous* columns (two categories only). We are counting N's, so it is our 'success' event. Convert everything that is not a success into one single category - e.g. convert all the A's into D's, e.g. using **Data>Code>Text to Text**. 'N' comes last alphabetically, so Minitab deems it a 'Success', and there is no need to change the **Value Order** as explained in (a). Now, run **Stat>Basic Statistics>1 Proportion: Samples in columns:** *C6*, and we get:

```
Test and CI for One Proportion: Time

Test of p = 0.5 vs p not = 0.5

Event = N

Variable    X    N   Sample p           95% CI          Z-Value   P-Value
Time       26   37   0.702703   (0.555428, 0.849978)    2.47      0.014
```

From this, we see that the proportion of suicides committed at night is between .56 and .85, with 95% confidence.

(c) Use **Stat>Basic Statistics>1-Sample t** (as in example 7.2 above)
(d) Use **Stat>Basic Statistics>1 Proportion** (as in part a. above)

Chapter 8 – Test of Hypothesis for Single Samples

In this chapter, we will execute one sample tests for the mean of a population, using
Stat>Basic Statistics>1-Sample Z
Stat>Basic Statistics>1-Sample t

Test for the population proportion, using
Stat>Basic Statistics>1 Proportion

Determine power, or type II error probability, with
Stat>Power and Sample Size>1-sample Z

And execute a chi-square test for the population variance, using
Calc>Column Statistics
Calc>Calculator
Calc>Probability Distributions>Chi-Square

Testing for a hypothesized mean

To test a hypothesis about μ, we use the **1-Sample Z** test when σ is known (quite unlikely!), or the **1-Sample t** test when σ is unknown. When σ is unknown and the sample size n is large, the resulting statistic (using *s*) has a *t* distribution with large degrees of freedom. The text approximates this with the standard normal distribution. Minitab's *t* calculation is more accurate, but the difference is slight.

Example 8.4 - M/S page 368

Open up the HOSPLOS data file. We want to test for evidence that the true mean length of stay (LOS) is less than 5. Since we have an assumed value for σ equal to 3.68, we may use the z-test here: **Stat>Basic Statistics>1-Sample Z**

In the dialog box, we have entered the null **test** value for the **mean** and the assumed value of the **Standard deviation**, and used the **Options** to choose the appropriate one-sided **alternative**: *less than*. The output is:

One-Sample Z: LOS

```
Test of mu = 5 vs < 5
The assumed standard deviation = 3.68
                                               95%
                                             Upper
Variable      N      Mean    StDev   SE Mean  Bound       Z      P
LOS         100   4.53000  3.67755   0.36800  5.13531  -1.28  0.101
```

The value of the z-statistic is shown, but all we really need is the P-value. Since $p = .101$ is greater than $\alpha=.05$, we do not have sufficient evidence to reject the null hypothesis.

What is that '**95% Upper Bound**' appearing in the output? Minitab assumes that you want a ***one-sided CI***, since you specified a one-sided alternative in your test. In this example, the one-sided CI for the mean is ($-\infty$, 5.13531). If you also ask for some **Graphs…**, a one-sided CI will be displayed. *For the more customary two-sided CI, you will have to redo the procedure, and change the alternative to two-sided.* The same goes for future procedures that we will discuss.

Examples 8.5 & 8.6 - M/S page 374

Look for the EMISSIONS data file on your CD, or open a new worksheet (**File>New: Minitab Worksheet**), and type the data *15.6, 16.2, 22.5, 20.5, 16.4, 19.4, 19.6, 17.9, 12.7, 14.9* into C1, and name this column *E-level*. To test the hypothesis that the true mean is *less than 20*, we perform a t-test via: **Stat>Basic Statistics>1-Sample t**

which produces:

One-Sample T: E-level

```
Test of mu = 20 vs < 20
                                              95%
                                           Upper
Variable    N     Mean    StDev   SE Mean  Bound      T       P
E-level    10   17.5700   2.9522   0.9336  19.2814  -2.60  0.014
```

There is some evidence to reject the null hypothesis, since the p-value is small, but the evidence is not quite sufficient at the requested 1% level (since $p > \alpha = .01$). To check the accuracy of this p-value calculation, ***don't forget to examine the data***:
You can use **Graph>Stem-and-Leaf**, to produce:

Stem-and-Leaf Display: E-level

```
Stem-and-leaf of E-level   N  = 10
Leaf Unit = 1.0

  1    1   2
  3    1   45
 (3)   1   667
  4    1   99
  2    2   0
  1    2   2
```

No indication of non-normality, so we can trust the p-value reported. Or when you run the **One-Sample t**-test, just remember to click on **Graphs** in the dialog box, and then choose some of the three available, e.g.:

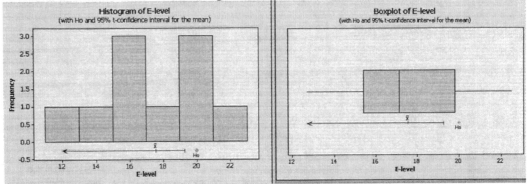

The arrow in each plot represents a *one-sided CI* (mentioned earlier), giving you a range of plausible values for the mean with 95% confidence (which also tells you that you can reject the null hypothesis at the 5% level, since the null value of *20* is outside of the interval).

Or, you might have a look at the normal probability plot with **Graph>Probability Plot-Single**:

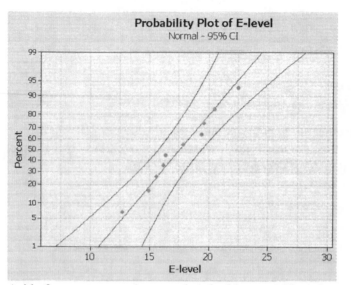

Aside from some erratic or random looking variation, the pattern is quite linear.

Exercise 8.70 - M/S page 379

Open up the GOBIANTS file. The relevant data on number of species is in C6. To test whether the true mean number of species differs from the 5, use **Stat>Basic Statistics>1-sample t: Samples in:** *C6*, **Test Mean:** *5*, **Options: Alternative:** *not equal*, and we get

One-Sample T: AntSpecies

```
Test of mu = 5 vs not = 5

Variable      N     Mean    StDev  SE Mean        95% CI         T      P
AntSpecies   11  12.8182  18.6752   5.6308  (0.2720, 25.3644)  1.39  0.195
```

There appears to be no evidence that the mean differs from 5, since *p* is large. But, is the p-value trustworthy?? Click the **Graphs...** option in the dialog box, and let's tick off all of them, producing:

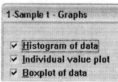

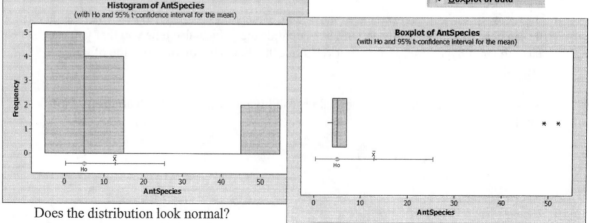

Does the distribution look normal?

The third plot available is called an **Individual Value Plot,** which sounds like it should just be a dotplot, eh? In fact it is, except for the introduction of *'jitter'* into the plot:

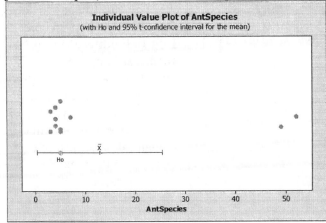

The actual data values correspond to the points projected down onto the horizontal axis, but they have been randomly perturbed or *jittered* in the vertical direction in this graph.

Why use *jitter*? Well, eliminate the jitter, and you get the dotplot below:

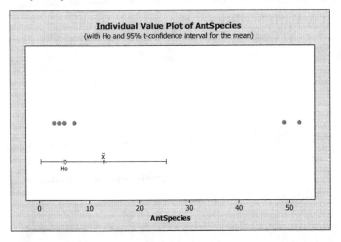

Now, we *cannot see the overlapped points* - potentially very misleading! You can reduce the magnitude of the jitter or reduce it to zero, by clicking on some points on the graph, and using the pop-up **Edit Individual Symbols** menu, shown at right. Note that 'X-direction' is actually the vertical direction in the graph above. I replaced *0.025* by *0.00* to get the last plot above.

Why randomly jiggle these points? Why not just pile them up neatly one on top of the other? *Well, indeed, Minitab release 15 added this option to the possibilities. In fact, the 'symmetrical offset' is the new default plot in release 15, and looks like:*

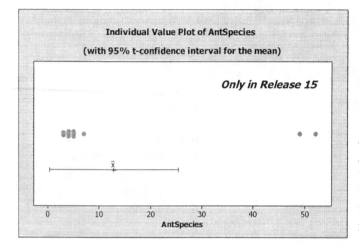

You can change this if you wish, by clicking on the dots, then using the pop-up box: **Edit Individual Symbols> Identical Points: Symbol Offset** Choose either *Offset symmetrically,* or *Jitter* or *None.* I prefer this release 15 version, over random jitter, though I'd rather just pile them up on the x-axis – maybe in release 16?

Testing for a hypothesized proportion

To test a hypothesis about p, use **Stat>Basic Statistics>1 Proportion**. To utilize the normal approximation (large n), select the **Option: Use test and interval based on normal distribution**. If n is not sufficiently large, de-select this option and Minitab will perform an *exact* test. Minitab will issue a warning if you use the normal approximation inappropriately.

Examples 8.7 & 8.8 - M/S page 381

We have 10 defectives in a sample of 300 batteries. Does this constitute strong evidence that in the entire shipment, we have less than 5% defective? Use **Stat>Basic Statistics>1 Proportion,** along with the **Options** provided, as below:

Note that here we have **Summarized data**, not raw data. We get:

Test and CI for One Proportion

```
Test of p = 0.05 vs p < 0.05
                                    95%
                                   Upper
    Sample    X     N   Sample p    Bound    Z-Value   P-Value
    1        10   300   0.033333  0.050380    -1.32     0.093
```

Since the P-value of .093 is high ($> \alpha = .01$), we do not have good support for the hypothesis that less than 5% of the entire lot are defective. Since no warning appears in the printout, the normal approximation should be adequate here. But, if you are not sure, de-select the **Option: Use test and interval based on normal approximation** and repeat.

Exercise 8.91 - M/S page 386

After taking a placebo, 70% of 7000 or 4900 patients reported improvement. If the placebo has no effect, then p = 0.5. Proceed with **Stat>Basic Stats>1 Proportion: Summarized data:** *7000* **trials,** *4900* **successes,** with **Options: Test proportion:** *0.5*, **Alternative:** *Not equal*, **Use test based on normal distribution**.

I have chosen a two-sided alternative, but if you wish to test only for evidence of beneficial effect (rather than for evidence of any type of effect), select **Alternative: Greater than**. It won't change the clear conclusion below:

Test and CI for One Proportion

```
Test of p = 0.5 vs p not = 0.5

    Sample      X      N   Sample p         95% CI          Z-Value  P-Value
    1        4900   7000  0.700000  (0.689265, 0.710735)     33.47    0.000
```

Since P is 0.000, such a large outcome cannot be due to chance. We have strong evidence of a (beneficial) placebo effect.

Type II Error Probability (β) and Power of a Test

We usually need to know how well the test performs when the null is not true, i.e. how often we make the right decision [the power] or the wrong decision [β = probability of a type II error]. Minitab's **Stat>Power and Sample Size** helps you calculate power (or β = 1 - power) for a variety of tests. It also helps you find the sample size required in order to achieve some specified power, at alternatives deemed to be of practical importance.

Example 8.9 - M/S page 389

Here, we are testing $H_0: \mu = 1.2$ versus $H_0: \mu \neq 1.2$, with $\alpha = .05$ (and repeat for $\alpha = .01$), where σ is assumed to equal .5. We want the power of the test (and $\beta = 1 -$ power) when $\mu = 1.1$. Enter this exact information, using **Stat>Power and Sample Size>1-sample Z:**

Power and Sample Size for 1-Sample Z	Power and Sample Size for 1-Sample Z - Options

Specify values for any two of the following:

Sample sizes: `100`

Differences: `-0.1`

Power values: ` `

Standard deviation: `0.5`   Options...

Help   OK   Cancel

Alternative Hypothesis
- Less than
- Not equal
- Greater than

Significance level: `0.05`

Store sample sizes in: ` `

Store differences in: ` `

Store power values in: ` `

Select

Help   OK   Cancel

Note that $\mu - \mu_0 = 1.1 - 1.2 = -0.1$, so we entered this under **Differences** above. We get

Power and Sample Size

```
1-Sample Z Test

Testing mean = null (versus not = null)
Calculating power for mean = null + difference
Alpha = 0.05  Assumed standard deviation = 0.5

              Sample
Difference     Size      Power
      -0.1      100   0.516005
```

So, the power is .516 and hence the type II error probability, β, is $1 - .516 = .484$.
To find the power and β, if α is set at .01, repeat the above, but change the value of the
Significance level in the **Options** box, to *.01*.

Exercise 8.100 - M/S page 393

We are testing H_0: $\mu=16$ vs H_a: $\mu<16$ with $\alpha=0.1$. We want the power of this test at
$\mu=15.5, 15, 14.5, 14, 13.5$, assuming $\sigma = 9.32$. Use **Stat>Power and Sample Size>
1-sample Z: Differences:** *-.5 –1 –1.5 –2 –2.5*, **Sample Size:** *33*, **Standard deviation:**
9.32, with **Options: Alternative Hypothesis:** *Less than*, **Significance level:** *0.10*, **Store
differences in:** *Diff*, **Store power values in:** *Power*. We get:

Power and Sample Size

```
1-Sample Z Test

Testing mean = null (versus < null)
Calculating power for mean = null + difference
Alpha = 0.1  Assumed standard deviation = 9.32

                Sample
Difference       Size       Power
      -0.5         33    0.165185
      -1.0         33    0.252967
      -1.5         33    0.360547
      -2.0         33    0.480534
      -2.5         33    0.602326
```

Use **Graph>Scatterplot-Simple: Y:***Power*, **X:***Diff*, and we get

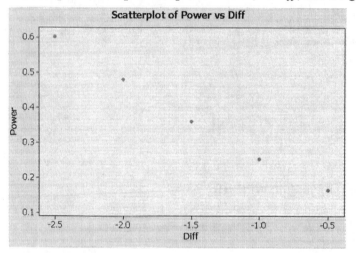

(c) We can estimate the power when $\mu = 14.25$, or Diff $= -1.75$, at about .41 or .42 from the plot above, or find the exact value by repeating part (a.) as above but with **Stat>Power and Sample Size: 1-Sample Z: Sample size:** *33*, **Differences:** *-1.75*

Test for a Population Variance (optional)

If you have Minitab release 15, you can perform a test for population variance simply via **Stat>Basic Statistics> 1 Variance.** This procedure is not included with Minitab 14, but you can still utilize Minitab 14, as you work your way step by step through the relevant calculations as follows:

(i) Find the sample standard deviation using **Calc>Column Statistics** or **Stat>Basic Statistics>Display Descriptive Stats**

(ii) Compute the test statistic: $(n-1) s^2 / \sigma^2$, using **Calc>Calculator**, or a hand calculator.

(iii) Use **Calc>Probability Distributions>Chi-Square: Cumulative Probability** to find the area to the left of your calculated statistic, from which the p-value can then be determined.

Example 8.10 - M/S page 395

Open up the data file FILLAMOUNTS or just type in the data. The null hypothesis here is that $\sigma^2 = 0.01$, with an alternative $\sigma^2 < 0.01$.

If you have Minitab release 15, select **Stat>Basic Statistics> 1 Variance** and fill in as shown below, including the **Options…**: Change *Enter standard deviation* to *Enter variance*, if you wish.

OK, OK, and we get:

Test and CI for One Standard Deviation: WEIGHT

```
Method

Null hypothesis        Sigma = 0.1
Alternative hypothesis  Sigma = < 0.1

The standard method is only for the normal distribution.
The adjusted method is for any continuous distribution.

Statistics
Variable   N   StDev  Variance
WEIGHT     10  0.0412  0.00169

Tests
Variable  Method    Chi-Square   DF  P-Value
WEIGHT    Standard        1.52  9.00   0.003
          Adjusted        1.13  6.67   0.010
```

Ignore the **Adjusted Method.** We are using the **Standard Method,** which is applicable for normally distributed data. The p-value is 0.003, so we can reject the null hypothesis rather conclusively.

Presumably you have already plotted the data as the starting point for you analysis, but if not, do so now. Recall that this test is quite sensitive to departures from normality (not very *robust*, in stat-talk). Your p-value above is only accurate if the data are nearly normally distributed. This is not easy to judge for such a small sample, but you should at least look for clear indications of non-normality, like obvious skewness or outliers.

Chapter 9 – Two Sample Inference

In this chapter, we will compare two means from independent samples, using
Stat>Basic Statistics>2-Sample t

Check normality assumptions with
Graph>Probability Plot

We will compare means, from a paired-data design, using
Stat>Basic Statistics>Paired-t

Compare proportions, for independent samples, with
Stat>Basic Statistics>2 Proportions

And compare variances from two independent samples using
Stat>Basic Statistics> 2 Variances

Comparing Two Means - Independent samples

To compare means of two independent samples, select Minitab's **Stat>Basic Statistics> 2-Sample t** procedure. Click on **'Assume equal variances'** if this is an appropriate assumption. If sample variances differ greatly or for large samples, be sure that **'Assume equal variances'** is *not selected* (the default).

Very unequal sample sizes is another sign that it might be best to stick with the safer default assumptions (unequal variances), as big differences in samples size reduces the *robustness* of the pooled t test.

For large samples, Minitab answers will differ very slightly from text answers due to use of a *t* distribution instead of standard normal.

You should also plot the data and look for severe departures from normality. If present, try a transformation (e.g. log) or non-parametric approach (see chap 14).

Examples 9.1- 9.3 - M/S pages 411 - 414

Open up the DIETSTUDY file. The data here are *stacked*, with all the weight losses in one column, C2, and the diet type (called the 'subscript' in the dialog box below) in another column, C1.

↓	C1-T	C2
	DIET	WTLOSS
1	LOWFAT	8
2	LOWFAT	10
3	LOWFAT	10
4	LOWFAT	12

This is generally the preferred layout, as opposed to having the data *unstacked*, where we would have 100 weight losses from the low-fat diet in one column, and the 100 weight losses from the regular diet in another column.

We wish to compare mean weight loss for the two diets. To execute the *large sample z-test*, we use Minitab's *unequal variances t-test* procedure, via **Stat>Basic Statistics>2-Sample t**, as follows:

Since the data are stacked, we choose **Samples in one column,** and indicate where the **Samples** of weight losses are stacked, and also where the diet indicator - referred to as **Subscripts** – are to be found. We have chosen the two-sided **Alternative**: *not equal* to the null hypothesis **Test difference** of *0.0*, and a *95%* **Confidence level** for the C.I. There is no reason to **Assume equal variances**, for the large sample test, so this is not ticked. We get:

Two-Sample T-Test and CI: WTLOSS, DIET

```
Two-sample T for WTLOSS

DIET       N    Mean   StDev   SE Mean
LOWFAT    100   9.31    4.67     0.47
REGULAR   100   7.40    4.04     0.40

Difference = mu (LOWFAT) - mu (REGULAR)
Estimate for difference:  1.91000
95% CI for difference:   (0.69298, 3.12702)
T-Test of difference = 0 (vs not =): T-Value = 3.10   P-Value = 0.002   DF = 193
```

The P-value for the test is a very small .002, indicating very strong evidence (far beyond the 5% level) of a difference, with the low-fat diet producing bigger losses, on average.

The displayed 95% C.I. is for μ_L - μ_R, rather than the other way around, due to alphabetical ordering (L before R). We see that the difference in average weight loss may be as small as 0.7 pounds or as much as 3.1 pounds, with 95% confidence.

Example 9.4 - M/S page 417

Open up the READING data file. The data are stacked with the scores in C1 = 'Scores' and the method in C2 = 'Method', where N = New method and S = Standard Method. We have two *small* samples to compare here. To execute the *equal variances* t-test or C.I., we use **Stat>Basic Statistics>2-Sample t** as follows:

Note that we have ticked **Assume equal variances** above. Click on **Graphs** and select some, as shown above. Also click on **Options** and check the **Confidence** level.

Even though we only want to construct a C.I. in this example, Minitab will execute both test and C.I. whenever invoking this procedure. We get:

Two-Sample T-Test and CI: SCORE, METHOD

```
Two-sample T for SCORE

METHOD   N   Mean  StDev  SE Mean
NEW     10  76.40   5.83      1.8
STD     12  72.33   6.34      1.8

Difference = mu (NEW) - mu (STD)
Estimate for difference:  4.06667
95% CI for difference:  (-1.39937, 9.53270)
T-Test of difference = 0 (vs not =): T-Value = 1.55  P-Value = 0.136  DF = 20
Both use Pooled StDev = 6.1199
```

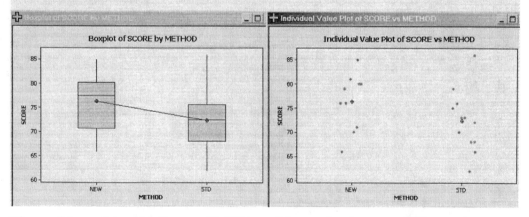

The confidence interval tells us, with 95% confidence, that the mean difference between the new method and the standard method may be anywhere from −1.4 to 9.5. Note that this interval contains '0', which indicates that there may well be no difference at all (at the 5% level of significance).

The plots show no signs of severe skewness or other sharp departures from normality, and spreads are fairly similar (look at the box widths in the boxplot). So conclusions from the equal variances *t* procedure should be trustworthy here.

NOTE - In the **Graphs…** sub-dialog box, Minitab offers you an *Individual value plot*, which is like a dotplot except that the points are randomly jiggled horizontally to avoid overlap of identical points. To reduce the amount of jitter, double-click on any data point, and use the pop-up **Edit Individual Symbols>Jitter** menu. In *Minitab release 15*, the new default plot is a *symmetrical offset* of the identical points, though you can change the display using **Edit Individual Symbols>Identical Points: Symbol Offset**.

To assess normality, it is very common to examine normal probability plots:
Graph>Probability Plots - Simple: Graph Variables: *Score***, Multiple Graphs: By Variables: By variables with groups in separate panels:** *Method*

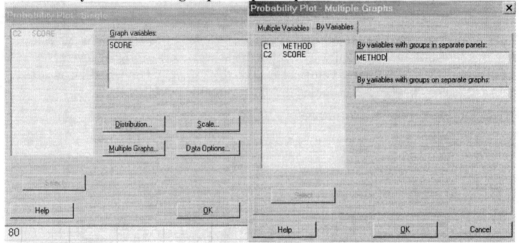

We need to ask for plots *by Method*, so that the data will be split up according to method. Otherwise, we would get one plot for all 200 scores.

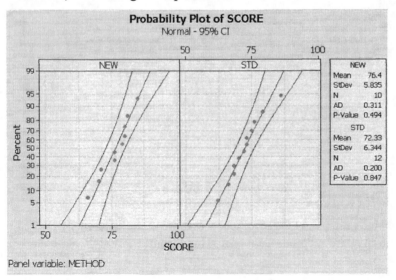

No severe departures from normality are evident above.

Exercise 9.16 - M/S page 426

Here we have the *summary statistics for each sample*, and not the raw data, so for a 90% confidence interval, proceed with:

```
2-Sample t Test and Confidence Interval            2-Sample t - Options                    ✕

                        ○ Samples in one column         Confidence      90

                          Samples:                      Test difference: 0.0

                          Subscripts

                        ○ Samples in different columns  Alternative:  not equal    ▼
                          First:
                          Second:                          Help            OK        Cancel

                        ● Summarized data
                            Sample size:  Mean:           Standard
                                                          deviation:
                          First:   127       39.08        6.73
                          Second: 114        38.79        6.94
                                                                      10    C11    C12    C13
                        ☐ Assume equal variances

       Select                          Graphs...        Options...
```

which produces:

Two-Sample T-Test and CI

```
Sample    N    Mean   StDev   SE Mean
1        127   39.08   6.73    0.60
2        114   38.79   6.94    0.65


Difference = mu (1) - mu (2)
Estimate for difference:  0.290000
90% CI for difference:  (-1.167651, 1.747651)
T-Test of difference = 0 (vs not =): T-Value = 0.33  P-Value = 0.743  DF = 234
```

Note that '0' is in the interval.

Exercise 9.27 - M/S page 429

Open up the SLI data file, where the DIQ scores are in C4 and the groups are in C3. To compare means for the two groups of children, use **Stat>Basic Statistics>2-Sample t: Samples in one column: Samples:** *C4*, **Subscripts:** *C3*, **Alternative:** *Not Equal*, **Confidence Level:** *95*, **Assume equal variances**. Also, click on **Graphs,** and select **Boxplot**. *Hit OK, and you get a message:*

`* ERROR * There must be exactly two distinct subscripts.`

The problem is that we have 3 instead of 2 distinct groups (subscripts) listed in C3. Select and delete the last 10 rows (OND group), rows 21 – 30. Repeat, and you get:

97

Two-Sample T-Test and CI: DIQ, Group

```
Two-sample T for DIQ

Group   N    Mean   StDev   SE Mean
SLI    10   93.60    9.08      2.9
YND    10   95.30    7.51      2.4


Difference = mu (SLI) - mu (YND)
Estimate for difference:  -1.70000
95% CI for difference:    (-9.53124, 6.13124)
T-Test of difference = 0 (vs not =): T-Value = -0.46  P-Value = 0.654  DF = 18
Both use Pooled StDev = 8.3350
```

Since the p-value is larger than $\alpha = .10$, we do not have evidence of a difference in mean DIQ, at the 10% level of significance.

Let's also check the assumptions required for the validity of this conclusion. The boxplots requested from **Graphs…** in the dialog box above are below:

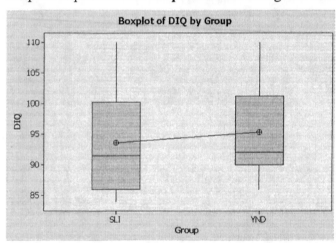

The line drawn from one box to the other in this graph connects the mean of one sample with the mean of the other sample.

Spreads are similar, and while there is some clear right-skewness, our results may still be tolerably accurate, considering the **robustness** of this test (insensitivity to moderate departures from assumptions). Also, see chapter 14 for alternative nonparametric tests.

You can also check normality by examining normal probability plots of these data, via **Graph>Probability Plot – Single: Graph:** *DIQ*, **Multiple Graphs: By Variables:** *Group*, as shown earlier in this section.

Comparing Means - Paired Data

When the observations in the two samples are paired (same subject used twice, paired subjects, etc.), we cannot use the two-sample t test discussed earlier, which assumes independent samples. Instead, we take the difference, for each set of paired observations, and perform a one-sample t test on the differences. No need to compute the differences yourself, as there is an explicit menu selection for this situation: **Stat>Basic Statistics> Paired-t**.

Example 9.5 - M/S page 434

Open up the GRADPAIRS data file.

↓	C1	C2	C3
	Pair	Male	Female
1	1	29300	28800
2	2	41500	41600
3	3	40400	39800
4	4	38500	38500

Each male has been paired with a female based on major and GPA, so the male and female salaries in each row of the table represent paired or correlated data.

Hence, we compare the males and female salaries, using the paired difference approach:
Stat>Basic Statistics>Paired t

Paired t (Test and Confidence Interval)

- Samples in columns
  First sample: Male
  Second: Female

Paired t - Options

Confidence: 95.0

Test mean: 0.0

Alternative: not equal ▾

- Summarized data (difference)
  Sample size:
  Mean:
  Standard

Paired t evaluates the first sample minus the second

Help | OK | Cancel

Select | Graphs... | Options...

Help | OK | Cancel

Click on **Options...** This procedure executes both a test (default null mean is '0') and a C.I. Set the **Confidence level** at '95', as requested. Click on **Graphs...** and select some of the choices, say the **Boxplot** and **Histogram**. We get:

Paired T-Test and CI: Male, Female

```
Paired T for Male - Female

                 N     Mean    StDev   SE Mean
Male            10  43930.0  11665.1    3688.8
Female          10  43530.0  11616.9    3673.6
Difference      10  400.000  434.613   137.437


95% CI for mean difference: (89.096, 710.904)
T-Test of mean difference = 0 (vs not = 0): T-Value = 2.91  P-Value = 0.017
```

The 95% C.I. is easily read off. It does not contain '0' - consistent with the test result indicating a significant difference in means ($p=.017 < .05$).

To confirm the accuracy of these conclusions, we should check the assumption of approximate normality (of the differences). Our **Graphs** choices yield:

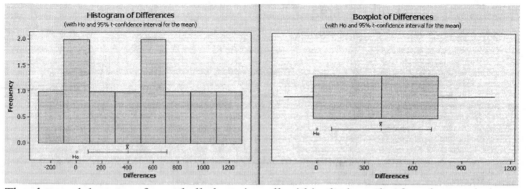

The observed departure from a bell-shape is well within the bounds of random variation, considering the small sample size.

If you prefer to examine a normal probability plot of the differences, then you have to compute the differences, place them in say C3 (**Calc>Calculator: Expression:** *C1 – C2*, **Store in:** *C3*), and proceed to examine C3 with **Graph>Probability Plot.**

Exercise 9.140 - M/S page 476

Open up the MSSTUDY data file. The paired subjects' ages are found in C3 and C6, named *MS-AGE* and *NON-AGE* respectively. To test for a difference in mean age, apply **Stat>Basic Statistics>Paired-t: First sample:** *MS-AGE*, **Second sample:** *NON-AGE*, with **Options: Mean:** *0*, **Alternative:** *Not Equal*. Also, click **Graphs...** and select some. We get:

Paired T-Test and CI: MS-AGE, NON-AGE

```
Paired T for MS-AGE - NON-AGE

                N      Mean      StDev    SE Mean
MS-AGE         10   38.7000     6.0194     1.9035
NON-AGE        10   37.9000     5.0431     1.5948
Difference     10   0.800000   3.881580   1.227464


95% CI for mean difference: (-1.976715, 3.576715)
T-Test of mean difference = 0 (vs not = 0): T-Value = 0.65   P-Value = 0.531
```

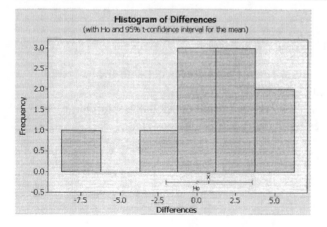

The P-value = 0.531, is large, so there is no significant difference in age between the two groups (i.e., the researchers did a good job of matching up the subjects age-wise).

You can repeat now this analysis for the other two variables, to check for successful matching.

Comparing Two Proportions

To compare proportions computed from two independent samples, enter the sample sizes and counts into **Stat>Basic Statistics>2 Proportions**. This produces both a CI and a test of significance for the difference between the true proportions, using a normal approximation for the sampling distribution of the estimated difference (assuming sufficiently large sample sizes).

Example 9.7 - M/S page 447

To assess whether the proportion of adults who smoke has decreased between 1995 and 2005, enter the observed sample counts of 555 and 578, along with the corresponding sample sizes of 1500 and 1750 respectively, into **Stat>Basic Statistics>2 Proportions**

Check the **Options...** Minitab states the hypothesis in terms of *First p − Second p*, so the **Test difference** is *0.0*, and the proper **Alternative** here is *greater than*, since this would indicate a decrease over time. And you should also select **Use pooled estimate of p for test**, to be consistent with the method in your textbook. **OK, OK,** and we get:

Test and CI for Two Proportions

```
Sample   X      N   Sample p
1       555   1500   0.370000
2       578   1750   0.330286


Difference = p (1) - p (2)
Estimate for difference:  0.0397143
95% lower bound for difference:  0.0121024
Test for difference = 0 (vs > 0):  Z = 2.37  P-Value = 0.009
```

The P-value is highly significant (and $< \alpha = .05$), indicating clear evidence of a decreased proportion of smokers.

In this example, we have **summarized** data, but if instead, we had the *raw* data, i.e. two columns (one for each year) of length 1500 and 1750, with categorical (smoker or non-smoker) observations in each, we would then select in the dialog box **Samples in Different Columns**. Or, if for each of the 3250 subjects, smoker status were in C1, and year of observation in C2, we would use **Samples in One Column** (subscripts in C2).

This procedure is based on the normal approximation to the distribution of a sample proportion, and is *not valid for small sample sizes*. Minitab tries to help, and where sample sizes warrant, will print out the following warning:

```
* NOTE * The normal approximation may be inaccurate for small samples.
```

accompanied by a more accurate P-value calculated from an exact test called *Fisher's exact test*. But you can interpret the P-value the usual way.

Exercise 9.61 - M/S page 451

To compare the two remission proportions, use **Stat>Basic Statistics>2 Proportions: Summarized Data: First:** *98* **trials,** *14* **events; Second:** *102* **trials,** *5* **events,** with **Options: Test Difference:** *0*, **Alternative:** *greater than*, **Use pooled estimate of p for test**. The alternative is 'greater than' since we only want to know if the treatment (sample 1) improves chances of remission, i.e. if $p_1 > p_2$ or $p_1 - p_2 > 0$.

Test and CI for Two Proportions

```
Sample   X    N   Sample p
1       14   98   0.142857
2        5  102   0.049020

Difference = p (1) - p (2)
Estimate for difference:  0.0938375
95% lower bound for difference:  0.0258888
Test for difference = 0 (vs > 0):  Z = 2.26  P-Value = 0.012
```

The two proportions are computed and printed out above: 0.142 for those using St. John's wort, and .049 for the placebo group. The P-value is .012 so we have ample evidence of a St. John's wort effect at the 10% level, but just miss significance at the more stringent 1% level.

Comparing Two Variances (optional)

Two procedures for comparing variances are produced by **Stat>Basic Statistics> 2 Variances** - the F-test discussed in the text, and another (non-parametric) test called Levene's test, which is more accurate if the normality assumption is questionable. *Note that a two-sided alternative is assumed, when calculating the p-value.*

Example 9.11 - M/S page 460

Open up the MICEWTS data file. The data are stacked with Weights in C2 and Supplier in C1. To compare the variability in weight of the supplied mice, use **Stat>Basic Statistics> 2 Variances** as follows:

```
2 Variances                                              ×

  C1    SUPPLIER      ⊙ Samples in one column
  C2    WEIGHT
                         Samples:   WEIGHT
                         Subscripts SUPPLIER

                       ○ Samples in different columns
                         First:
                         Second:

                       ○ Summarized data
                                   Sample size:  Variance:
                         First:
                         Second:

       Select              Options...        Storage...

     Help                    OK              Cancel
```

Results pop up in both a graphical window and in the session window. In the latter:

Test for Equal Variances: WEIGHT versus SUPPLIER

95% Bonferroni confidence intervals for standard deviations

SUPPLIER	N	Lower	StDev	Upper
1	13	0.138579	0.202133	0.361467
2	18	0.070860	0.098193	0.156836

F-Test (normal distribution)
Test statistic = 4.24, p-value = 0.007

Levene's Test (any continuous distribution)
Test statistic = 4.31, p-value = 0.047

We see that the F-test statistic = 4.24 (ratio of bigger sample variance over smaller sample variance), and its p-value = 0.007, assuming a two-sided alternative, which is the case here. The p-value is less than the specified $\alpha = .10$, so we have adequate evidence that the variability in weights is not the same for the two suppliers. If normality were questionable, we would put more trust in Levene's test, whose P-value of 0.047, would lead us to the same conclusion here, since it is also less than the specified α.

What if the alternative had been one-sided? Since .007 is a two-tailed probability, the probability that F > 4.24 is .007/2. Hence, the P-value would be either 0.007/2, if the data were consistent with the alternative hypothesis, or 1 - .007/2, if the data pointed the other way.

Chapter 10 – Analysis of Variance

In this chapter, we will construct ANOVA tables for various designs, using
Stat>ANOVA>One-way (or **One-way (Unstacked)**)
Stat>ANOVA>Two-way

Plot the relevant means with
Stat>ANOVA>Main Effects Plot
Stat>ANOVA>Interactions Plot

Split one worksheet into two, according to the distinct values in one column, with
Data>Split Worksheet

Enter a simple pattern of numbers into a column with
Calc>Make Patterned Data

One-way ANOVA

If there is one factor present in a completely randomized design, we may test the null
hypothesis that the factor has no effect on the response (i.e. the treatment means are all
equal), using **Stat>ANOVA** and selecting **One-way.**

Example 10.3 - M/S page 488

To randomly assign the 15 customers entered in C1 below to the three brands, you can
just use **Calc>Random Data>Sample From Columns** as below, to randomly scramble
the 15 customers in C1, producing the second C1 shown below. Then, the first 5 in the
list (# 4, 13, 9, 10, 3) get brand A, the next 5 get brand B, and the last 5 get brand C. The
5 A's, 5 B's, 5 C's entered in C2 weren't needed, but are there for clarity.

Or, you can start with the C1, C2 as on the left, and then scramble C2, instead of C1, to
match each customer in your list with a randomly selected brand.

Examples 10.4 & 10.5 - M/S pages 493 & 496

Open the GOLFCRD data file. In the first two columns, you will see all the data *stacked*, with the distance of each drive in C2, and the corresponding brand in C1.

↓	C1-T	C2
	Brand	Distance
1	A	251.2
2	A	245.1
3	A	248.0
4	A	251.1

To test the null hypothesis that (mean) distance does not depend on brand, we use **Stat>ANOVA>One-way**. The other selection **ANOVA>One-way (Unstacked)** would be the right choice if the distances were placed in 4 separate columns, one corresponding to each brand, like the *unstacked* layout in Table 10.1 in the text. The stacked format is preferable. **STAT>ANOVA>One-way** produces the dialog box on the left:

Enter the **Response** and **Factor**, as shown, click on **Graphs...**, and tick off **Boxplots**. **OK, OK** produces

One-way ANOVA: Distance versus Brand

```
Source  DF    SS     MS      F      P
Brand    3  2794.4  931.5  43.99  0.000
Error   36   762.3   21.2
Total   39  3556.7

S = 4.602   R-Sq = 78.57%   R-Sq(adj) = 76.78%

                            Individual 95% CIs For Mean Based on
                            Pooled StDev
Level   N   Mean   StDev   --------+---------+---------+---------+-
A      10  250.78   4.74   (---*---)
B      10  261.06   3.87                  (---*---)
C      10  269.95   4.50                              (----*---)
D      10  249.32   5.20   (---*---)
                           --------+---------+---------+---------+-
                            252.0     259.0     266.0     273.0

Pooled StDev = 4.60
```

Above, we see the standard ANOVA table, and additional output showing sample means, standard deviations, and confidence intervals for each of the four Brand means. The p-value of .000 in the ANOVA table indicates overwhelming evidence that the brands are not all equivalent. But is the p-value (and 95% confidence levels) trustworthy? Check the **Graphs** that you requested:

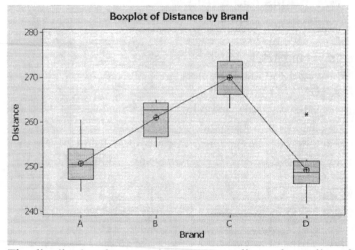

The distributional assumptions are normality and equality of variance. The plots above show little difference in dispersion among the samples, and no severe departures from normality (considering the small sample sizes). The p-value can be trusted.

If you prefer to examine histograms, like those shown in the text, use **Graph>Histogram-Simple: Graph variables:** *Distance*, **Multiple Graphs: By Variables: By:** *Brand*.   To include fitted normal curves, also select **Data View> Distribution: Fit distribution:** *Normal*.
   Or you can examine normal probability plots via **Graph>Probability Plot-Single: Graph:** *Distance*, **Multiple Graphs: By Variables: By:** *Brand*.

Multiple Comparisons

There are a plethora of multiple comparison procedures that may be used to execute pair-wise comparisons among the means. The text mentions e.g. the Bonferroni and Tukey methods. Tukey's is generally preferable to Bonferroni's, and is included with **ANOVA>One-way** under **Comparisons…**   Minitab does not require that the sample sizes be equal.
   You can apply the Bonferroni method too, using **Comparisons…: Fisher's** where you would enter for the requested **individual error rate** the experimentwise error rate divided by the number of comparisons.

Example 10.6 - M/S page 506

Open up the worksheet again. We have already concluded that differences exist among the brands. To apply Tukey's method, select **Stat>ANOVA: One-way,** again, but this time, also click on **Comparisons…**

Select **Tukey's** and enter either '5' or '.05' for the family (experiment-wise) error rate, because we want an overall confidence level of 95% for our batch of confidence intervals. **OK, OK**, gives us the same basic ANOVA output as shown earlier in Example 10.4, but additionally, we see the following:

```
Tukey 95% Simultaneous Confidence Intervals
All Pairwise Comparisons among Levels of Brand

Individual confidence level = 98.93%

Brand = A subtracted from:

Brand   Lower   Center   Upper   -------+---------+---------+---------+--
B       4.736   10.280   15.824                      (---*---)
C      13.626   19.170   24.714                            (---*--)
D      -7.004   -1.460    4.084              (---*---)
                                -------+---------+---------+---------+--
                                     -15        0        15        30


Brand = B subtracted from:

Brand    Lower   Center   Upper   -------+---------+---------+---------+--
C        3.346    8.890   14.434                    (---*---)
D      -17.284  -11.740   -6.196        (---*---)
                                 -------+---------+---------+---------+--
                                      -15        0        15        30


Brand = C subtracted from:

Brand    Lower   Center    Upper   -------+---------+---------+---------+--
D      -26.174  -20.630  -15.086   (--*---)
                                  -------+---------+---------+---------+--
                                       -15        0        15        30
```

Above, we see first brand A compared with B, and with C, and with D, followed by brand B compared with C, and with D, and finally brand C compared with D, so we have all the six possible comparisons. E.g., (4.736, 15.824) is the C.I. for μ_B-μ_A. Since this interval does not contain '0', we may declare a significant difference between brand A and brand B. Likewise, we see that every comparison above is significant except for the comparison of brand D and brand A (the only interval containing '0').

In order to rank the brands from highest mean distance to lowest, go back to the following portion of the ANOVA output:

```
                             Individual 95% CIs For Mean Based on
                             Pooled StDev
    Level   N    Mean   StDev  --------+---------+---------+---------+-
    A      10   250.78   4.74   (---*---)
    B      10   261.06   3.87                    (---*---)
    C      10   269.95   4.50                                  (----*---)
    D      10   249.32   5.20  (---*---)
                               --------+---------+---------+---------+-
                                  252.0     259.0     266.0     273.0

    Pooled StDev = 4.60
```

Read off the means, rank and list them. Connect those that do not differ significantly with a solid line. We get:

Mean: $\overline{249.32 \quad 250.78}$ 261.06 269.95
Brand: D   A   B   C

To construct a confidence interval for any one of the treatment means, read off approximate limits from the output above, or work them out more precisely using:

$$\textbf{Sample Mean} \pm t_{\alpha/2}\, \textbf{s}\, /\sqrt{\textbf{n}}$$

where **s** is the 'Pooled StDev' given in the output: 4.60 ($= \sqrt{MSE}$), with 36 df.

Exercise 10.115 - M/S page 551

(a) Open up the FACES data file, where dominance ratings are in C1 and facial expressions are in C2. To compare ratings among the different facial expressions, run **Stat>ANOVA>One-Way: Response:** *RATING*, **Factor:** *FACE*, **Graphs:** *Boxplots*. We get:

One-way ANOVA: RATING versus FACE

```
Source  DF     SS    MS     F      P
FACE     5   23.09  4.62  3.96  0.007
Error   30   34.99  1.17
Total   35   58.07

S = 1.080    R-Sq = 39.75%    R-Sq(adj) = 29.71%

                        Individual 95% CIs For Mean Based on
                        Pooled StDev
Level   N    Mean   StDev  --------+---------+---------+---------+
A       6   0.853   0.829                    (--------*--------)
D       6   0.605   0.785                  (--------*--------)
F       6  -0.738   1.436    (--------*--------)
H       6   1.018   0.646                     (--------*--------)
N       6  -0.320   1.262       (--------*-------)
S       6  -1.045   1.273  (--------*--------)
                           --------+---------+---------+---------+
                              -1.0       0.0       1.0       2.0

Pooled StDev = 1.080
```

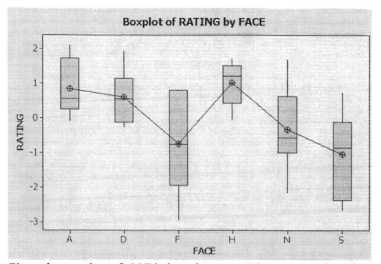

Boxplot of RATING by FACE

Since the p-value of .007 is less than $\alpha = .10$, we can reject the null hypothesis at the 10% level of significance, and conclude that facial expression does matter.

The graph allows a check on the assumptions. We see no substantial non-normality (e.g. severe skewness), and while some samples appear more dispersed than others, the samples are very small, so this can easily be explained by the random variation present. As well, the sample sizes are equal, improving the procedure's ***robustness***. Our conclusions should hold water.

(b) We will compare mean ratings for the 6 facial expressions using Tukey's method with an error rate of 5%, i.e. **Stat>ANOVA>One-way: Response:** *Rating,* **Factor:** *Face,* **Comparisons: Tukey's, family error rate:** *5.* We get the one-way ANOVA table shown in part (a), followed by the additional output below:

```
Tukey 95% Simultaneous Confidence Intervals
All Pairwise Comparisons among Levels of FACE

Individual confidence level = 99.51%

FACE = A subtracted from:

FACE   Lower   Center   Upper     +---------+---------+---------+---------
D     -2.144  -0.248   1.647                (---------*--------)
F     -3.487  -1.592   0.304          (--------*---------)
H     -1.731   0.165   2.061                  (---------*--------)
N     -3.069  -1.173   0.722             (--------*---------)
S     -3.794  -1.898  -0.003        (---------*--------)
                                +---------+---------+---------+---------
                              -4.0      -2.0      0.0       2.0

FACE = D subtracted from:

FACE   Lower   Center   Upper     +---------+---------+---------+---------
F     -3.239  -1.343   0.552             (--------*---------)
H     -1.482   0.413   2.309                    (--------*---------)
N     -2.821  -0.925   0.971               (--------*---------)
S     -3.546  -1.650   0.246          (---------*--------)
                                +---------+---------+---------+---------
                              -4.0      -2.0      0.0       2.0
```

```
FACE = F subtracted from:

FACE    Lower   Center   Upper    +---------+---------+---------+---------
H       -0.139   1.757   3.652                       (---------*--------)
N       -1.477   0.418   2.314              (--------*---------)
S       -2.202  -0.307   1.589          (--------*---------)
                                         +---------+---------+---------+---------
                                      -4.0      -2.0      0.0       2.0

FACE = H subtracted from:

FACE    Lower   Center   Upper    +---------+---------+---------+---------
N       -3.234  -1.338   0.557              (---------*---------)
S       -3.959  -2.063  -0.168          (---------*--------)
                                         +---------+---------+---------+---------
                                      -4.0      -2.0      0.0       2.0

FACE = N subtracted from:

FACE    Lower   Center   Upper    +---------+---------+---------+---------
S       -2.621  -0.725   1.171              (--------*---------)
                                         +---------+---------+---------+---------
                                      -4.0      -2.0      0.0       2.0
```

Rank the facial expressions, using the list of means up above, and then underline or otherwise connect the means that do not differ significantly (intervals containing '0'):

$$\underline{\text{S} \qquad \text{F} \qquad \text{N}} \qquad \underline{\text{D} \qquad \text{A} \qquad \text{H}}$$

-1.045 -.738 -.320 .605 .853 1.018

In summary, Happy and Angry faces both rate above Sad (by finding the pairs of treatments that are not connected by any bars).

Randomized Block Design

If we group our experimental units into blocks, before assigning treatments, we need to amend the ANOVA to include an extra SS for the block-to-block variation. Since blocks are now a second source of variation in the design, use **ANOVA>Two-way**, to calculate the proper ANOVA table.

   Two-way does not have an option for multiple comparison procedures. You can still request if you apply **Stat>ANOVA>General Linear Model,** but this procedure is only included with the *professional version of Minitab*. If you have the professional version, try using **Help**, to figure out how to handle the dialog boxes.

   To check assumptions, you can use the *residual plots* listed under **Graphs…,** but you should first study the section on residual plots in chapter 12.

Examples 10.8 & 10.9 - M/S pages 516 & 518

Open up the GOLFRBD data file, containing data from a randomized block design. *Note how the data are entered:*

Every row corresponds to one drive of the ball. In column 3, we see the distance. In column 2 we identify the brand used, and in column 1, we identify the golfer.

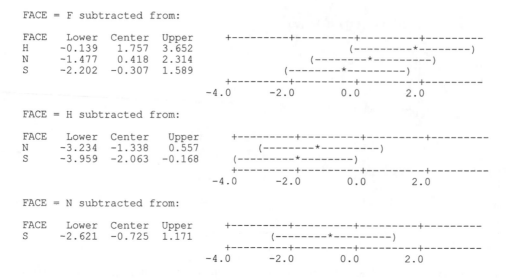

✦	C1	C2-T	C3
	Golfer	Brand	Distance
1	1	A	202.4
2	1	B	203.2
3	1	C	223.7
4	1	D	203.6
5	2	A	242.0
6	2	B	248.7

We are interested in how C3 (distance) depends on C2 (brand) and C1 (golfer). For answers, use the F-ratios printed in the two-way ANOVA output obtained from: **Stat>ANOVA>Two-way,** where we enter *Distance* as the **Response,** and *Brand* and *Golfer* in any order next to **Row factor** and **Column factor:**

I also ticked off **Display means** for *Brand*, since we will also want to compare these means pairwise, if the global test is significant, to learn more about where the differences lie. We get

Two-way ANOVA: Distance versus Brand, Golfer

```
Source   DF      SS       MS      F      P
Brand     3   3298.7  1099.55  54.31  0.000
Golfer    9  12073.9  1341.54  66.26  0.000
Error    27    546.6    20.25
Total    39  15919.2

S = 4.499   R-Sq = 96.57%   R-Sq(adj) = 95.04%

                   Individual 95% CIs For Mean Based on
                   Pooled StDev
Brand   Mean    --------+---------+---------+---------+-
A      227.05          (---*--)
B      233.18              (--*---)
C      245.33                            (---*--)
D      220.73   (---*---)
                --------+---------+---------+---------+-
                   224.0     232.0     240.0     248.0
```

Since each p-value is 0.000, we have strong evidence for differences in mean among brands, and also for differences among golfers (suggesting that blocking was effective)..

You can plot the brand means with **Stat>ANOVA>Main Effects Plot: Response:** Distance, **Factor:** Brand

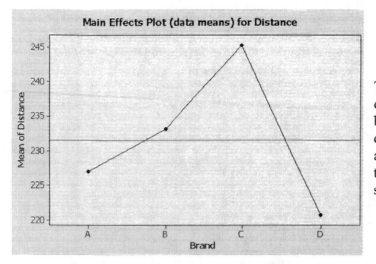

The greatest mean distance is achieved with brand C, and indeed, it differs significantly from all the other brands (see the Bonferroni analysis shown in your text).

Minitab's **Two-Way** command does not include multiple comparisons options, but if you have the professional version of Minitab, select **Stat>ANOVA>General Linear Model**, with **Responses:** *Distance*, **Model:** *Brand Golfer*. Select **Comparisons…** and then choose **Pairwise Comparisons** with **Terms:** *Brand*. Select the **Bonferroni** or **Tukey** method. Choose **Test** and/or **Confidence Interval**.

Unfortunately, you cannot use the individual mean 95% CI's shown in the output to precisely test for differences between pairs of means (though non-overlap does suggest *some* evidence of a difference).

Exercise 10.69 - M/S page 525

Open up the PLANTS data file, where in columns C1 - C3, you will find the student number, plant setting (1 = Live, 2 = Photo, 3 = None) and maximum finger temperature, respectively. Does the type of plant setting affect relaxation level (measured by finger temperature)? Students are the blocks. Run **Stat>ANOVA>Two-way: Response:** *Temperature*, **Row Factor:** *Plant*, **Column Factor:** *Student*

Two-way ANOVA: Temperature versus Plant, Student

Source	DF	SS	MS	F	P
Plant	2	0.122	0.06100	0.02	0.981
Student	9	18.415	2.04611	0.63	0.754
Error	18	58.038	3.22433		
Total	29	76.575			

$S = 1.796$ R-Sq = 24.21%   R-Sq(adj) = 0.00%

The large p-value of .981 for the factor Plant ($> \alpha = .10$) indicates no evidence of a difference in mean relaxation level among the three experimental plant settings.

Factorial Experiments

When the design incorporates two factors, we put the response values in one column, factor 1 settings in a second column, factor 2 settings in a third column (similar to a RBD design), and then use **ANOVA>Two-Way** for the number crunching. With replicates at each factor-level combination, the ANOVA will now include SS's for the *main effect* of factor A, the *main effect* of factor B, and the *interaction* between the factors.

Two-Way has no option for multiple comparison procedures, but if you have the professional version of Minitab, you may use **Comparisons...** in **Stat>ANOVA> General Linear Model**. Use Minitab's **Help** for more details.

If you wish to do an initial test of significance on the treatment combinations, you have to enter distinct treatment numbers into a single column and run **ANOVA>One-way**. And if you discover a significant interaction, the **One-way>Comparisons...** may be useful if you wish to execute multiple comparisons among all the treatments.

To check assumptions, you can use the **Graphs...** sub-menu, to examine individual samples via **Boxplots** or **Individual value plots** or, better yet, examine the **Residual Plots** (after reading up on these types of plots in chapter 12).

Example 10.10 - M/S page 531

Open up the GOLFFAC1 data file. Note that the response Distance is in C3, the factor Club in C1, the factor Brand in C2. Write treatment numbers (1-8) in C4, where 1 denotes the DRIVER/A, 2 denotes DRIVER/B, etc. Name this column 'Treatment'. Or you can use **Calc>Make Patterned Data** (see exercise below).

↓	C1-T	C2-T	C3	C4
	Club	Brand	Distance	Treatment
1	DRIVER	A	226.4	1
2	DRIVER	A	232.6	1
3	DRIVER	A	234.0	1
4	DRIVER	A	220.7	1
5	DRIVER	B	238.3	2
6	DRIVER	B	231.7	2

To test for any difference among the 8 treatments, run a **One-Way ANOVA**:

One-Way Analysis of Variance

C1	Club	**Response** Distance
C2	Brand	
C3	Distance	
C4	Treatment	**Factor:** Treatment

One-way ANOVA: Distance versus Treatment

```
Source      DF      SS      MS       F       P
Treatment    7  33659.8  4808.5  140.35  0.000
Error       24    822.2    34.3
Total       31  34482.0
```

The small p-value tells us that treatments are not all the same. To study the main effects and interaction, proceed with **Stat>ANOVA>Two-Way**

Two-Way Analysis of Variance　　　　　　　　　　　　　　　　　　✕

C1	Club
C2	Brand
C3	Distance
C4	Treatment

Response: Distance

Row factor: Club　　　　　☐ Display means

Column factor: Brand　　　☐ Display means

which produces

Two-way ANOVA: Distance versus Club, Brand

Source	DF	SS	MS	F	P
Club	1	32093.1	32093.1	936.75	0.000
Brand	3	800.7	266.9	7.79	0.001
Interaction	3	766.0	255.3	7.45	0.001
Error	24	822.2	34.3		
Total	31	34482.0			

The interaction is significant ($p = .001 < \alpha = .10$), so the relationship between the brands is not the same for drivers and for 5-irons. Maybe the best brand for drivers will not be the best brand for 5-irons. We can examine the nature of the interaction, by plotting the means (like Fig 10.24 in text), using **Stat>ANOVA>Interactions Plot**

Interactions Plot　　　　　　　　　　　　　　　　　　✕

C1	Club
C2	Brand
C3	Distance
C4	Treatment

Responses:

Distance

Factors:

Club Brand

☐ Display full interaction plot matrix

which produces the following plot of means (to put *Club* on the x-axis, just reverse the order of listing of the factors in the box above):

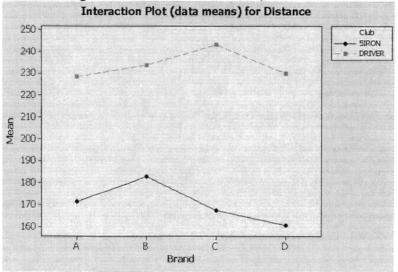

With 5-irons, brand B appears best, while with drivers, brand C appears best. To confirm these impressions, one simple way to proceed is to split the worksheet according to Club with **Data>Split Worksheet: By:** *Club*. Then, for each type of club (i.e. each new worksheet), run **Stat>ANOVA>One-way** with **Comparisons: Tukey's, family error rate:** *10*. For the Driver, e.g. we get:

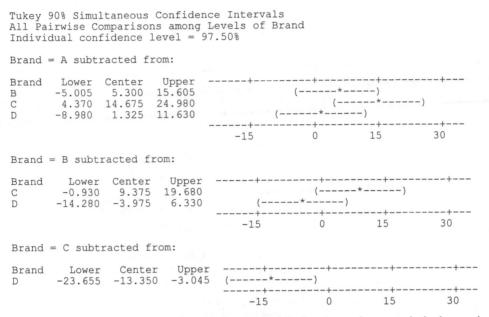

```
Tukey 90% Simultaneous Confidence Intervals
All Pairwise Comparisons among Levels of Brand
Individual confidence level = 97.50%

Brand = A subtracted from:

Brand   Lower   Center   Upper    ------+---------+---------+---------+---
B       -5.005   5.300   15.605                   (------*-----)
C        4.370  14.675   24.980                     (------*------)
D       -8.980   1.325   11.630            (------*------)
                                  ------+---------+---------+---------+---
                                      -15        0        15        30

Brand = B subtracted from:

Brand   Lower   Center   Upper    ------+---------+---------+---------+---
C       -0.930   9.375   19.680                 (------*------)
D      -14.280  -3.975    6.330        (------*------)
                                  ------+---------+---------+---------+---
                                      -15        0        15        30

Brand = C subtracted from:

Brand   Lower   Center   Upper    ------+---------+---------+---------+---
D      -23.655 -13.350   -3.045   (------*------)
                                  ------+---------+---------+---------+---
                                      -15        0        15        30
```

We see that C is significantly better than A and D, but just misses statistical superiority over B (the interval for $\mu_C - \mu_B$ just catches '0').

Repeat for the 5-iron (worksheet), and you will find that brand B is significantly better than the other three brands.

Example 10.11 - M/S page 535

This is similar to Example 10.10. Open up GOLFFAC2, and run a two-way ANOVA: **Stat>ANOVA>Two-Way** with the same entries in the dialog box as before (I omit the test on Treatments). The output is:

Two-way ANOVA: Distance versus Club, Brand

Source	DF	SS	MS	F	P
Club	1	46443.9	46443.9	1887.94	0.000
Brand	3	3410.3	1136.8	46.21	0.000
Interaction	3	105.2	35.1	1.42	0.260
Error	24	590.4	24.6		
Total	31	50549.8			

S = 4.960   R-Sq = 98.83%   R-Sq(adj) = 98.49%

There is no evidence of an interaction (p=.260). The main effects of Brand and Club are both highly significant. The club effect is no surprise, as we already know that drivers carry the ball further than 5-irons. Distance also depends on brand, and we should examine this further.

We can plot all the 8 treatment means, like Fig. 10.27 in the text, using
Stat>ANOVA>Interactions Plot: Response: *Distance*, **Factors:** *Club Brand*

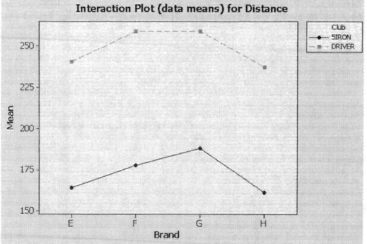

The plot shows little indication of an interaction (i.e. the brand differences look quite similar for 5-irons and drivers), quite consistent with our test result.

In the absence of interaction, it makes good sense to plot the *overall* means for each brand, or for each type of club, using **Stat>ANOVA>Main Effects Plot: Response:** *Distance*, **Factors:** *Club Brand*. We get:

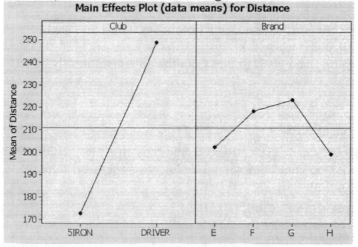

The significant Brand effect appears to be mostly due to the big difference between the high distances achieved by brands F and G, and the low distances achieved by brands E and H. This impression is confirmed in the multiple comparisons analysis shown in the text (Fig 10.26), which can be replicated, using the *professional version of Minitab*, by selecting **Stat>ANOVA>General Linear Model: Response:** *Distance*, **Model:** *Club Brand Club*Brand*, and then using the **Comparisons…: Pairwise Comparisons, Terms:** *Brand*, **Method:** *Tukey*.

Exercise 10.93 - M/S page 543

Open up the VITAMINB data file. *Kidney weight* in C1 is the response. The factor *Rat size* is in C2 and the second factor *Diet* is in C3. To test first for treatment differences, we need to enter some treatment numbers into, say, C4 (name it 'Treatment'). The first 7 rows of data are from treatment 1, the next 7 from treatment 2, etc., so either enter these numbers into C4 directly, or use **Calc>Make Patterned Data>Simple Set of Numbers: Store in:** *Treatment*, **From first value:** *1*, **To last value:** *4*, **In steps of:** *1*, **List each value:** *7* **times**.

We test for treatment differences via **Stat>ANOVA>One-way: Response:** *C1*, **Factor:** *C4*, and also select **Graphs:** *Boxplots,* so that you can check model assumptions. We get:

One-way ANOVA: Kidney Weight versus Treatment

Source	DF	SS	MS	F	P
Treatment	3	8.1168	2.7056	47.34	0.000
Error	24	1.3715	0.0571		
Total	27	9.4883			

From the low p-value, we are confident that treatment differences exist. You can do a rough check on your assumptions by examining the boxplots below, or by using other data plots selected from the main menu's **Graph** choices (like normal probability plots), where you must remember to specify **Multiple Graphs>By Variables**.

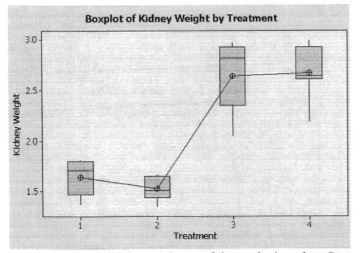

For the more useful factorial part of the analysis, select **Stat>ANOVA>Two-Way: Response:** *C1*, **Row Factor:** *C2*, **Column Factor:** *C3*. We see

Two-way ANOVA: Kidney Weight versus Rat Size, Diet

Source	DF	SS	MS	F	P
Rat Size	1	8.06789	8.06789	141.18	0.000
Diet	1	0.01243	0.01243	0.22	0.645
Interaction	1	0.03643	0.03643	0.64	0.432
Error	24	1.37151	0.05715		
Total	27	9.48827			

There is no evidence of interaction, no evidence of a Diet effect, but strong evidence of an effect from Rat Size.

A picture helps clarify, so use **Stat>ANOVA>Interactions Plot: Response:** *C1*, **Factors:** *C2 C3*. You will get the plot on the left below. Or you can reverse the order of the factors listed, and you will get the plot on the right.

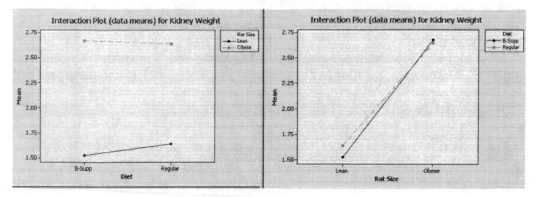

Either plot shows the large effect of rat size, and the comparatively minor (and non-significant) interaction and diet effects.

Chapter 11 – Simple Linear Regression

In this chapter, we will plot the response variable versus the predictor, using
Graph>Scatterplot

Add the least-squares line to the data plot, with
Stat>Regression>Fitted Line Plot

Make inferences and predictions with a regression model, using
Stat>Regression>Regression

Calculate the correlation coefficient, with
Stat>Basic Statistics>Correlation

Fitting by Least-Squares

The method of *least-squares* or *regression* may be used to fit a model showing how a *response variable* depends on another variable called the *predictor*. The simplest model is a straight line, but least squares may also be used to fit a variety of different models, including e.g. quadratic models and models with many predictors (see chapter 12). The fitted line may be calculated and drawn on the plot using **Stat>Regression>Fitted Line Plot**, but **Stat>Regression>Regression** will provide more information and useful options.

Figures 11.3, 11.5, 11.6c - M/S pages 564, 567, 569

Take a moment to type in the data into C1 and C2 as shown (or look for the STIMULUS data file):

↓	C1	C2
	Drug	Time
1	1	1
2	2	1
3	3	2
4	4	2
5	5	4

We can produce a scattergram (scatterplot) showing reaction time versus drug amount, using
Graph>Scatterplot - Simple as follows:

Scatterplot - Simple			✕

C1	Drug		Y variables	X variables	▲
C2	Time	1	Time	Drug	
		2			
		3			

We get

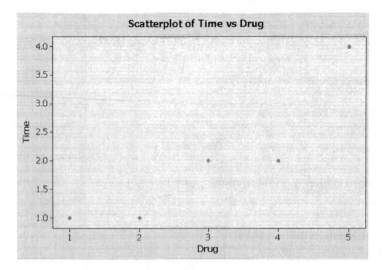

To find the equation of the regression line, and draw it on the plot, we may use
Stat>Regression>Fitted Line Plot

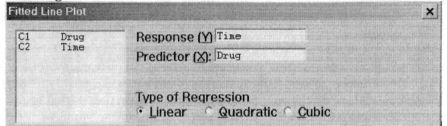

which gives us

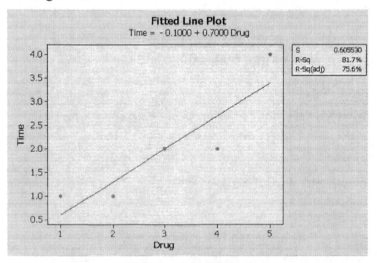

Alternatively, to calculate the least-squares equation, along with much additional useful information, to be discussed later, use **Stat>Regression>Regression**

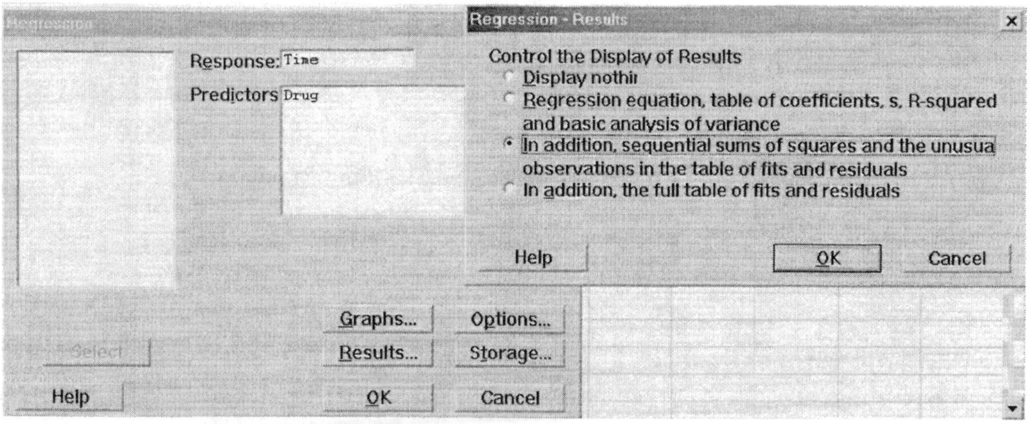

Fill in the **Response** and **Predictors** as above. Click on **Results** and it should appear as above, with the third level of **Display** ticked. You can choose to have less displayed, if you prefer, but I recommend this level, as all of the information displayed will eventually become useful to us. **OK, OK,** and in the session window, we see

```
Regression Analysis: Time versus Drug

The regression equation is
Time = - 0.100 + 0.700 Drug

Predictor      Coef    SE Coef      T       P
Constant    -0.1000     0.6351   -0.16   0.885
Drug         0.7000     0.1915    3.66   0.035

S = 0.605530   R-Sq = 81.7%   R-Sq(adj) = 75.6%

Analysis of Variance

Source         DF      SS       MS       F       P
Regression      1   4.9000   4.9000   13.36   0.035
Residual Error  3   1.1000   0.3667
Total           4   6.0000
```

At the top, the fitted equation is clearly displayed. Additional significant figures for the slope and intercept appear below **Coef**. The Sum of Squared deviations minimized by the regression line appears next to **Residual Error** (1.1000) in the **Analysis of Variance** table. The rest of the information in the table will be discussed later.

Exercise 11.28 - M/S page 577

Open up the OJUICE data file. Note carefully from the question which variable is the predictor and which is the response. Regress with **Stat>Regression>Regression: Response:** *Sweetness,* **Predictors:** *Pectin.* Also, click on **Options** and fill in: **Prediction intervals for new observations:** *300.*

At the top, we get the regression equation:

Regression Analysis: Sweetness versus Pectin

```
The regression equation is
Sweetness = 6.25 - 0.00231 Pectin
```

So, sweetness decreases by .231 for every increase of 100ppm in pectin.

At the bottom, we have:

```
Predicted Values for New Observations
New
Obs     Fit  SE Fit        95% CI              95% PI
  1  5.5589  0.0587  (5.4372, 5.6806)   (5.0967, 6.0211)

Values of Predictors for New Observations
New
Obs  Pectin
  1     300
```

The **Fit** or prediction when the pectin measures 300ppm, is **5.5589** (ignore the other entries on this line, for now). You can check this out by plugging '300' into the regression equation above.

Model Assumptions and Estimating the Variance

It is important to check the assumptions of our model. A scattergram helps, but the best approach is through the use of *residual plots*, which we defer until section 12.13. Our model assumes a constant variance σ^2 for the random errors. The estimate of σ, denoted *s*, is clearly displayed in Minitab's **Stat>Regression>Regression** output or in the **Stat>Regression>Fitted Line Plot** output.

Example 11.2 - M/S page 580

Open up again the worksheet used earlier in this chapter (STIMULUS), or just type in the data on drug amount and reaction time. As before, we use
Stat>Regression>Regression: Response: *Time*, **Predictor:** *Drug*, and obtain:

Regression Analysis: Time versus Drug

```
The regression equation is  Time = - 0.100 + 0.700 Drug

Predictor       Coef  SE Coef       T       P
Constant     -0.1000   0.6351   -0.16   0.885
Drug          0.7000   0.1915    3.66   0.035

S = 0.605530   R-Sq = 81.7%   R-Sq(adj) = 75.6%

Analysis of Variance
Source          DF      SS       MS       F       P
Regression       1  4.9000   4.9000   13.36   0.035
Residual Error   3  1.1000   0.3667
Total            4  6.0000
```

We see that the estimated standard deviation of the errors is **s = 0.605530**. We may also find the value of s^2 by examining the displayed Analysis of Variance table. s^2 is exactly the same as the **Residual Error MS** of **0.3667**, which is just the Residual Error SS (SSE) = 1.1000 divided by the Residual Error DF = n − 2 = 3.

Exercise 11.46 - M/S page 583

Open up the CUTTOOL data file, where C1 − C3 contain the data on Cutting Speed, Useful Life (hours) for brand A, and Useful Life (hours) for brand B, respectively. We can examine the data with fitted lines, using **Stat>Regression>Fitted Line Plot: Response:** *HoursA*, **Predictor:** *Speed*, **Type of Regression:** *Linear*, and then repeat changing the **Response** to *HoursB*.

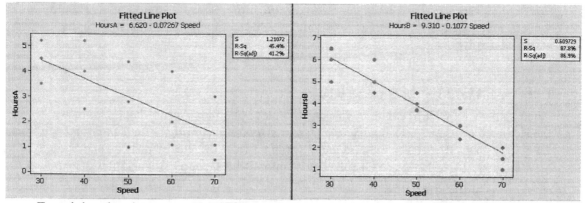

Examining the plots, we see considerably less scatter about the regression line, when predicting brand B lifetimes, which is also evident from comparing the regression standard errors (*s*), printed in the boxes to the right of each graph: **1.211** for brand A, but only **.6097** for brand B. [We could also compare the **R-Sq** values printed in the boxes, but this will be discussed later]

Inferences about the slope

The **Stat>Regression>Regression** output provides an estimate of the slope (Coef), and its standard error (SE Coef). Using these, we can then construct a CI or execute a t-test. The *t* value for a test of H: β = 0 is also displayed in the output with corresponding p-value (two-tailed).

Example 11.3 - M/S page 584

Open up our first simple regression data worksheet again (STIMULUS). Regress reaction time on the amount of drug, i.e. **Stat>Regression>Regression: Response:** *Time*, **Predictor:** *Drug*. Key information about the individual parameters appears right below the regression equation:

Regression Analysis: Time versus Drug

```
The regression equation is
Time = - 0.100 + 0.700 Drug

Predictor     Coef   SE Coef      T      P
Constant   -0.1000    0.6351  -0.16  0.885
Drug        0.7000    0.1915   3.66  0.035
```

The slope is the **Coef** of **Drug** in the equation, and equals .7000. Its standard error, **SE Coef** equals 0.1915. Hence a 95% C.I. for the true slope would be .7000 ± 3.182(.1915), where 3.182 is the *t* value chopping off .025 to the right in a *t* (3df) distribution.

The estimated slope over its standard error, or Coef/StDev above, equals 3.66, and the corresponding p-value is **0.035**, from which we may conclude, at the 5% level, that the amount of drug does help to predict reaction time, via a straight-line model. ***Note that this p-value assumes a two-sided alternative to the null.*** Adjust it appropriately should you encounter a one-sided alternative (divide by 2, if the sample is consistent with the stated alternative).

Exercise 11.120 - M/S page 619

Open up the ORGCHEM data file. We want to plot Maximum Absorption (C2) versus the Hammett Constant (C3), but ***using different symbols according to the compound***. In C1 = COMPOUND, replace the compound variants '1a', '1b', etc, by just '1', and '2a', '2b', etc by just '2'. Now, use **Graph>Scatterplot-With Groups: Y variables:** *MAXABS*, **X variables:** *HAMMETT*, **Categorical variables:** *COMPOUND*, and we see

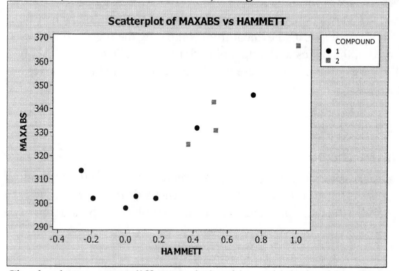

Clearly, there are two different relationships, which should be fitted separately. *For compound 1, though, the relationship appears curved and hence a straight line may not be the most appropriate model choice.*

For (b), (c) and (d), use **Data>Split Worksheet: By variables:** *COMPOUND*, to break up the worksheet into two, according to the compound. Then select **Stat>Regression>Regression** or **Stat>Regression>Fitted Line Plot**, enter *MAXABS* and

HAMMETT for the response and predictor, respectively. Repeat for each worksheet, to obtain the following two outputs:

Results for: ORGCHEM.MTP(COMPOUND = 1)

Regression Analysis: MAXABS versus HAMMETT

```
The regression equation is  MAXABS = 308 + 41.7 HAMMETT

Predictor      Coef  SE Coef       T      P
Constant    308.137    4.896   62.94  0.000
HAMMETT       41.71    13.82    3.02  0.029

S = 11.9432   R-Sq = 64.6%   R-Sq(adj) = 57.5%

Analysis of Variance
Source           DF       SS      MS     F      P
Regression        1   1299.7  1299.7  9.11  0.029
Residual Error    5    713.2   142.6
Total             6   2012.9
```

Results for: ORGCHEM.MTP(COMPOUND = 2)

Regression Analysis: MAXABS versus HAMMETT

```
The regression equation is  MAXABS = 303 + 64.1 HAMMETT

Predictor      Coef  SE Coef       T      P
Constant    302.588    8.732   34.65  0.001
HAMMETT       64.05    13.36    4.79  0.041

S = 6.43656   R-Sq = 92.0%   R-Sq(adj) = 88.0%

Analysis of Variance
Source           DF       SS      MS      F      P
Regression        1   952.14  952.14  22.98  0.041
Residual Error    2    82.86   41.43
Total             3  1035.00
```

The two fitted equations appear to indicate that the absorption rate increases at a much faster rate, as the Hammett constant increases, when using compound 2 (though this disparity may also be attributed to the influence of the two negative Hammett Constant points for compound 1). Both models are statistically significant at the 5% level but not at the 1% level (p-values of .029 and .041).

The Correlation Coefficient and the Coefficient of Determination

To find the correlation between two variables, r, use **Stat>Basic Statistics>Correlation**. The square of r, called the *coefficient of determination*, is a key measure of the predictive power of the fitted model and is always included in the regression output, written as '*R-Sq*'.

Example 11.4 - M/S page 593

Open up the CASINO data file, containing the number of employees in C1 and the crime rate in C2. . To calculate the correlation between the number of employees and the crime rate, we use **Stat>Basic Statistics>Correlation**:

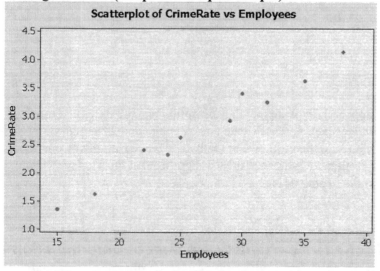

By default, **Display p-values**, is selected, which give you an assessment of the statistical significance of the observed association (same as testing the slope of the fitted straight line).

We get

Correlations: Employees, CrimeRate

```
Pearson correlation of Employees and CrimeRate = 0.987
P-Value = 0.000
```

The very high (and very significant) correlation is no surprise, if we look at the scattergram below (**Graph>Scatterplot-Simple**):

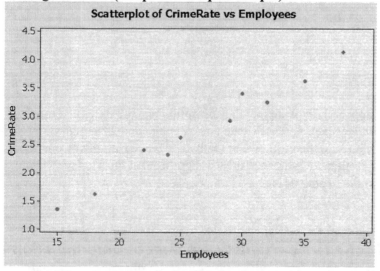

Example 11.5 - M/S page 596

Open up once again the STIMULUS worksheet. To find r^2, just look for **R-Sq** in the output from: **Stat>Regression>Regression: Response:** *Time*, **Predictor:** *Drug* (or from that little box included in the graph produced by **Stat>Regression>Fitted Line Plot**)

Regression Analysis: Time versus Drug

```
The regression equation is
Time = - 0.100 + 0.700 Drug

Predictor     Coef   SE Coef      T      P
Constant   -0.1000    0.6351  -0.16  0.885
Drug        0.7000    0.1915   3.66  0.035

S = 0.605530   R-Sq = 81.7%   R-Sq(adj) = 75.6%
```

Ignore the R-Sq(adj) *value in the output for now, but it will be useful when we discuss multiple regression*

We see that the amount of drug explains (linearly) 81.7% of the variation among subjects' reaction times, leaving only 18% of this variation unaccounted for (presumably due to other differences among the subjects).

Exercise 11.85 - M/S page 601

Open up the BOXING2 data file, containing blood lactate levels in C1, and perceived recovery in C2. Let's run the regression, to find r^2: **Stat>Regression>Regression: Response:** *Lactate*, **Predictor:** *Recovery*, and at the top of the output, we see:

Regression Analysis: Lactate versus Recovery

```
The regression equation is  Lactate = 2.97 + 0.127 Recovery

Predictor     Coef   SE Coef     T      P
Constant    2.9696   0.7896   3.76  0.002
Recovery   0.12667   0.04878   2.60  0.021

S = 0.950722   R-Sq = 32.5%   R-Sq(adj) = 27.7%
```

So, a modest *32.5%* of the variation on lactate levels is explained by the linear relationship with perceived recovery. If you square root 0.325, you get ±0.57, so r must equal +0.57, considering the positive slope of the fitted regression equation above.

 Of course, we can find the correlation directly (and square it to determine the coefficient of determination) using **Stat>Basic Statistics>Correlation: Variables:** *Lactate Recovery*

Correlations: Lactate, Recovery

```
Pearson correlation of Lactate and Recovery = 0.570
P-Value = 0.021
```

Confidence and Prediction Intervals

To find a confidence interval for the mean response at some setting of the predictor, or to obtain a prediction interval for a random future observation on the response, use the available **Options** when running your **Regression**. *Both* intervals are printed out when you request **Prediction intervals**. Choose the one that you need.

Examples 11.6 & 11.7 - M/S page 603

Once again, open up the simple data set (STIMULUS) with which we started this chapter. To find either the confidence interval for the mean, or the prediction interval, when we have a 4% drug concentration, run the regression: **Stat>Regression>Regression,** and use **Options…: Prediction intervals for…** Plug in *'4'* as below:

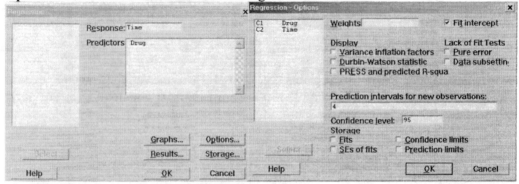

The *default* confidence level of 95% is what we want. Leave **Fit intercept** checked, as un-checking it will fit a line with an intercept of '0'. **OK, OK,** and in the output, in addition to all the stuff that we have seen before, way down at the bottom, you can see:

```
Predicted Values for New Observations
New
Obs    Fit    SE Fit       95% CI          95% PI
  1   2.700   0.332   (1.645, 3.755)   (0.503, 4.897)

Values of Predictors for New Observations
New
Obs   Drug
  1   4.00
```

You can read off the prediction interval, *PI,* or the confidence interval, *CI,* as desired. The *Fit* of 2.700 results from plugging 4.00 for the drug amount in the regression equation. *'SE Fit'* is precisely what your text refers to as the 'estimated standard error of $\hat{y}$', so the CI displayed above is simply constructed as: $2.700 \pm t_{.025} (0.332)$.

Exercise 11.103 - M/S page 609

Open up the CUTTOOL data file again, showing 'Speed', 'HoursA', 'HoursB' in C1 – C3 respectively. For a CI for mean lifetime for brand A when the cutting speed is 45, we use **Stat>Regression>Regression: Response:** *HoursA*, **Predictor:** *Speed*, with **Options: Prediction intervals …:** *45*, **Confidence level:** *90*, which yields the usual regression output, and, at the bottom:

```
Predicted Values for New Observations
New
Obs    Fit   SE Fit       90% CI          90% PI
  1   3.350  0.332   (2.763, 3.937)   (1.127, 5.573)

Values of Predictors for New Observations
New
Obs   Speed
  1   45.0
```

Repeat, changing the **Response** to '*HoursB*', and we see, at the bottom:

```
Predicted Values for New Observations
New
Obs    Fit  SE Fit       90% CI            90% PI
  1  4.465   0.167  (4.169, 4.761)   (3.345, 5.585)

Values of Predictors for New Observations
New
Obs  Speed
  1   45.0
```

In (a), we are estimating means, not predicting future random observations, so it is the CI's in the outputs above that we use. It is no surprise that the CI for brand A mean lifetime is the wider one, since *s* is bigger there, indicating larger random errors or greater dispersion about the regression line. For (b), use the PI's from the outputs.

Section 11.7 - A Complete Example - M/S page 609

Open up the FIREDAM data file. We want to model the relationship between fire damage (C2) and distance from nearest fire station (C1). Start by examining the data and fitted line, using **Stat>Regression>Fitted Line Plot: Response:** *Damage*, **Predictor:** *Distance*, **Type of Regression:** *Linear*. We get

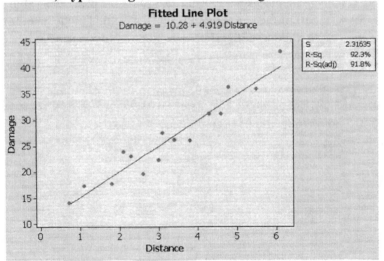

Things looks suitable for regression analysis, i.e. the pattern is linear and the relationship appears to be fairly strong. Also, the scatter is fairly constant about the line, suggesting constant error variance. We also want (Step 5 in text) a prediction of fire damage at a distance of 3.5 miles. Run **Stat>Regression>Regression: Response:** *Damage*, **Predictor:** *Distance*, with **Options: Prediction intervals for new observations:** *3.5*, and we get:

Regression Analysis: Damage versus Distance

```
The regression equation is    Damage = 10.3 + 4.92 Distance

Predictor     Coef   SE Coef      T       P
Constant    10.278     1.420    7.24   0.000
Distance    4.9193    0.3927   12.53   0.000

S = 2.31635    R-Sq = 92.3%    R-Sq(adj) = 91.8%

Analysis of Variance
Source          DF       SS       MS        F       P
Regression       1   841.77   841.77   156.89   0.000
Residual Error  13    69.75     5.37
Total           14   911.52

Predicted Values for New Observations
New
Obs      Fit   SE Fit        95% CI              95% PI
  1   27.496    0.604  (26.190, 28.801)  (22.324, 32.667)

Values of Predictors for New Observations
New
Obs   Distance
  1       3.50
```

Distance is quite significant as a predictor of Damage, since the *p*-value for the coefficient of Distance is only 0.000 (to 3 decimal places). If you only wish to consider the one-sided alternative that Damage increases with Distance, divide the printed p-value by 2. The model explains 92% (R-Sq) of the variation in Damages, so this model should be quite useful for prediction. Our prediction of the Damages in a burning house 3.5 miles away from the fire station, is anywhere from $22,324 to $32,667, with 95% confidence (from the PI).

From the slope, we can estimate that for every extra mile away from the station, average damages increase by $4,919, or more precisely by between $4.9193 \pm t_{.025,\ 13DF} (.3927)$ thousands of dollars, i.e. $4070 to $5768, with 95% confidence.

Chapter 12 – Multiple Regression and Model Building

In this chapter, we will, fit a multiple regression model, with
Stat>Regression>Regression

Construct data plots, using
Graph>Scatterplot
Graph>Matrix Plot

Superimpose a fitted quadratic model on a data plot, with
Stat>Regression>Fitted Line Plot

Calculate higher order predictor terms for the model, or log-transform the response, using
Calc>Calculator

Calculate dummy variables with
Calc>Make Indicator Variables

Find p-values for the F distribution, with
Calc>Probability Distributions>F

Run a step-wise regression with
Stat>Regression>Stepwise

Construct a batch of four useful residual plots, with
Stat>Regression>Residual Plots

Check for multicollinearity of predictors, using
Stat>Basic Statistics>Correlation

Fitting & Inference for the Model and Parameters

To fit a first order model in k predictors, we again may apply the *least squares criterion*, i.e., choose the plane which minimizes the sum of the squared differences between the observed responses and the predicted or fitted responses. The math is messy. Let Minitab's **Stat>Regression>Regression** do the hard work.

The coefficient of determination, *R-Sq*, and its adjusted version *R-Sq (adj)*, estimate how well the predictors jointly explain changes in the response. The *F* statistic and *p*-value from the Analysis of Variance (ANOVA) table test the predictive utility of the model as a whole.

Next to each *Predictor*, we see the fitted *Coef*ficient and its estimated standard error (*SE*). Their ratio, denoted '*T*' may be used to test the null hypothesis that the true coefficient (β) equals zero, and the corresponding *p*-value is also shown (assuming a two-sided alternative).

Examples 12.1 & 12.3 & 12.4 - M/S pages 631, 636, 641

Open up the GFCLOCKS data file showing a sample of
32 auction prices of grandfather clocks, along with their
age and the number of bidders. We want to investigate
the dependence of price on age and #bidders.

↓	C1 AGE	C2 NUMBIDS	C3 PRICE
1	127	13	1235
2	115	12	1080
3	127	7	845
4	150	9	1522

We can plot *Sale Price* versus each possible predictor
using **Graph>Scatterplot-Simple** with
Multiple Graphs: Multiple Variables: *In separate panels*, as below:

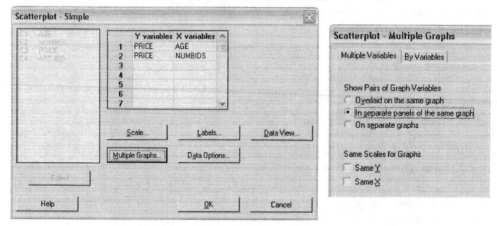

We get

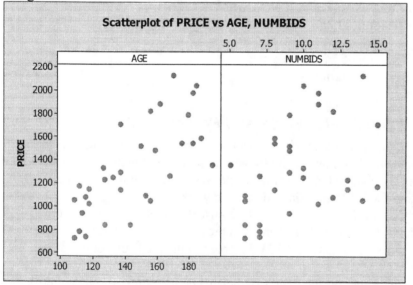

Alternatively, you can produce the same display using **Graph>Matrix Plot**:

Price has a positive relationship with each predictor, with Age being the stronger single predictor. But perhaps a model with both predictors will explain Price better than any one predictor alone. To run a multiple regression on all predictors, simply use **Stat>Regression>Regression**, as before, but now enter both variables next to **Predictors:**

Part of the output is below:

Regression Analysis: PRICE versus AGE, NUMBIDS

```
The regression equation is
PRICE = - 1339 + 12.7 AGE + 86.0 NUMBIDS

Predictor      Coef   SE Coef       T      P
Constant    -1339.0     173.8   -7.70  0.000
AGE         12.7406    0.9047   14.08  0.000
NUMBIDS      85.953     8.729    9.85  0.000

S = 133.485   R-Sq = 89.2%   R-Sq(adj) = 88.5%

Analysis of Variance
Source          DF        SS       MS       F      P
Regression       2   4283063  2141531  120.19  0.000
Residual Error  29    516727    17818
Total           31   4799790
```

At the top is the fitted regression equation in both predictors. The parameter estimates also appear under **Coef**. The sum of squared errors minimized by this fit is equal to the **Residual Error SS = 516727**, and the estimated standard deviation of the errors is $s =$ **133.485** (or square root the Residual Error MS).

To test the hypothesis that mean auction price increases with the number of bidders, controlling for age, we take the Coef of Numbids, and divide by its estimated SE, producing the displayed T value of **9.85**. Assuming a two-sided alternative, Minitab calculates p as $P(|t|>9.85) =$ **0.000**. Divide by 2 for the correct p-value here.

To find a 90% CI for the true coefficient of Age, take the Coef shown above, and wrap $t_{.05,29df}$ (SE Coef) around it: $\mathbf{12.7406} \pm \mathbf{1.699(0.9047)}$.

We see that this model explains about **89.2%** (*R-Sq*) of the variation in prices (or 88.5%, adjusted for sample size and # predictors). The **F**-value of **120.19** is highly significant with $p = \mathbf{.000}$ ($< \alpha = .05$), so the model as a whole is useful for predicting prices.

Exercise 12.21 - M/S page 645

Open up the SNOWGEESE data file, and regress Weight Change (%) in C3 on Digestive Efficiency (%) in C4 and Acid-Detergent Fibre (%) in C5, using **Stat>Regression>Regression: Response:** *C2*, **Predictors:** *C3 C4*. We get:

Regression Analysis: WtChange versus DigEff, ADFiber

```
The regression equation is
WtChange = 12.2 - 0.0265 DigEff - 0.458 ADFiber

Predictor       Coef  SE Coef       T      P
Constant      12.180    4.402    2.77  0.009
DigEff      -0.02654  0.05349   -0.50  0.623
ADFiber      -0.4578   0.1283   -3.57  0.001

S = 3.51948   R-Sq = 52.9%   R-Sq(adj) = 50.5%

Analysis of Variance

Source          DF       SS       MS       F      P
Regression       2   542.03   271.02   21.88  0.000
Residual Error  39   483.08    12.39
Total           41  1025.12
```

The model is highly significant, since $p = 0.000$ for the regression model, and the model explains *52.9%* of the variation in earnings (*50.5%*, adjusted). Digestive efficiency is not a significant predictor in the model since the *p*-value for its coefficient is quite large (.623), whereas A-D fibre is significant (*p*=.001), and we can estimate with 99% confidence that for every additional 1% fibre, weight goes down by $0.458 \pm t_{.005,39} (0.128)$ percent.

Using the Model for Estimation and Prediction

To find a confidence interval for the mean response, or a prediction interval for a future random observation on the response, run the **Regression** procedure with **Options…: Prediction Intervals for new observations**, and specify the desired settings for the predictors.

Example 12.5 - M/S page 649

Open up the GFCLOCKS data file again. (a) For a 95% CI for the average auction price for 150 year old clocks with 10 bidders, run **Stat>Regression>Regression** and plug in as follows:

Note that you have to enter the values of the predictors in the exact same order that you entered the predictors in the main dialog box. E.g. if we had entered "10 150", Minitab would assume 10 is the age, with 150 bidders. **OK, OK**, and at the bottom of the session window output, we see:

```
Predicted Values for New Observations
New
Obs     Fit   SE Fit        95% CI              95% PI
  1  1431.7     24.6  (1381.4, 1481.9)  (1154.1, 1709.3)


Values of Predictors for New Observations
New
Obs  Age  Bidders
  1  150     10.0
```

Using the 95% CI above, we can say with 95% confidence that the mean auction price for all 150 year old clocks with 10 bidders is between $1381.40 and $1481.90.

(b) The above output also gives you the prediction interval (95% PI) for the price of a single clock being sold with these same characteristics. With 95% confidence, its price should fall between $1154 and $1709.

(c) To predict the auction price for one 50 year old clock that has 2 bidders, proceed as above, but now plug in "*50 2*" for the **new observations**, and read off the *PI* instead of the *CI* in the output. However, the interval would not be of much use, since this would constitute *extrapolation* well beyond the range of the observed data (always plot the data in various ways before trusting any statistical analysis!). But, even if you proceed naively, Minitab tries to help! You get the requested output, but with a warning (also note the ridiculous negative intervals):

```
Predicted Values for New Observations
New
Obs    Fit  SE Fit        95% CI              95% PI
  1  -530.0   123.0  (-781.5, -278.5)  (-901.2, -158.8)XX
XX denotes a point that is an extreme outlier in the predictors.


Values of Predictors for New Observations
New
Obs   Age  Bidders
  1  50.0     2.00
```

Interaction and Higher Order Models

A first order model in the predictors may not be the correct model. There may be interaction in the way the predictors affect the response, or the predictors may affect the response in a non-linear fashion. Use **Calc>Calculator** to calculate higher order terms such as a product like x_2x_3 or a squared term like x_4^2. Then include these terms among the predictors when you regress, so that you can examine and test their contribution to the model.

Example 12.6 - M/S page 655

Open up again the grandfather clock data. Do age and #bidders interact in their effect on clock price? To answer this, enter an interaction term as an additional predictor in the model, and then test its significance. Use **Calc>Calculator**, to calculate the product of *Age* and *Bidders*, and store the result in C4 named '*AgeBid*':

```
Calculator                                            ×

 C1   Age          Store result in variable: AgeBid
 C2   Bidders
 C3   Price        Expression:
                   'Age' * 'Bidders'
```

Next, regress on all three predictors, **Stat>Regression>Regression**

```
Regression                                            ×

 C1   Age          Response: Price
 C2   Bidders
 C3   Price        Predictors Age Bidders AgeBid
 C4   AgeBid
```

And we get

Regression Analysis: Price versus Age, Bidders, AgeBid

```
The regression equation is  Price = 320 + 0.88 Age - 93.3 Bidders + 1.30 AgeBid

Predictor    Coef   SE Coef      T      P
Constant    320.5     295.1   1.09  0.287
Age         0.878     2.032   0.43  0.669
Bidders    -93.26     29.89  -3.12  0.004
AgeBid     1.2978    0.2123   6.11  0.000

S = 88.9145   R-Sq = 95.4%   R-Sq(adj) = 94.9%

Analysis of Variance
Source           DF        SS        MS       F      P
Regression        3   4578427   1526142  193.04  0.000
Residual Error   28    221362      7906
Total            31   4799790
```

The model as a whole is significant ($p=0.000$). The interaction term *AgeBid* has a *t*-ratio of *6.11* with displayed p-value of *0.000*, but since the alternative tested here is only for a 'positive' interaction (i.e. price increases more rapidly with age, when there are more bidders), the true p-value is half the posted value or 0.000/2 = .000. There is strong evidence of interaction.

Example 12.7 - M/S page 661

Open up the ELECTRIC data file. Examine the relationship between electrical usage and home size with **Graph>Scatterplot-Simple: Y:** *Usage,* **X:** *Size*

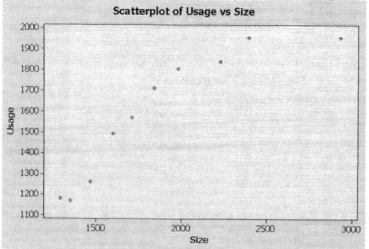

This is quite suggestive of a curved or quadratic type of relation. Let's visually examine first a linear fit, and then a quadratic fit, to the data, using **Stat>Regression>Fitted Line Plot: Type of Regression:** *Linear*

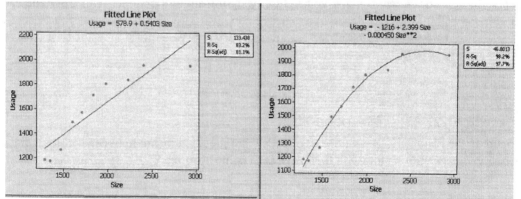

Repeat, but change to **Regression:** *Quadratic.* Here are the two resulting fitted line plots:

What a difference a square makes! The errors or deviations about the line are much smaller, and also more random in nature, for the quadratic fit. Also, note the difference in the *R-Sq* values displayed.

Define a new Size^2 predictor term to include in the model, using **Calc>Calculator**

```
Calculator                                                          ×
 C1     Size
 C2     Usage        Store result in variable:  Size-Sq

                     Expression:
                     'Size'**2
```

You will see the squares of the home sizes show up in a column named 'Size-Sq'. Now, regress on Size and Size^2, with **Stat>Regression>Regression**

```
Regression                                                          ×
 C1     Size
 C2     Usage        Response: Usage
 C3     Size-Sq
                     Predictors Size 'Size-Sq'
```

This gives us:

```
The regression equation is  Usage = - 1216 + 2.40 Size - 0.000450 Size-Sq

Predictor            Coef      SE Coef       T       P
Constant          -1216.1        242.8   -5.01   0.002
Size               2.3989       0.2458    9.76   0.000
Size-Sq        -0.00045004   0.00005908   -7.62   0.000

S = 46.8013    R-Sq = 98.2%    R-Sq(adj) = 97.7%

Analysis of Variance
Source            DF        SS        MS        F       P
Regression         2    831070    415535   189.71   0.000
Residual Error     7     15333      2190
Total              9    846402
```

We see that the model as a whole is highly significant in explaining variation in usage, with F = 189.71 and p-value = 0.000. And the very high R-Sq(adj) value of 97.7 % indicates a very close fit to the data, indeed.

  The Size-Sq coefficient is negative and highly significant in the model with a p-value of 0.000, indicating strong evidence of a concave downward trend, i.e., the rate of increase in electrical usage slows as home size increases (correct p-value for a one sided alternative is .000/2).

Exercise 12.161 - M/S page 734

Open up the PHOSPHOR data file. To investigate the relationship between soil loss and dissolved phosphorus, start with **Graph>Scatterplot-Simple: Y:** *Phosphorus*, **X:** *SoilLoss*

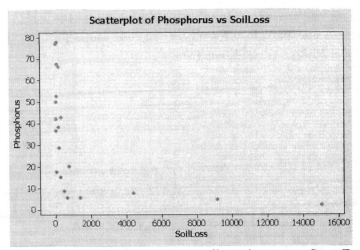

The relationship appears quite non-linear in nature. **Stat>Regression>Fitted Line Plot:**
Y: *Phosphorus*, **X:** *SoilLoss*, **Regression Model:** *Quadratic*, gives us:

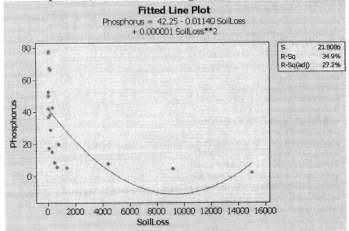

Again, we see that the relationship is clearly non-linear, but even a quadratic model
appears to be questionable. Form the predictor $SoilLoss^2$, using **Calc>Calculator:**
Expression: *SoilLoss **2*, **Store in:** *SoilLoss-Sq*, and then regress with **Stat>Regression**
>Regression: Response: *Phosphorus*, **Predictors:** *SoilLoss SoilLoss-Sq*, and we get:

```
The regression equation is
Phosphorus = 42.2 - 0.0114 SoilLoss + 0.000001 SoilLoss-Sq

Predictor            Coef     SE Coef        T       P
Constant           42.247       5.712     7.40   0.000
SoilLoss         -0.011404    0.005053    -2.26   0.037
SoilLoss-Sq     0.00000061  0.00000037     1.66   0.115

S = 21.8086   R-Sq = 34.9%   R-Sq(adj) = 27.2%

Analysis of Variance
Source           DF       SS       MS       F       P
Regression        2   4325.4   2162.7    4.55   0.026
Residual Error   17   8085.5    475.6
Total            19  12410.9
```

We examine the SoilLoss-Sq term, and since its p-value of 0.115 exceeds $\alpha = .05$, we find the evidence for quadratic curvature to be insufficient. While this result appears mysterious, look at the fitted line plot for a straight-line model:

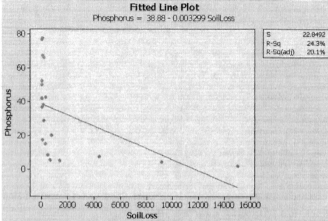

Fitted Line Plot
Phosphorus = 38.88 - 0.003299 SoilLoss

S	22.8492
R-Sq	24.3%
R-Sq(adj)	20.1%

Did the quadratic model above reduce the (vertical) deviations about the fitted line? Not by much!

The real problem here is that neither a simple straight line nor quadratic curve fits well. A transformation of one or both of the variables might help, or perhaps the range of values for soil loss is too great for us to fit just one simple model.

Qualitative Predictors – Using dummies

The regression model can only be fit using quantitative predictors, so a special coding is needed to include qualitative variables as predictors. Minitab's **Calc>Indicator Variables** creates k dummy variables for a qualitative variable with k levels, where each dummy distinguishes a particular setting or level of the variable (= '1' for that level, = '0' otherwise). Choose only k-1 of them to use as predictors.

Example 12.9 - M/S page 673

Open the GOLFCRD data file. Notice the stacked arrangement of the data with brand in C1 and distance in C2. *Delete the unstacked data you will see in columns C3-C6, so that we can reuse these columns, below.*

↓	C1-T	C2
	Brand	Distance
1	A	251.2
2	A	245.1
3	A	248.0
4	A	251.1

We want to create dummy (indicator) variables to replace the qualitative predictor Brand, in C1.
Use **Calc>Make Indicator Variables:**

Make Indicator Variables   ✕

| C1 | Brand |
| C2 | Distance |

Indicator variables for:
Brand

Store results in:
c3-c6

Note that since there are **four brands**, we need to name **four columns** for the indicator variables, with each column being an *indicator* of one of the brands. A portion of the new worksheet is shown below (Minitab's new release also adds useful names to

↓	C1-T	C2	C3	C4	C5	C6	C7
	Brand	Distance					
1	A	251.2	1	0	0	0	
2	A	245.1	1	0	0	0	
3	A	248.0	1	0	0	0	
4	A	251.1	1	0	0	0	
5	A	260.5	1	0	0	0	
6	A	250.0	1	0	0	0	
7	A	253.9	1	0	0	0	
8	A	244.6	1	0	0	0	
9	A	254.6	1	0	0	0	
10	A	248.8	1	0	0	0	
11	B	263.2	0	1	0	0	
12	B	262.9	0	1	0	0	
13	B	265.0	0	1	0	0	
14	B	254.5	0	1	0	0	
15	B	264.3	0	1	0	0	
16	B	257.0	0	1	0	0	
17	B	262.8	0	1	0	0	
18	B	264.4	0	1	0	0	
19	B	260.6	0	1	0	0	
20	B	255.9	0	1	0	0	
21	C	269.7	0	0	1	0	
22	C	263.2	0	0	1	0	

In C3 is the indicator for Brand A ('1' if Brand A, '0' otherwise), in C4, the indicator for Brand B ('1' if Brand B, '0' otherwise), etc. Note Minitab works alphabetically, assigning the first dummy (C3) to the category first in alphabetical order ('A'), etc. You should also add some helpful names to the new columns, like 'Brand_A', 'Brand_B', etc.

   Now, we regress on the indicators, but we cannot use all of them. Pick any three that you wish, depending on which comparisons you wish to set up via the coefficients of the dummies. To follow the text, we omit the first dummy above, and then

Stat>Regression>Regression: Response: *C2*, **Predictors:** *C4-C6*   gives us:

```
The regression equation is
Distance = 251 + 10.3 C4 + 19.2 C5 - 1.46 C6

Predictor      Coef   SE Coef       T       P
Constant    250.780     1.455  172.34   0.000
C4           10.280     2.058    5.00   0.000
C5           19.170     2.058    9.32   0.000
C6           -1.460     2.058   -0.71   0.483

S = 4.60163    R-Sq = 78.6%    R-Sq(adj) = 76.8%

Analysis of Variance
Source            DF        SS      MS       F       P
Regression         3   2794.39  931.46   43.99   0.000
Residual Error    36    762.30   21.18
Total             39   3556.69
```

The F value = *43.99* and p-value = *0.000* above are *exactly* the same as those obtained when running a one-way ANOVA before. Again we conclude that brand does affect distance.

Exercise 12.79 - M/S page 679

(b), (f), (g) Open up the NZBIRDS data file, where in each row, we have data on one of the species, with species body mass in c9 and species diet type in c7. Since diet type is qualitative, with four categories, we need to form dummy variables. Use **Calc>Make Indicator Variables: Indicator Variables for:** *Diet*, **Store in:** *C11-C14*. Note we need to specify 4 columns for dummies since there are 4 diets. Now regress on any 3 of the indicators, say the last three: **Stat>Regression>Regression: Response:** *'Body Mass'*, **Predictors:** *C12-C14*, gives us:

Regression Analysis: Body Mass versus C12, C13, C14

```
The regression equation is
Body Mass = 903 + 26206 C12 - 660 C13 + 2997 C14

Predictor    Coef   SE Coef      T      P
Constant      903      4171   0.22  0.829
C12         26206      6090   4.30  0.000
C13          -660      5772  -0.11  0.909
C14          2997     14298   0.21  0.834

S = 27352.1   R-Sq = 16.5%   R-Sq(adj) = 14.5%

Analysis of Variance

Source          DF          SS          MS     F      P
Regression       3  18924196564  6308065521  8.43  0.000
Residual Error 128  95761566077   748137235
Total          131  1.14686E+11
```

The low p-value for the F statistic in the ANOVA table gives good evidence that mean body mass does depend on the type of diet of the species.

Each dummy's **Coef** above estimates the difference in mean body weight between species with some particular type of diet ('indicated' by that dummy) and species with a fish diet, since we omitted c11 and c11=1 if Fish diet (check the worksheet). E.g. since c13 = 1 if diet is invertebrates, we estimate that such species have a mean biomass **660** grams less than those eating fish, on average. Look at the worksheet, to be sure you understand the dummies. It might be helpful to give your dummies more suggestive names, like "Fish" for C11, etc.

Note that you can produce the same ANOVA table by running a one-way ANOVA: **Stat>ANOVA>One-way: Response:** *'Body Mass'*, **Factor:** *Diet*. We get:

One-way ANOVA: Body Mass versus Diet

```
Source  DF          SS          MS     F      P
Diet     3  18924196564  6308065521  8.43  0.000
Error  128  95761566077   748137235
Total  131  1.14686E+11

S = 27352   R-Sq = 16.50%   R-Sq(adj) = 14.54%
```

Using Quantitative and Qualitative Variables in the Model

The regression model (also called the *general linear model*) allows you to fit by least-squares one coherent model with quantitative variables, qualitative variables, and higher order terms for curvature and interaction. Just **Calc** the terms desired and **Regress**.

Example 12.11 - M/S page 681

Open up the CASTING data file. We want to fit a model relating productivity (#castings) to incentive and type of plant, allowing for interaction in the effects of these two variables. First, form dummies for the type of plant. Since there are two types of plant, two dummy variables will be created by Minitab, using **Calc>Make Indicator Variables:**

Minitab operates alphabetically, so looking at the worksheet, you can see that C4 is the indicator for Nonunion, and C5 is the indicator for Union. We only need one, so following the text, we will use the indicator for Nonunion, in C4, and let's name C4 'Dummy'.

Next, we need an interaction term, which is formed by taking the product of Incentive in C3 with the dummy for Nonunion: **Calc>Calculator:**

Now, regress on three predictors: **Stat>Regression>Regression:**

[After double-clicking on C6 to copy it over, its name appears in parentheses. These parentheses are necessary for clarity when using a *column name containing spaces or some special symbols like the '*' that I used*]

We get

```
The regression equation is
Castings = 1366 + 6.22 Incentive + 47.8 Dummy + 0.03 Inc*Dummy

Predictor      Coef   SE Coef      T      P
Constant    1365.83     51.84  26.35  0.000
Incentive     6.217     1.667   3.73  0.002
Dummy         47.78     73.31   0.65  0.525
Inc*Dummy     0.033     2.358   0.01  0.989

S = 40.8387   R-Sq = 71.1%    R-Sq(adj) = 64.9%

Analysis of Variance
Source          DF      SS      MS      F      P
Regression       3   57332   19111  11.46  0.000
Residual Error  14   23349    1668
Total           17   80682
```

The fitted equation appears above. Plug in '1' for Dummy (whence Inc*Dummy = Incentive), to obtain the fitted relation for the non-union plants and plug in '0' for Dummy (whence Inc*Dummy = '0') to obtain the relation for the union plants.

Does the relation between number of castings produced and incentive depend on the type of plant? The coefficient of the product term has a large p-value of **0.989**, so there is no evidence of such a dependence, or interaction.

You can plot the data with superimposed regression lines separately fitted for each group (i.e. plant type) by using: **Scatterplots-With Regression and Groups: Y:** *Castings,* **X:** *Incentives,* **Categorical variables:** *Plant*

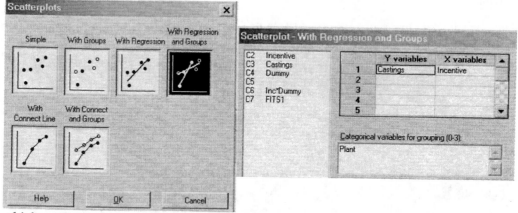

which produces:

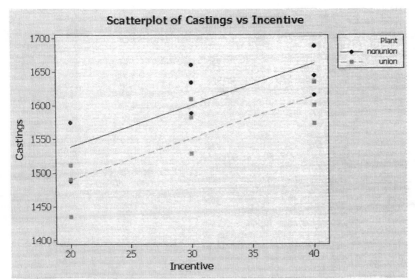

Note how nearly parallel the fitted lines are, consistent with our finding of no significant interaction.

Comparing Nested Models

We can test for significance a portion of the predictors in any model, by running two regressions, one for the *complete model* and one for the *reduced model*, then comparing SSE's. Or you may use the *Sequential SS* shown in the output for the complete model.

Example 12.12 - M/S page 688

Open the CARNATIONS data file. We wish to fit a complete second order model to the data, and assess the contribution of the 2nd order terms. First, calculate the three 2nd order terms (if not already calculated for you in the saved worksheet), using **Calc>Calculator:**

Now, regress Height on all predictors with **Stat>Regression>Regression**

```
Regression                                                           ×
  C1    Temp          Response: Height
  C2    Fertilizer
  C3    Height        Predictors  Temp Fertilizer 'Temp-Sq'
  C4    Temp-Sq                   'Fert-Sq' TempFert
  C5    Fert-Sq
  C6    TempFert
```

We get

```
The regression equation is
Height = - 5128 + 31.1 Temp + 140 Fertilizer - 0.133 Temp-Sq - 1.14 Fert-Sq
         - 0.145 TempFert

Predictor         Coef    SE Coef        T       P
Constant       -5127.9      110.3   -46.49   0.000
Temp            31.096      1.344    23.13   0.000
Fertilizer     139.747      3.140    44.50   0.000
Temp-Sq      -0.133389   0.006853   -19.46   0.000
Fert-Sq       -1.14422    0.02741   -41.74   0.000
TempFert     -0.145500   0.009692   -15.01   0.000


S = 1.67870   R-Sq = 99.3%   R-Sq(adj) = 99.1%


Analysis of Variance
Source          DF      SS       MS        F       P
Regression       5  8402.3   1680.5   596.32   0.000
Residual Error  21    59.2      2.8
Total           26  8461.4

Source         DF   Seq SS
Temp            1   1510.7
Fertilizer      1    279.3
Temp-Sq         1   1067.6
Fert-Sq         1   4909.7
TempFert        1    635.1
```

The fitted equation appears above. To assess the contribution of the second order terms, regress again, using only the predictors Temp and Fertilizer (the reduced model):

```
The regression equation is
Height = 106 - 0.916 Temp + 0.788 Fertilizer

Predictor       Coef   SE Coef        T       P
Constant      106.09     55.95     1.90   0.070
Temp         -0.9161    0.3930    -2.33   0.028
Fertilizer    0.7878    0.7860     1.00   0.326


S = 16.6727   R-Sq = 21.2%   R-Sq(adj) = 14.6%

Analysis of Variance
Source          DF      SS      MS      F       P
Regression       2  1789.9   895.0   3.22   0.058
Residual Error  24  6671.5   278.0
Total           26  8461.4
```

Take the difference of the SSE's $= SSE_R - SSE_C = 6671.5 - 59.2 = 6612.3$, plug it into the F statistic $= 6612.3/3 \div (59.2/21) = 781.6$, and then find the P-value, either approximately from your F-tables, or by using **Calc>Probability Distributions> F:** *Cumulative Probability,* **Numerator DF: 3, Denominator DF: 21, Input Constant:** *781.6*

Cumulative Distribution Function

```
F distribution with 3 DF in numerator and 21 DF in denominator

     x        P( X <= x )
  781.6             1
```

Hence, the p-value is $P(F > 781.6) = 1 - 1 = 0$ (to 5 decimal places), providing very strong evidence that the second order model fits the data better than a first order model.

Alternatively you can find $SSE_R - SSE_C$ by using the *Seq SS*** display in the output for the complete model and adding up the 3 Seq SS's for the 3 second order terms, which equals *1067.6 + 4909.7 + 635.1* = 6612.4, the same as $SSE_R - SSE_C$, aside from rounding error. This approach works **only** if we list the predictors being tested last in the *Seq SS* table (so list them last in the Regression dialog box list of predictors). No need to separately fit the reduced model.

** The *Seq SS* for a predictor gives the drop in the SSE caused by adding that predictor to a model with all the predictors listed earlier (above) in the table.

Stepwise Regression

Often we have many candidate predictors available. A common screening procedure to help select a useful subset of the candidate predictors is called stepwise regression. Use Minitab's **Stat>Regression>Stepwise**, and enter all the predictors being considered.

Example 12.13 - M/S page 699

Open the EXECSAL data file. We have 10 candidate predictors X1 – X10 (see text for their descriptions) to consider for a model to be used to predict log executive salary. Run Minitab's stepwise regression procedure: **Stat>Regression>Stepwise**, and fill in the **Predictors** and **Response**. Also, click on **Methods…**

Review the choice of **Methods** and change as desired. The default choices shown above are to enter or remove a variable based on a *15%* significance test. You may choose an **initial** set of predictors to start with, if you wish, or force some **predictors to be included in every model**. Here, we begin screening with no predictors and then add one at a time (with possible removals along the way). The results are:

```
Stepwise Regression: Y versus X1, X2, X3, X4, X5, X6, X7, X8, X9, X10

   Alpha-to-Enter: 0.15   Alpha-to-Remove: 0.15

Response is Y on 10 predictors, with N = 100

Step              1        2        3        4        5
Constant     11.091   10.968   10.783   10.278    9.962

X1           0.0278   0.0273   0.0273   0.0273   0.0273
T-Value       12.62    15.13    18.80    24.68    26.50
P-Value       0.000    0.000    0.000    0.000    0.000

X3                     0.197    0.233    0.232    0.225
T-Value                 7.10    10.17    13.30    13.74
P-Value                 0.000    0.000    0.000    0.000

X4                             0.00048  0.00055  0.00052
T-Value                           7.32    10.92    11.06
P-Value                          0.000    0.000    0.000

X2                                      0.0300   0.0291
T-Value                                   8.38     8.72
P-Value                                  0.000    0.000

X5                                               0.00196
T-Value                                             3.95
P-Value                                           0.000

S            0.161    0.131    0.106   0.0807   0.0751
R-Sq         61.90    74.92    83.91    90.75    92.06
R-Sq(adj)    61.51    74.40    83.41    90.36    91.64
Mallows C-p  343.9    195.5     93.8     16.8      3.6
```

There were five steps, and then the process stopped. At **Step 1**, X1 (Experience) was added to the model. The resulting equation (11.091 + 0.0278 X1), t-value for the coefficient of X1 (12.62), s (0.161), and R-squared (61.90%), are all shown in the column underneath Step 1. At **Step 2**, X3 was added to the model, and the new coefficients, t-values, s, and R-squared are printed, in the next column. At **Step 5**, X5 was added to the model. No variables can be removed (all are significant at 15% level). The remaining candidates are assessed and none of them contribute significantly. The process stops.

Residual Analysis - Checking the Assumptions

Residual plots help us check our assumptions and gain insight into possible revisions of the model. They can easily be created using the **Graphs** sub-dialog box when running Minitab's **Stat>Regression>Regression** *or* **Fitted Line Plot**.

Example 12.14 - M/S page 704

Open the ELECTRIC data file. Select **Stat>Regression>Regression,** and plug in the response (Usage) and predictor (Size). To display all the residuals, you need to increase the amount of information displayed, so choose **Results**, and go for the maximal display including *the full table of fits and residuals*:

Regression

Response: USAGE

Predictors: SIZE

Regression - Results

Control the Display of Results

○ _D_isplay nothing

○ _R_egression equation, table of coefficients, s, R-squared, and basic analysis of variance

○ _I_n addition, sequential sums of squares and the unusual observations in the table of fits and residuals

● _I_n addition, the full table of fits and residuals

Graphs...

Results...

Select

Help   OK   Help   Cancel   Help   OK

We get:

Regression Analysis: Usage versus Size

The regression equation is Usage = 579 + 0.540 Size

Predictor	Coef	SE Coef	T	P
Constant	578.9	167.0	3.47	0.008
Size	0.54030	0.08593	6.29	0.000

S = 133.438   R-Sq = 83.2%   R-Sq(adj) = 81.1%

Analysis of Variance

Source	DF	SS	MS	F	P
Regression	1	703957	703957	39.54	0.000
Residual Error	8	142445	17806		
Total	9	846402			

Obs	Size	Usage	Fit	SE Fit	Residual	St Resid
1	1290	1182.0	1275.9	66.0	-93.9	-0.81
2	1350	1172.0	1308.3	62.1	-136.3	-1.15
3	1470	1264.0	1373.2	55.0	-109.2	-0.90
4	1600	1493.0	1443.4	48.6	49.6	0.40
5	1710	1571.0	1502.8	44.7	68.2	0.54
6	1840	1711.0	1573.1	42.3	137.9	1.09
7	1980	1804.0	1648.7	43.1	155.3	1.23
8	2230	1840.0	1783.8	51.8	56.2	0.46
9	2400	1956.0	1875.7	61.5	80.3	0.68
10	2930	1954.0	2162.0	99.6	-208.0	-2.34R

R denotes an observation with a large standardized residual.

To graph the residuals, repeat the above, but this time, also select **Graphs...** from the **Regression** dialog box above, and fill in:

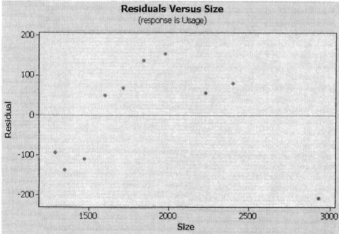

OK. OK, and we now get the regression output again, along with the following plot of **Residuals versus:** *Size*:

The points show a clear curving pattern, indicating a lack of fit of the straight-line model, and a need to amend it to allow for curvature. You can also easily produce other types of useful residual plots by ticking off your desired choices in the **Regression - Graphs** box.

Next let's try on for size, a quadratic model. We could repeat all of the above, after calculating and adding a 'Size-squared' term to the predictors, or more quickly, let's just select **Stat>Regression>Fitted Line Plot: Type of Regression:** *Quadratic,* and just like above, select and request **Graphs…: Residuals versus:** *Size,* in the dialog box. We see:

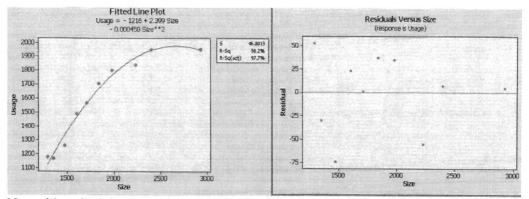

Now, things look better, with no systematic patterns in the residual plot - just a random bobbing about '0'.

Examples 12.15 & 12.16 - M/S pages 707 & 710

Open up the GFCLOCKALT data file. We wish to regress auction price on two predictors and their product, and then perform a residual analysis. First calculate the product term with **Calc>Calculator: Store in:** *AgeBid*, **Expression:** *Age * Bidders* . Then run **Stat>Regression>Regression,** click on **Graphs...,** and request the plot of *Residuals* versus *Bidders*, and the two plots that will help you check normality of the residuals:

If you want to print out all of the residuals, like shown in Figure 12.36 in the text, use the sub-dialog box for **Results...** and click on *In addition, the full table of fits and residuals*. Usually, though, there is no need to print out all the residuals. **OK, OK** produces:

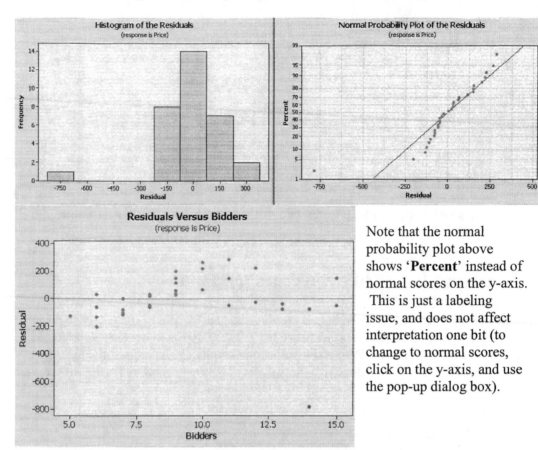

Note that the normal probability plot above shows '**Percent**' instead of normal scores on the y-axis. This is just a labeling issue, and does not affect interpretation one bit (to change to normal scores, click on the y-axis, and use the pop-up dialog box).

If you use the 3rd or 4th control level of the display of **Regression - Results…**, Minitab will flag any *unusual observations*, with symbols 'R' or 'X':

```
Unusual Observations
Obs   Age    Price      Fit   SE Fit   Residual   St Resid
 17   170   1131.0   1914.4    116.7     -783.4     -4.80R
 31   194   1356.0   1480.7    133.6     -124.7     -0.83 X

R denotes an observation with a large standardized residual.
X denotes an observation whose X value gives it large influence.
```

'*R*' indicates an observation with a large residual (*St*andardized *Resid*ual bigger than 2 in magnitude). '*X*' indicates an observation with unusual values for the predictors. The latter may have a large influence on the fit, while possessing a small residual.

Minitab has flagged the observation that was altered here, which has an extremely large residual of –783.4, or –4.80 standardized. Standardized residuals are handy for assessing relative size, since standardized values are not scale-dependent. [The standardizing operation does not use 's' as the standard deviation of a residual. It's a bit more complicated.]

The outlier, observation #17, stands out clearly in all the plots (and judging by the curvature in the plot of residuals versus Bidders, we should also consider adding Bidders^2 to the model). Outliers should be investigated and handled with considerable care, as one sole outlier can badly distort the fit and hence your understanding of the true relationship for the general population.

I recommend routine use of: **Stat>Regression>Regression: Graphs: Residual Plots:** *Four in one,* which produces a nice batch of four useful plots:

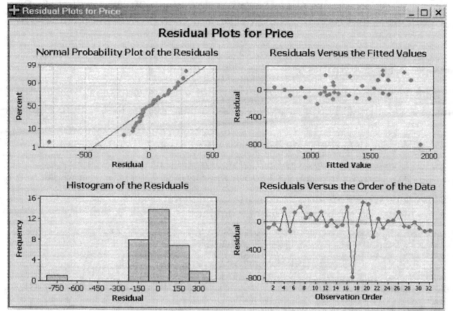

We get a normal probability plot of the residuals, a plot of residuals versus row number (Order of the Data), a histogram of residuals, and residuals versus the fits or predicted values (same as residuals versus predictor, except for labeling, when there is only one predictor). In all of these, that one outlier stands out clearly. We can also use these plots to check for various problems with the model (e.g. non-normality, non-constant variance).

If we delete the outlier (row 17), re-run the regression, and use **Graphs: Residual Plots:** *Four in one*, we now get:

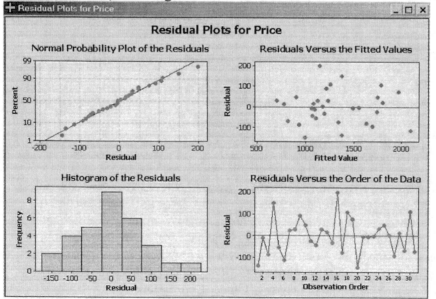

The error distribution now is quite close to normal. Variance is quite constant. You can also verify from the rest of the output that *s* and *R-squared* are much improved too.

The lack of pattern in 'Residuals vs. Fitted Values' suggests there is no obvious problem with the model we fitted, but you can also plot the residuals vs. each individual predictor (like #bidders). 'Residuals vs. Fits' contains some overlapping information with each of the latter plots, since the fits (predictions) are linear combinations of all the predictors.

Example 12.17 - M/S page 712

Sometimes, a transformation of the data, like a logarithm, will serve to remedy one or more problems, e.g. stabilizing the variance and/or symmetrizing the distribution of the errors.

Open up the SOCWORK data file, containing data on *salary* in C2 and years of *experience* in C1 for a sample of social workers. Regress salary on experience with **Stat>Regression>Regression: Response:** *Salary*, **Predictors:** *Experience*, and use **Graphs: Residual Plots:** *Four in one*

In the session window, you get the usual output, but the residual plots look like:

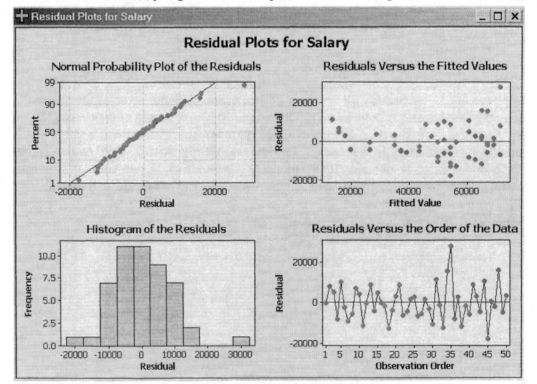

No point in drawing any conclusions from the regression output in the session window, since we have a big problem here: non-constant variance. We see this in the plot of residuals versus fitted values. The good news here is that it is a fairly steady trend, with dispersion increasing gradually as (fitted) salaries increase. Often in this situation, a transformation will remedy the problem. Let's try a log transform of the salary, using **Calc>Calculator.** Under **Functions**, find '*Natural log*' and double click it to copy it (denoted *LOGE*, which stands for 'log base e') under **Expression:**

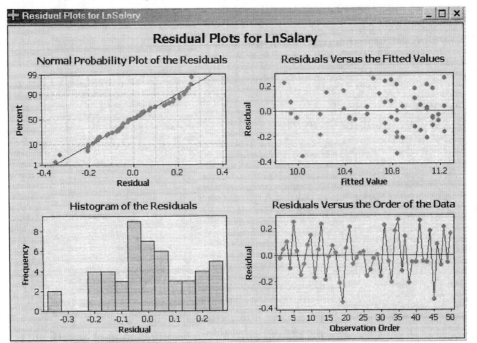

Now double click on '*C2 Salary*' on the left, to replace the highlighted '*number*' by '*Salary*' as the argument for the LOGE function. **Store result in:** *LnSalary*. Hit **OK** and the logs appear in a column named *LnSalary*.

Now, regress with the new response variable: **Stat>Regression>Regression: Response:** *LnSalary*, **Predictors:** *Experience*, with **Graphs: Residual Plots:** *Four in one*. **OK, OK**, and we get

The trend in dispersion is gone! And the distribution is still fairly close to normal. We can now use the regression output in the session window to make reliable inferences with this model.

Multicollinearity of Predictors

Use **Stat**>**Basic Statistics**>**Correlation** to check the correlations among all the predictors being considered for inclusion in the model, so as to spot possible problems of multicollinearity with the fitted model.

Example 12.18 - M/S page 720

Open up the FTC data. We are interested in regressing CO on Tar, Nicotine, and Weight. To look for possible multicollinearity problems, we find the correlations among all three predictors, using **Stat**>**Basic Statistics**>**Correlation**:

We get a two-way table of correlations:

```
Correlations: Tar, Nicotine, Weight

                  Tar   Nicotine
Nicotine        0.977
                0.000

Weight          0.491      0.500
                0.013      0.011

Cell Contents: Pearson correlation
               P-Value
```

Each entry in the table is a correlation between the variable directly above and the variable directly to the side. Underneath the correlation is its *p-value* (since all three p-values shown are small, all correlations are highly statistically significant).

We see a very high correlation of *0.977* between Tar and Nicotine. It may be difficult to disentangle the effects of these two predictors in a multiple regression. Weight has a moderate positive correlation with Tar (*0.491*) and with Nicotine (*0.500*).

If CO is regressed on any of these predictors individually, we get a significant positive relationship, as expected, but what happens if we regress CO on all three correlated predictors?   Use **Stat>Regression>Regression: Response:** *CO*, **Predictors:** *Tar Nicotine Weight*, and we get:

```
The regression equation is  CO = 3.20 + 0.963 Tar - 2.63 Nicotine - 0.13 Weight

Predictor    Coef   SE Coef       T      P
Constant    3.202     3.462    0.93  0.365
Tar        0.9626    0.2422    3.97  0.001
Nicotine   -2.632     3.901   -0.67  0.507
Weight     -0.130     3.885   -0.03  0.974

S = 1.44573   R-Sq = 91.9%   R-Sq(adj) = 90.7%

Analysis of Variance
Source          DF      SS      MS      F      P
Regression       3  495.26  165.09  78.98  0.000
Residual Error  21   43.89    2.09
Total           24  539.15
```

We see a couple of surprising negative coefficients, for *Nicotine* and *Weight*, but neither coefficient is significant. Only the t-test for *Tar* is significant ($p = 0.001$). There appears to be much redundancy (overlapping information) in these predictors.

Exercise 12.163 - M/S page 735

Open up the MNSALES data, with data on apartment building sales in Minneapolis. We are interesting in examining the relationship between sales price (C2) and various characteristics of the apartment buildings (C3-C8).

(a) Write a model relating SalePrice to Apartments and Condition, using dummy variables for the latter.

(b), (d) Use **Graph>Scatterplot-With Regression and Groups>Y:** *SalePrice,* **X:** *Apartments,* **Categorical:** *Condition*

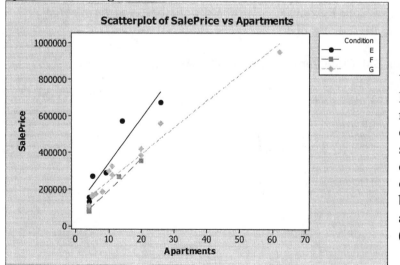

What do you think? Does the relationship different significantly depending on the condition of the building? You will answer this in part (e).

(c) Create the dummies with **Calc>Make Indicator Variables: For:** *Condition,* **Store results in:** *'Cond_E' 'Cond_F' 'Cond_G'*

   Then, run the regression: **Stat>Regression>Regression: Response:** *SalePrice,* **Predictors:** *Apartments 'Cond_F' 'Cond_G'* and you get:

Regression Analysis: SalePrice versus Apartments, Cond_F, Cond_G

```
The regression equation is
SalePrice = 188875 + 15617 Apartments - 152487 Cond_F - 103046 Cond_G

Predictor       Coef   SE Coef       T       P
Constant      188875     28588    6.61   0.000
Apartments     15617      1066   14.66   0.000
Cond_F       -152487     39157   -3.89   0.001
Cond_G       -103046     31784   -3.24   0.004
```

Now, you can plug in for the dummies, to get equations for each condition (e.g. Cond_F = 0, Cond_G = 0, for the excellent condition buildings)

(e) Use **Calc>Calculator** to form the products of the two dummies above with *Apartments* and then regress again, with these two additional predictors in the model. Test the two product terms for significance, using the approach of section 12.9. Significance means that the relationship is not the same for each condition status.

(f) To check the for multicollinearity, compute the correlations via **Stat>Basic Statistics>Correlation: Variables:** *Apartments Age LotSize ParkSpaces BldArea* and you will get:

Correlations: Apartments, Age, LotSize, ParkSpaces, BldArea

```
            Apartments       Age   LotSize   ParkSpaces
Age             -0.014
LotSize          0.800    -0.191
ParkSpaces       0.224    -0.363     0.167
BldArea          0.878     0.027     0.673        0.089

Cell Contents: Pearson correlation
```

There are some high correlations present. Let's run the regression: **Stat>Regression>Regression: Response:** *SalePrice* **Predictors:** *Apartments Age LotSize ParkSpaces BldArea* and we get:

Regression Analysis: SalePrice versus Apartments, Age, ...

```
The regression equation is
SalePrice = 92788 + 4140 Apartments - 853 Age + 0.96 LotSize + 2696 ParkSpaces
            + 15.5 BldArea

Predictor      Coef   SE Coef        T       P
Constant      92788     28688     3.23   0.004
Apartments     4140      1490     2.78   0.012
Age          -853.2     298.8    -2.86   0.010
LotSize       0.962     2.869     0.34   0.741
ParkSpaces     2696      1577     1.71   0.104
BldArea      15.544     1.462    10.63   0.000

S = 33217.9   R-Sq = 98.0%   R-Sq(adj) = 97.5%

Analysis of Variance
Source           DF           SS            MS        F       P
Regression        5   1.05290E+12   2.10581E+11   190.84   0.000
Residual Error   19   20965196398    1103431389
Total            24   1.07387E+12
```

```
Unusual Observations
Obs  Apartments  SalePrice     Fit  SE Fit  Residual  St Resid
 10        62.0     950000  930390   30651     19610    1.53 X
 23        14.0     573200  521446   21889     51754    2.07R
 25         5.0     272000  199065   12661     72935    2.37R
R denotes an observation with a large standardized residual.
X denotes an observation whose X value gives it large influence.
```

The multicollinearity likely explains the non-significant test of *LotSize* wheras *ParkSpaces* may simply be a weak predictor of sales price.

Also, from the list of *Unusual Observations,* we see that observation #10 may have a big *influence* on the fit of the regression model, due to its unusual characteristics (e.g. very large number of apartments). Remove it and refit to examine its influence. If large, consider excluding it from your analysis, with explanation.

(g) We want to regress the sale prices of apartments (C2) on the five quantitative characteristics of the apartments in C3-C7, and then conduct a residual analysis to check our assumptions. Regress using **Stat>Regression> Regression: Response:** *C2*, **Predictors:** *C3-C7*, with **Graphs: Residual Plots:** *Four in one,* and we get the usual session window regression output, along with the following residual plots:

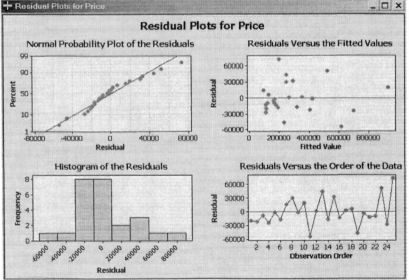

The distribution is close enough to normal, with no clear outliers, and there is no strong pattern when we look at residuals versus fits (well, maybe somewhat smaller dispersion for the very cheap houses).

We may also wish to plot the residuals versus each individual predictor, which can be done very conveniently using the same sub-dialog box, with **Regression - Graphs: Residual Plots: Residuals versus the variables:** *C3-C7*:

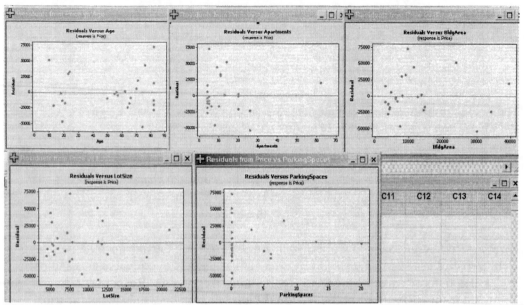

None of these show e.g. any curvature, which might indicate a need for higher order terms in that particular predictor. However, some trends in dispersion are present.

Chapter 13 – Categorical Data Analysis

In this chapter, we perform a chi-square test on a one-way table, using
Calc>Calculator
Calc>Column Statistics
Calc>Probability Distributions>Chi-Square

Perform a test of association for two categorical variables, with
Stat>Tables>Cross Tabulation and Chi-Square
Stat>Tables>Chi-Square Test (Table in Worksheet)

And graphically compare the row (or column) distributions, using
Graph>Bar Chart

One categorical variable

We can test a null hypothesis about the distribution of a single categorical variable, by calculating expected counts under the hypothesis, comparing expected counts and observed counts in a chi-square statistic, and then using the chi-square distribution to find the p-value. Unfortunately, there is no shortcut with Minitab student release 14, so you will have to work through the calculations step by step, as demonstrated below.

However, if you have the *professional version of Minitab,* you can use **Stat>Tables> Chi-Square Goodness-of-Fit Test (One variable)**

Example 13.2 - M/S page 747

Enter the four categories of opinion about marijuana possession in C1, their respective counts in C2, and the hypothesized proportions in C3.

↓	C1-T	C2	C3
	Opinion	Count	Proportion
1	Legalize	39	0.07
2	Decriminalize	99	0.18
3	ExistingLaws	336	0.65
4	None	26	0.10

We will work out the chi-square statistic, step by step, using **Calc>Calculator**.

First, find the expected counts and store them in a C4::

Now, calculate the four individual terms in the chi-square statistic, and store them in C5, named 'Terms':

Calculator

C1	Opinion
C2	Count
C3	Proportion
C4	Expected

Store result in variable: Terms

Expression:
('Count' - 'Expected')**2 / 'Expected'

This produces

↓	C1-T	C2	C3	C4	C5
	Opinion	Count	Proportion	Expected	Terms
1	Legalize	39	0.07	35	0.4571
2	Decriminalize	99	0.18	90	0.9000
3	ExistingLaws	336	0.65	325	0.3723
4	None	26	0.10	50	11.5200

Now add up the four terms, to get the value of the chi-square statistic, and store the result in a *stored constant* named 'ChiSq', using **Calc>Column Statistics:**

Column Statistics

C2	Count
C3	Proportion
C4	Expected
C5	Terms

Statistic
- Sum
- Mean
- Standard deviation
- Minimum
- Maximum
- Range
- Median
- Sum of squares
- N total
- N nonmissing
- N missing

Input variable Terms

Store result in ChiSq (Optional)

In the session window, we see: Sum of Terms = 13.2495

To evaluate significance, find the p-value or $P(\chi^2 > 13.2495) = 1 - P(\chi^2 < 13.249)$, using **Calc>Probability Distributions>Chi-Square:**

Chi-Square Distribution

| K1 | ChiSq |

- Probability density
- Cumulative probability
  Noncentrality parameter: 0.0
- Inverse cumulative probability
  Noncentrality parameter: 0.0

Degrees of freedom: 3

- Input column:
  Optional storage:

- Input constant: ChiSq
  Optional storage:

Select

We get:

```
Chi-Square with 3 DF

       x      P( X <= x )
13.2495      0.995873
```

So, the p-value is 1 - .995873 = .004, and we have strong evidence (p < α =.01) that the hypothesized proportions are not correct.

If you have the professional version of Minitab, just use: **Stat>Tables> Chi-Square Goodness-of-Fit Test (One variable): Observed counts:** *C2,* **Test: Specific Proportions:** *C3*

Exercise 13.55 - M/S page 774

To test the hypothesized probabilities, enter the counts into C1 named 'count', and the probabilities into C2 named 'prob'. Now, use **Calc>Calculator** successively to calculate the expected counts **Expression:** *85 * C2,* **Store in :** *C3,* and then **Expression:** *(C1-C3)**2/C3,* **Store in:** *C4.* Then **Calc>Column Statistics: Statistic:** *Sum,* **Input:** *C4,* **Store:** *ChiSq .* Then, **Calc>Probability Distributions> Chi-Square:** *Cumulative Probability,* **DF:** *4,* **Input constant:** *ChiSq,* and we get:

```
Chi-Square with 4 DF

       x   P( X <= x )
17.1631      0.998203
```

So, p = 1 − 0.998203 = .002, which gives strong evidence that the placebo percentages do not apply to the patients given the drug.

If you have the professional version of Minitab, use: **Stat>Tables> Chi-Square Goodness-of-Fit Test (One variable): Observed counts:** *C1,* **Test: Specific Proportions:** *C2*

[For (b) and (c), use **Stat>Basic Statistics>1 Proportion**]

Two way tables- testing for association

To test for an association between two categorical variables, where summarized data (counts) are available, go to **Stat>Tables** and use either **Chi-Square Test (Table in Worksheet)** or **Cross Tabulation and Chi-Square** depending on how the observed counts (frequencies) have been entered in the worksheet. [For raw data, you must select **Cross Tabulation**]

Example 13.3 - M/S page 759

Here we have summary counts from a two-way classification of 500 men with respect to marital status and religion. There are two possible ways to enter such data in a worksheet. One way is to enter the data just as it appears in TABLE 13.8, i.e. as shown below (you can label the columns, but not the rows):

↓	C1	C2	C3	C4	C5
	A	B	C	D	None
1	39	19	12	28	18
2	172	61	44	70	37

For a chi-square analysis, enter the columns containing the table in **Stat>Tables>Chi-Square test (Table in Worksheet)**:

Chi-Square Test (Table in Worksheet)	☒

C1	A	Columns containing the table:
C2	B	A B C D None
C3	C	
C4	D	
C5	None	

We get

Chi-Square Test: A, B, C, D, None

```
Expected counts are printed below observed counts
Chi-Square contributions are printed below expected counts

            A       B       C       D    None   Total
    1      39      19      12      28      18     116
        48.95   18.56   12.99   22.74   12.76
        2.023   0.010   0.076   1.219   2.152


    2     172      61      44      70      37     384
       162.05   61.44   43.01   75.26   42.24
        0.611   0.003   0.023   0.368   0.650


Total     211      80      56      98      55     500

Chi-Sq = 7.135, DF = 4, P-Value = 0.129
```

Since the *p-value = 0.129* is big (> α), there is insufficient evidence of a relationship between marital status and religion.

A much better way to enter the data, which gives us more flexibility, and better graphing possibilities, is to enter the data as follows:

↓	C1-T	C2-T	C3
	Religion	MaritalStatus	Counts
1	A	Divorced	39
2	B	Divorced	19
3	C	Divorced	12
4	D	Divorced	28
5	None	Divorced	18
6	A	NeverDivorced	172
7	B	NeverDivorced	61
8	C	NeverDivorced	44
9	D	NeverDivorced	70
10	None	NeverDivorced	37

Note that all the 10 counts are in C3, with religious status in C1, and marital status in C2.

For a chi-square analysis, and display of relevant percentages, use
Stat>Tables>Cross Tabulation and Chi-Square

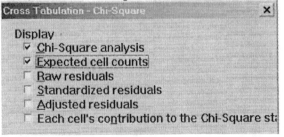

The two categorical variables under analysis must be entered under **Categorical variables** (the first listed variable forms the rows, the second forms the columns), and next to **Frequencies are in,** we indicate the column with the *summary counts* - If you do not indicate where the summary counts may be found, Minitab will think that each row in the table reflects the outcome of the two variables for *just one subject* (appropriate if the worksheet contained raw rather than summarized data). I have also chosen to display the **Counts** and the **Column percents** in the output.

Now, click on **Chi-Square...** in the box, and tick:

OK, OK, and we get the following output:

Tabulated statistics: MaritalStatus, Religion

```
Using frequencies in Counts

Rows: MaritalStatus   Columns: Religion

                    A        B        C        D     None      All

Divorced           39       19       12       28       18      116
                18.48    23.75    21.43    28.57    32.73    23.20
                48.95    18.56    12.99    22.74    12.76   116.00

NeverDivorced     172       61       44       70       37      384
                81.52    76.25    78.57    71.43    67.27    76.80
               162.05    61.44    43.01    75.26    42.24   384.00

All               211       80       56       98       55      500
               100.00   100.00   100.00   100.00   100.00   100.00
               211.00    80.00    56.00    98.00    55.00   500.00


Cell Contents:        Count
                      % of Column
                      Expected count

Pearson Chi-Square = 7.135, DF = 4, P-Value = 0.129
Likelihood Ratio Chi-Square = 6.985, DF = 4, P-Value = 0.137
```

Again, we see the *p-value of 0.0129* displayed above (ignore the *p-value=0.137* from the *Likelihood Ratio* test, given at the bottom). In the table, we calculated the relevant percentages, i.e. column rather than row percentages, so that we may see how marital status varies from religion to religion, and not the other way around.

To display, these percentages in a graph, we use **Graph>Bar Chart,** with choices shown below:

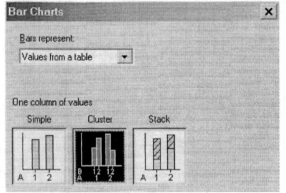

Hit **OK,** and continue as below, filling in the main dialog box, and some **Options…**

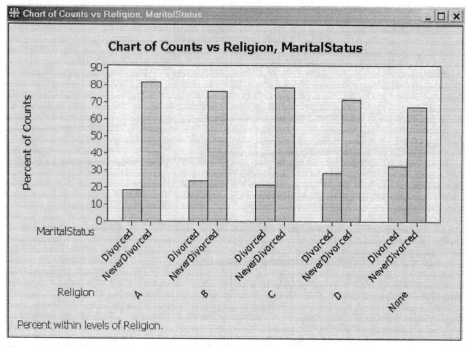

Note we made *Religion* the **outermost** variable in the plot below, and asked to plot **Y as Percent** (rather than the counts), so that these will add to 100% **within** each religion (**categories at level 1**). We get:

From the above, we can readily compare marital status from religion to religion, e.g. note that religion A has the smallest percent divorced.

Exercise 13.33 - M/S page 766

Open the ETHNIC file, containing Region, Strategy and Count in C1, C2, C3 resp. Run
Stat>Tables>Cross Tabulation and Chi-Square:
Categorical Variables: *Region, Strategy,*
Display: *Counts, Row Percents,*

Frequencies are in: *Count*;
Chi-Square… Display: *Chi-Square analysis, Expected cell counts.*

OK, OK produces the following output:

```
Tabulated statistics: Region, Strategy

Using frequencies in Count

Rows: Region   Columns: Strategy

              Mobil    None     War     All

Africa           36      39      20      95
              37.89   41.05   21.05  100.00
              38.69   36.62   19.69   95.00

LatinAmer        31      24       7      62
              50.00   38.71   11.29  100.00
              25.25   23.90   12.85   62.00

PostComm         23      32       4      59
              38.98   54.24    6.78  100.00
              24.03   22.74   12.23   59.00

SEAsia           22      11      26      59
              37.29   18.64   44.07  100.00
              24.03   22.74   12.23   59.00

All             112     106      57     275
              40.73   38.55   20.73  100.00
             112.00  106.00   57.00  275.00

Cell Contents:        Count
                      % of Row
                      Expected count

Pearson Chi-Square = 35.412, DF = 6, P-Value = 0.000
Likelihood Ratio Chi-Square = 35.031, DF = 6, P-Value = 0.000
```

Note that the categories for each variable were ordered alphabetically in the display. You can change this if desired, using **Editor>Column>Value Order,** after clicking anywhere on that column in the worksheet. Re-ordering the political strategies e.g. here makes sense, since they can be placed in a natural order from least extreme to most extreme (reordered for the graph below).

The tiny *p-value* tells us that strategy does vary according to region. For a picture of the row percentages displayed above, use
Graph>Bar Chart-Cluster, One column of values, Bars represent: *Values from a table*   with
Graph variables: *Count*, **Categorical variables:** *Region Strategy*,
Bar Chart Options…: *Show Y as Percent,* **Take percent:** *Within categories at level 1.*

OK, OK, and we get

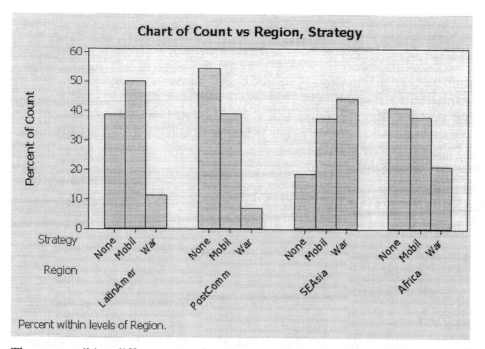

The most striking differences are between South/Southeast/East Asia, and the other three regions. We see terrorism/rebellion/war at comparatively high levels and non-action at comparatively low levels in S./S.E./E. Asia.

Chapter 14 – Nonparametric Statistics

In this chapter, we execute various non-parametric procedures, including
Stat>Nonparametrics>1-Sample Sign
Stat>Nonparametrics>Mann-Whitney (a.k.a. Wilcoxon Rank Sum Test)
Stat>Nonparametrics>1-Sample Wilcoxon (a.k.a. Wilcoxon Signed Rank Test)
Stat>Nonparametrics>Kruskal-Wallace
Stat>Nonparametrics>Friedman

We will rank data, using
Data>Rank

Unstack samples of data using
Data>Unstack Columns

The Sign Test

An alternative to the one-sample *t* test, the sign test is used to test a hypothesis about the *median* of a population (or to find a CI for the median). Look under
Stat>Nonparametrics.

Figure 14.3 - M/S page 14- 6

Open SUBABUSE or just type the data into C1 yourself, from Table 14.1. We wish to test for evidence that the median is less than *1.00*. Go to **Stat>Nonparametrics>1-Sample Sign,** enter your **Test median**, and select the correct **Alternative**:

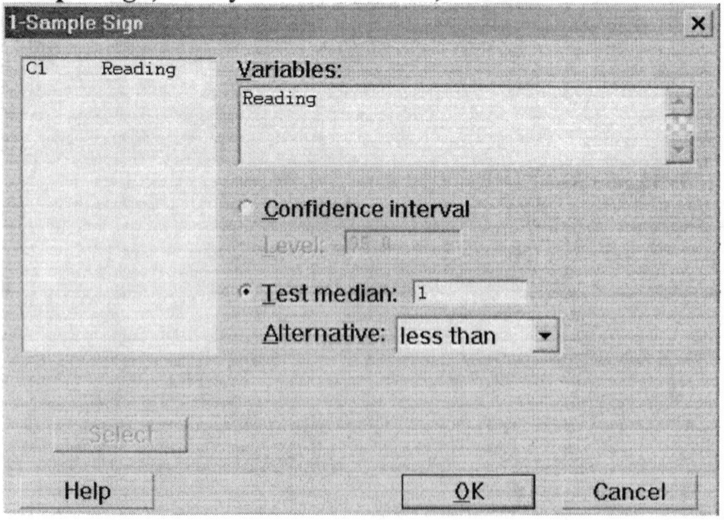

We get

Sign Test for Median: Reading

```
Sign test of median =  1.000 versus <  1.000

           N  Below  Equal  Above      P  Median
Reading    8      7      0      1  0.0352  0.8350
```

The reported *p*-value is *0.0352*, less than α = .05, so we have sufficient evidence that the median is less than 1.0.

Exercise 14.9 - M/S page 14-10

Open up the AMMONIA data file, containing the ammonia levels in C1. To test if the median exceeds 1.5, use **Stat>Nonparametrics>1-Sample Sign: Variables:** *C1*, **Test Median:** *1.5*, **Alternative:** *greater than*. We get

Sign Test for Median: Level

```
Sign test of median =  1.500 versus >  1.500

         N  Below  Equal  Above      P  Median
Level    8      4      1      3  0.7734   1.490
```

Since the sample median is *1.490*, which is less than the hypothesized median of 1.50, we cannot possibly have evidence for a median higher than 1.50. The high *p*-value just confirms the obvious.

Wilcoxon Rank Sum Test – for 2 independent samples

This procedure is a *rank-transformation* based alternative to the 2-sample *t* test for two independent samples. It doesn't appear to be on the menu, but Minitab lists another procedure called the Mann-Whitney test, which turns out to be 100% equivalent, though differing in some details of the calculations. So, we will execute the *Wilcoxon rank sum test* by applying **Stat>Nonparametrics>Mann-Whitney**. Minitab calculates the p-value using a normal approximation with continuity correction.

Example 14.2 - M/S page 14-14

Open the DRUGS data file. The reaction times are all stacked in C2 with the drug in C1 (see below). You can tell from the dialog box here (see below) that Minitab requires *unstacked data* for this procedure. So, first, we have to *unstack* the data using **Data>Unstack Columns**

Unstack Columns

| C1 | Drug |
| C2 | Reactime |

Unstack the data `Reactime`

Using subscripts `Drug`

☐ Include missing as a subscript value

Store unstacked data:
  ○ In new worksheet
    Name: _____ (Optional)
  ● After last column in use
  ☑ Name the columns containing the unstacked

[Select]

[Help]   [OK]   [Cancel]

Hit **OK** and now you see the unstacked reaction times in *C3* and *C4*:

↓	C1-T	C2	C3	C4
	Drug	Reactime	Reactime_A	Reactime_B
1	A	1.96	1.96	2.11
2	A	2.24	2.24	2.43
3	A	1.71	1.71	2.07
4	A	2.41	2.41	2.71
5	A	1.62	1.62	2.50
6	A	1.93	1.93	2.84
7	B	2.11		2.88
8	B	2.43		

To test for a difference in reaction time between the two drugs, click on
Stat>Nonparametrics>Mann-Whitney: and indicate where the two samples may be found:

Mann-Whitney

C2	Reactime
C3	Reactime_A
C4	Reactime_B

First Sample: `'Reactime_A'`

Second Samp `'Reactime_B'`

Confidence level: `95.0`

Alternative `not equal` ▾

We get:

Mann-Whitney Test and CI: Reactime_A, Reactime_B

```
             N   Median
Reactime_A   6   1.9450
Reactime_B   7   2.5000

Point estimate for ETA1-ETA2 is -0.4950
96.2 Percent CI for ETA1-ETA2 is (-0.9497,-0.1099)
W = 25.0
Test of ETA1 = ETA2 vs ETA1 not = ETA2 is significant at 0.0184
```

The p-value of *0.0184* is less than α=.05, so we have evidence of a difference between the drugs at the specified 5% level of significance. ['*ETA*' in the output is the symbol for a population median]

The normal approximation: Minitab estimates the p-value using the *normal approximation* discussed in the text, rather than doing an exact calculation. The p-value of .0184 here differs slightly from the normal approximation based p-value of .0152 shown in the SAS output in Fig 14.6 in the text, because Minitab includes a *continuity correction* of 0.5 in the top of the z statistic. With this correction and a correction for ties, Minitab's normal approximation will be fairly accurate even for samples sizes smaller than 10.

The Kruskal-Wallace test, discussed later, may also be applied here (it generalizes this test), and will give you the same z-value & p-value as displayed in the SAS output.

An exact test: To perform an *exact test*, find the rank sum of the smaller sample and assess significance from a table of exact critical values. '*W*' in the output gives the sum of the ranks for the **First Sample** in the dialog box. Make sure this is the smaller sample. In the example above, *W= 25.0* is the rank sum for drug A, which has the smaller sample size, so compare it with the critical values found in the table in the back of the text: *28* and *56*. Hence, our rank sum is significant at the 5% level.

Exercise 14.27 - M/S page 14-19

Open up the EYEMOVE data file. We wish to compare visual acuity of deaf and hearing children. Here we find the acuity values for the deaf children in *C1* and the acuity values for the hearing children in *C2*, i.e. *unstacked data*, just what is required by Minitab for this test, so we do not have to bother unstacking the data as in example 14.2 above.

↓	C1	C2
	DEAF	HEARING
1	2.75	1.15
2	3.14	1.65
3	3.23	1.43

Run **Stat>Nonparametrics>Mann-Whitney: First Sample:** *C1,* **Second Sample:** *C2,* **Alternative:** *greater than.* We get

Mann-Whitney Test and CI: DEAF, HEARING

```
          N   Median
DEAF     10   2.3750
HEARING  10   1.6450


Point estimate for ETA1-ETA2 is 0.8050
95.5 Percent CI for ETA1-ETA2 is (0.4600,1.2700)
W = 150.5
Test of ETA1 = ETA2 vs ETA1 > ETA2 is significant at 0.0003
The test is significant at 0.0003 (adjusted for ties)
```

(a) $W = 150.5$ is the rank sum for the first sample that you enter in the dialog box (the deaf children). For an exact test, use the table of critical values in the text, which gives **127** as the upper critical value for a one-sided alternative at $\alpha = .05$. Since W exceeds 127, we have evidence at the 5% level that visual acuity tends to be higher for deaf children.

(b) The very small, normal approximation based, p-value of 0.0003 ($< .05$) displayed in the output also supports the conclusion of superior acuity for deaf children. [If you wish to see the actual numerical value of the z statistic, use **Stat>Nonparametrics>Kruskal-Wallace** after stacking the data with **Data>Stack>Columns**]

Wilcoxon Signed Rank Test – for paired data

This is a rank-based nonparametric alternative to the paired difference t test, which can be applied to paired data. Minitab applies the approximate test based on a normal approximation, but because a continuity correction is also applied, the results will be quite accurate even for rather small samples. You have to form the differences yourself, and then apply the **1-Sample Wilcoxon** test to the single column of differences.

Example 14.3 - M/S page 14-23

Open up the CRIMEPLAN data file. The data are paired, with each expert's rating of plan A in one column and rating of plan B in another column.

↓	C1	C2	C3
	Expert	A	B
1	1	7	9
2	2	4	5
3	3	8	8
4	4	9	8

Find the difference in ratings for each expert, and place these in a column named *Difference*:

```
Calculator                                              x
 C1    Expert    Store result in variable: Difference
 C2    A
 C3    B         Expression:
                 'A' - 'B'
```

We want to know if there is evidence that plan A has lower ratings, in general, than plan B, or that the 'A – B' differences *tend to fall below zero*, so proceed with **Stat>Nonparametrics>1-Sample Wilcoxon**, as follows:

which gives

Wilcoxon Signed Rank Test: Difference

```
Test of median = 0.000000 versus median < 0.000000

                        N
                      for    Wilcoxon              Estimated
              N      Test   Statistic        P       Median
Difference   10         9        15.5    0.221       -1.000
```

The high p-value of *0.221* (> α =.05) is consistent with the null hypothesis of no difference in effectiveness.

Exact test: If you wanted to perform an exact test, use the **Wilcoxon Statistic** displayed in the output. However, Minitab always computes this as the sum of the positive ranks, i.e. T_+. Since $T_+ = 15.5$ is bigger than the exact critical value of **8**, obtained from the table of critical values in your text, there is insufficient evidence of a difference in effectiveness, at the 5% level of significance. [If you needed to find T_-, use the fact that $T_+ + T_- = n(n+1)/2$]

Exercise 14.48 - M/S page 14- 28

Open the POWVEP data file containing the 157 day VEP measurements in C2 and the 379 day VEP measurements in C3. Calculate the differences using **Calc>Calculator: Expression:** *C3 – C2*, **Store in:** *Diff*. Test for a significant increase in the VEP scores with **Stat>Nonparametrics>1-Sample Wilcoxon: Variables:** *Diff*, **Test Median:** *0.0*, **Alternative:** *greater than*. We get

Wilcoxon Signed Rank Test: Diff

```
Test of median = 0.000000 versus median > 0.000000

              N
            for    Wilcoxon              Estimated
       N   Test   Statistic        P       Median
Diff  11     11        57.0    0.018       0.9450
```

Since $p = 0.018$ is less than $\alpha=.05$, we have sufficient evidence at the 5% level, to conclude that neurological impairment tends to be greater at the later date. [If you want to check significance using the exact critical value tables in the text, note that $T =$ $(11)(12)/2 - 57.0 = 9$]

Kruskal-Wallace Test - comparing independent samples

This nonparametric alternative to the one-way ANOVA F-test allows you to compare location for any number of independent samples. Minitab's procedure generalizes the normal approximated Wilcoxon Rank Sum Test, hence, if you apply it to only two samples, the results will be identical to those obtained using the Wilcoxon Test (aside from a slight difference due to continuity correction of latter).

Figure 14.11 - M/S page 14- 31

Open the HOSPBEDS data file. The data on number of unoccupied beds per hospital are stacked in C2, with the hospital number (#1, 2 or 3) in C1.

↓	C1	C2
	Hospital	Beds
1	1	6
2	1	38
3	1	3
4	1	17

To compare numbers of unoccupied beds at the three hospitals, run **Stat>Nonparametrics>Kruskal-Wallace**:

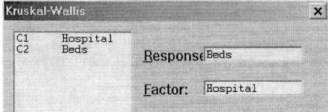

which produces

Kruskal-Wallis Test: Beds versus Hospital

```
Kruskal-Wallis Test on Beds

Hospital   N   Median   Ave Rank      Z
1         10    15.50     12.0    -1.54
2         10    31.50     21.1     2.44
3         10    18.50     13.5    -0.90
Overall   30             15.5

H = 6.10   DF = 2   P = 0.047
H = 6.10   DF = 2   P = 0.047   (adjusted for ties)
```

We have attained significance at the 5% level, since $p = 0.047$, and may conclude that not all three hospitals are equivalent in the distribution of numbers of empty beds.

Exercise 14.59 - M/S page 14- 34

Open up the DRINKERS data file, containing the GROUP in C1 and the task SCORE in C2. To compare scores across the groups, use **Stat>Nonparametric>Kruskal-Wallace:** **Response:** *C2*, **Factor:** *C1*

Kruskal-Wallis Test: SCORE versus GROUP

```
Kruskal-Wallis Test on SCORE

GROUP    N   Median   Ave Rank       Z
A       11   0.1600       10.2   -3.66
AC      11   0.2200       19.5   -0.91
AR      11   0.5000       32.2    2.89
P       11   0.4300       28.1    1.68
Overall 44                22.5

H = 19.03  DF = 3   P = 0.000
H = 19.04  DF = 3   P = 0.000  (adjusted for ties)
```

The small *p*-value indicates strong evidence or differences among groups.

Friedman Test for a Randomized Block Design

If the assumptions necessary to run an ANOVA F-test for a randomized block design are questionable, we may instead choose to apply a nonparametric procedure named after Milton **Friedman**, which again utilizes a rank transformation of the data.

Figure 14.13 - M/S page 14- 37

Open the REACTION2 data file, partially displayed to the side. Note how the data are in *stacked* format, with all the reaction times in one column, C3, drug in another column, C2, and subject number (block) in a third column, C1. This is precisely the format required by Minitab in order to run this procedure (if formatted differently, do some cutting and pasting, or try **Data>Stack**).

↓	C1 SUBJECT	C2-T DRUG	C3 TIME
1	1	A	1.21
2	2	A	1.63
3	3	A	1.42
4	4	A	2.43
5	5	A	1.16
6	6	A	1.94
7	1	B	1.48
8	2	B	1.85
9	3	B	2.06

Now run **Stat>Nonparametrics>Friedman**

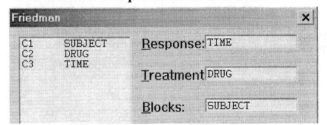

which produces:

Friedman Test: TIME versus DRUG blocked by SUBJECT

```
S = 8.33   DF = 2   P = 0.016

                        Sum
                        of
DRUG   N   Est Median   Ranks
A      6       1.5283    7.0
B      6       1.7417   12.0
C      6       1.8950   17.0

Grand median = 1.7217
```

The small *p*-value of *0.016* (< α) indicates evidence of a difference in reaction time among the drugs.

Exercise 14.74 - M/S page 14- 40

Open up the PLANTS data file, where in columns C1 - C3, you will find the student number, plant setting (1 = Live, 2 = Photo, 3 = None) and maximum finger temperature, respectively. Does the type of plant setting affect relaxation level (measured by finger temperature)? Students are the blocks. Run **Stat>Nonparametrics>Friedman:** **Response:** *Temp*, **Treatment:** *Condition*, **Blocks:** *Student*

Friedman Test: Temperature versus Plant blocked by Student

```
S = 0.20   DF = 2   P = 0.905

                        Sum
                        of
Plant   N   Est Median   Ranks
1      10      95.600    19.0
2      10      95.683    20.0
3      10      95.817    21.0

Grand median = 95.700
```

With the large *p*-value, there clearly is no evidence that finger temperature depends on the experimental condition.

Spearman's Rank Correlation

The correlation coefficient discussed in earlier chapters is called *Pearson's (product-moment) correlation coefficient*. An alternative way of examining correlation is to transform the data to ranks, and then calculate Pearson's correlation, as usual, *but on the ranks*. This produces *Spearman's correlation coefficient*.

Example 14.4 - M/S page 14- 43

Open the NEWBORN data file. To find Spearman's rank correlation between cigarettes smoked (C1) and baby's weight (C2), we first rank the data in each column of data using

↓	C1 Cigarettes	C2 Weight
1	12	7.7
2	15	8.1
3	35	6.9
4	21	8.2

Data>Rank

Or use **Calc>Calculator:**
Expression: *Rank*

Now, correlate (Pearson) the two columns of ranks with
Stat>Basic Statistics>Correlation:

which produces

Correlations: CigRanks, WeightRanks

```
Pearson correlation of CigRanks and WeightRanks = -0.425
P-Value = 0.115
```

So, Spearman's correlation coefficient is *-0.425* here, indicative of a weak negative association, with lighter babies associated with heavier smoking mothers. However, with a *p*-value of *0.115* (divide by 2 for a one-sided alternative), the association fails to attain statistically significance at the 5% level.

Note: The p-value produced this way is an approximation and will be inaccurate if the sample size does not exceed 10, in which case you should compare the reported correlation with a table of critical values (like the one in the back of your text).

Exercise 14.85 - M/S page 14- 49

Open the BOXING2 data file, containing Blood Lactate Level in C1 and Perceived Recovery in C2. To find Spearman's correlation for these two variables, compute and store, in a new column, the ranks of the lactate levels using **Data>Rank: Rank data in:** *Lactate*, **Store ranks in:** *LactateRank*. Similarly compute and store the ranks of the perceived recovery values with **Data>Rank: Rank data in:** *Recovery*, **Store ranks in:** *RecoveryRank*. Apply **Stat>Basic Statistics>Correlation** to the two columns of *ranks*, and we get

Correlations: LactateRank, RecoveryRank

```
Pearson correlation of LactateRank and RecoveryRank = 0.713
P-Value = 0.002
```

We see a moderately strong positive association, which is also highly significant with a *p*-value of only *0.002* ($< \alpha = .10$). Higher blood lactate levels tend to accompany higher perceived recovery.

Note: You can also use **Stat>Tables>Cross Tabulation and Chi-Square,** enter the variables, then click on **Other Stats…** and select the last listed **Measure of Association**:

OK, OK, and at the bottom of the output, you will see:

```
Pearson's r       0.708760
Spearman's rho    0.713453
```

This gives us the correct value of 0.713 for Spearman's correlation, but unfortunately, we do not get a p-value, so you would now have to use your table of critical values for a test of significance.

Though the output above displays something called 'Pearson's r', this is not in fact our familiar Pearson's correlation for the raw or unranked data. Ignore this number. If you wanted Pearson's correlation for the actual numerical data here, you would instead use **Stat>Basic Statistics>Correlation: Variables:** *Lactate Recovery*, and you would get 0.570.

Index

THE TI-84 PLUS & SILVER MANUAL

Singh Kelly

Pensacola Junior College

Statistics ELEVENTH EDITION

McClave | Sincich

Upper Saddle River, NJ 07458

Getting Started
The TI-84 & TI-83+

Important Note: If you purchased a **used** calculator your first step is to return the calculator to its original default settings. Follow these steps: Turn **ON** the calculator. Then to reset to original settings as follows: For the **TI-83 Plus, Silver & TI-84 Plus, Silver**

Press: **2nd + 4 ENTER** This will clear all Lists

Press: **2nd + 7 2 2 ENTER** This will reset the calculator

These will help you avoid some unwanted **"ERROR"** messages.

NOTE: In this new addition I have included a special section devoted to **"ERROR"** messages. It is located at the end of this handbook, pp 82-86.

After you finish each problem *Press*: **CLEAR** to return to the home screen. **It may take more than one time**. It is a great idea to begin each new problem from the home screen.☺

Another note here: Sometimes,

▼ Is used for **arrow down**. ► Is used for **arrow right**.

▲ Is used for **arrow up**. ◄ Is used for **arrow left**.

One feature of the TI-84 Plus & Silver Addition is the use of Scientific Notation when displayed values are of extreme quantities.
(*i.e.* *2.33454332E-7 is the calculator's notation for the value 0.000000233454332*)

When reading a displayed value take care to **read across the entire line.** The calculator will **always** use Scientific Notation for extreme values whether it is set in Scientific MODE or not.

Lastly, the calculator can convert fractions to decimals and decimals to fractions. For example: if you have the fraction **7/11** on your screen and you want to convert it to a decimal
press: **MATH ▼ ENTER ENTER** .6363636364 will be displayed. (This is the decimal approximation for 7/11)
To convert it to a fraction
press: **MATH ENTER ENTER** 7/11 will be displayed.

(Some decimal approximations represent irrational numbers and of course they cannot be converted into fractions)

Contents: McClave/Sincich Text Manual

	Calculator

Entering Data, Using Lists & Graphical Methods

To create a histogram, or any other type of statistical plot on the calculator you must first enter the data into one of the 6 lists found on the calculator.

Example 2.2   *Percentage of Student Loans in Default.*
To enter the data *press*: **STAT ENTER** you will see these two screens:

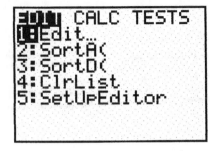
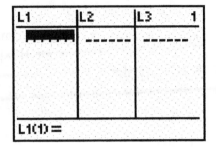

This is the LISTS Menu. You will be using this screen every time you enter data
(*The only exception is when entering data for a χ2 test. You will then use the matrix menu*)

Type each number followed by pressing **ENTER**
After typing the first 6 numbers you will see this screen:

Note: *The bottom of the TI-84 screen indicates which LIST and which entry is active.(LIST1 & it is waiting for the 7th entry)*

Once you have entered all the data *press*: **2ⁿᵈ   Y=**

You will see this screen.
*This is the **Statistical Plots** Menu. Whenever you want to plot a graph you will need to adjust the parameters within this menu. You may graph up to 3 things on the same screen by turning on up to 3 Plots.*

For now we only want 1 plot (a histogram) of the data we just entered so we will only turn on Plot1. We will enter the parameters section for Plot1 by pressing **ENTER**

After pressing the enter key you will see the highlighted plot.
(Plot1 in this case)

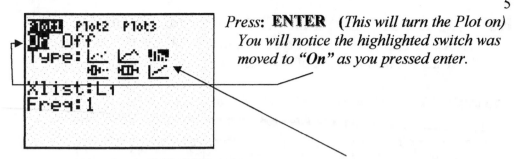

Press: **ENTER** (*This will turn the Plot on*) *You will notice the highlighted switch was moved to "On" as you pressed enter.*

Here you have six choices for graph **type.** They are, in order, Scatter-plot, Line-graph, Histogram, Boxplot with outliers, Boxplot without outliers, Normal Probability plot.

For our current problem we will highlight Histogram by following these steps.

Press: ▼ ► ► *This will highlight the Histogram.*

Press: **ENTER**

Press: ▼ *Beside the "Xlist :"* <u>you</u> *must enter the list where* <u>you</u> *put your data. (in my case I used List1)*

Press: **2nd 1** *All lists are located as 2nd functions on the corresponding numeral key.(i.e. 2nd 1 gives List1, 2nd 2 gives List2, 2nd 3 gives List3, etc.)*

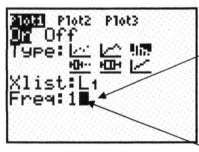

Press: **ALPHA 1** *This step is only needed if your screen does not show a 1 beside* **Freq:**.

In some cases the data comes to you in a frequency table, in which case, you will enter the List that contains that frequency here; otherwise, it should be **1.**

ZOOM 9 is a built-in option of the calculator. By pressing **ZOOM 9** you will see this Histogram:

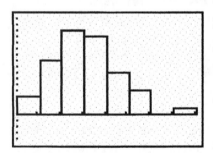

ZOOM 9 *will give you a graph, but that graph may not be appropriate for your problem, as is the case here. By pressing* **TRACE** *the calculator will display several characteristics of the graph.*

You can see on this next display the min=2.7 and the max<5.1285714. This is the class width of the first class as chosen by the calculator.

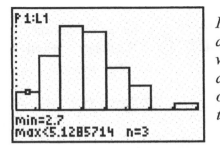

*For our problem we should set our display differently. It is appropriate to set the class width at **1** and the classes should begin at **2.5** and end with **20.5.** Using these setting we will obtain the same display as the SPSS display in the text.*

To adjust the window settings *press:* **WINDOW**

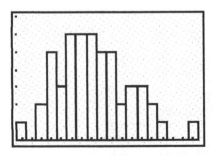

This is the menu where you will adjust the window settings. *(It will be necessary to re-adjust these settings for each new problem)* For our current problem we need to match the settings seen on the next screen. *From inspection of the SPSS display found in the text you can discern why we choose these settings.*

```
WINDOW
  Xmin=2.5          ← Lower limit
  Xmax=20.5         ← Upper limit
  Xscl=1            ← Class width
  Ymin=0            ← Lowest possible frequency
  Ymax=7            ← Highest frequency
  Yscl=1            ← Convenience
  Xres=1            ← Contrast for display
```

To see the Histogram *press:* **GRAPH**

By pressing **TRACE** *you can arrow across to have each class boundary and frequency displayed.*
Note: *the calculator does not display titles units or numerical scale. If you are printing any graph you must add those items yourself using your favorite word processor.*

Now we see the exact display seen in the text. (except for title, numerical scales, and legend) **Remember:** you will need to adjust the window settings for each new problem. The calculator is a wonderful machine but it does not "know" what is appropriate for your problem.☺

Exercise 2.26

To graph a histogram for this problem we must first locate the class mark for each of the classes. Recall the class mark is the midpoint of the class. We will enter the class marks into List1 and the corresponding frequency into List2.

Class Mark (L1)	Measurement Class	Relative Frequency (L2)
1.5	.5 – 2.5	.10
3.5	2.5 – 4.5	.15
5.5	4.5 – 6.5	.25
7.5	6.5 – 8.5	.20
9.5	8.5 – 10.5	.05
11.5	10.5 – 12.5	.10
13.5	12.5 – 14.5	.10
15.5	14.5 – 16.5	.05

Begin by typing the data into the Lists.
Press: **STAT ENTER**

To clear the entire list, highlight the **L1** *all the way at the top of the screen. And press*: **CLEAR ENTER**

To delete any individual entry within a given list highlight that entry and press: **DEL**

Always enter data for each new problem into an empty list.

After you have entered the class marks into **L1** and the frequencies into **L2** you should be looking at the upper right screen.

Remember to enter the decimal points as they appear in the text.

We now need to adjust the window and then change some settings within the **STAT PLOT** *MENU.*

Press: **WINDOW**

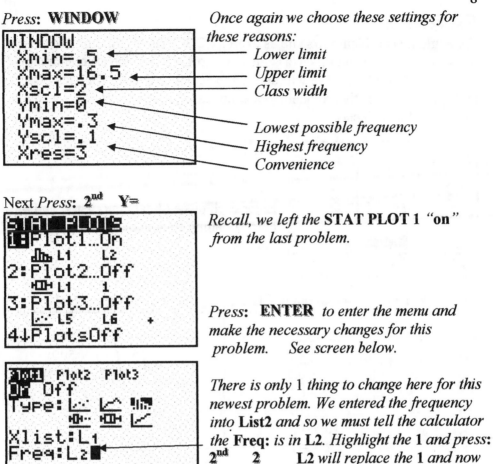

Once again we choose these settings for these reasons:

Xmin=.5 ← Lower limit
Xmax=16.5 ← Upper limit
Xscl=2 ← Class width
Ymin=0 ← Lowest possible frequency
Ymax=.3 ← Highest frequency
Xres=3 ← Convenience

Next *Press*: **2ⁿᵈ Y=**

*Recall, we left the **STAT PLOT 1** "on" from the last problem.*

Press: **ENTER** *to enter the menu and make the necessary changes for this problem.   See screen below.*

*There is only 1 thing to change here for this newest problem. We entered the frequency into **List2** and so we must tell the calculator the **Freq**: is in **L2**. Highlight the 1 and press:* **2ⁿᵈ   2**   *L2 will replace the 1 and now we can plot the graph.*

Press: **GRAPH**   you will see the following Histogram:

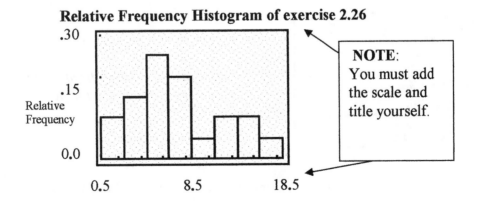

Relative Frequency Histogram of exercise 2.26

.30

.15

Relative
Frequency

0.0

0.5   8.5   18.5

NOTE:
You must add
the scale and
title yourself.

Exercise 2.27

We begin this problem the same as exercise 2.26, namely finding the class marks. We need to enter the **class marks** into **List1** and the corresponding **frequencies** into **List2.** We will derive the actual frequency for each class by multiplying the relative frequency of each class by 500 to achieve the actual frequency for each class.

Press: **STAT ENTER**

Enter each piece of data into list 1, and the corresponding frequency that you calculated into list 2.

Class Mark (L1)	Measurement Class	Derived Frequency (L2)
1.5	.5 – 2.5	50 (.10 X 500)
3.5	2.5 – 4.5	75 (.15 X 500)
5.5	4.5 – 6.5	125 (.25 X 500)
7.5	6.5 – 8.5	100 etc.
9.5	8.5 – 10.5	25
11.5	10.5 – 12.5	50
13.5	12.5 – 14.5	50
15.5	14.5 – 16.5	25

Once again we must adjust the window to fit our current data.
Press: **WINDOW** to view the old settings.
You will see the following screen:

```
WINDOW
 Xmin=.5
 Xmax=16.5
 Xscl=2
 Ymin=0
 Ymax=.3
 Yscl=.1
 Xres=3
```

Recall: these are the settings from the last problem. The new settings must match our new problem.

The next screen shows the menu <u>after</u> all adjustments have been made.

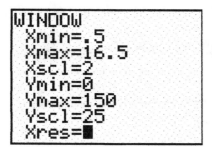

Settings after you have made the changes.

Since the last problem used List1 for class marks and List2 for relative frequency and we graphed a Histogram there are no changes needed within the STAT PLOT Menu.

Press: **GRAPH**

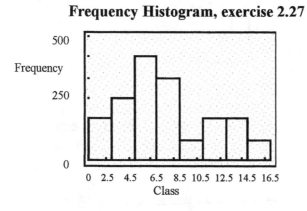

Frequency Histogram, exercise 2.27

Recall: *you must add the title, scales, and any necessary units needed for your problem.*

Example 2.3

To calculate the mean you must first enter the data into the list.

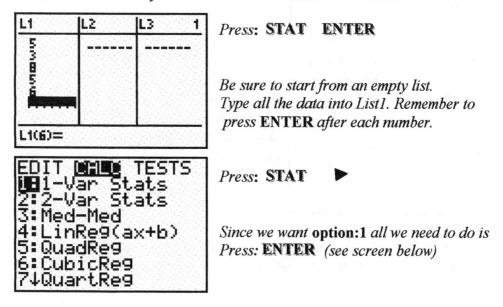

Press: **STAT   ENTER**

Be sure to start from an empty list.
Type all the data into List1. Remember to press **ENTER** *after each number.*

Press: **STAT   ▶**

Since we want **option:1** *all we need to do is*
Press: **ENTER** *(see screen below)*

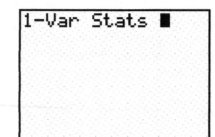

You must enter the List where you typed the data.
Press: **2ⁿᵈ 1 ENTER**

The screen below left will appear.
Press: ▼ *a few times, and you will see the screen below right.*

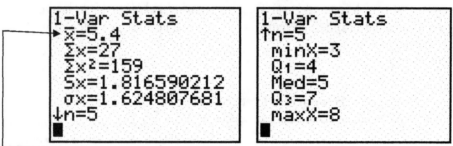

As can be seen from the screen the mean is **5.4**

Exercise 2.53

To calculate the mean and median of the data given we must first enter the data into the calculator. I will use list 1.

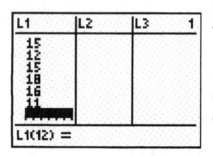

Press: **STAT ENTER**

This will take you to the Lists Menu.

Clear any old data from the list as it may interfere with this new problem.

After you have entered all the data *Press:* **STAT ▶ ENTER**

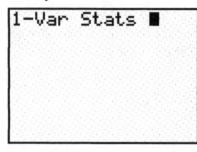

At this time you must enter the list designation where you placed the data. Since I put the data in List 1, I will press: **2ⁿᵈ 1**

Next Press: **ENTER**

The TI-84 will display a lot of Statistics. Some you will need. Some you will not need. I will point out the basic descriptive statistics most often asked for.

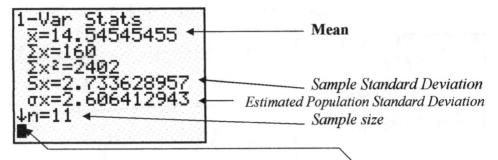

Mean

Sample Standard Deviation
Estimated Population Standard Deviation
Sample size

Notice the arrow pointing downward in front of the n=11. This tells us there is more data that can be seen by pressing the arrow down key.

Press: ▼ a few times and you will see this new screen.

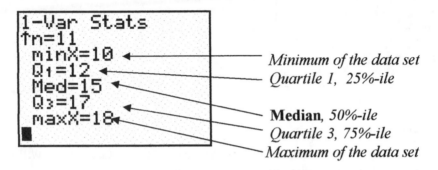

Minimum of the data set
Quartile 1, 25%-ile

Median, *50%-ile*
Quartile 3, 75%-ile
Maximum of the data set

Together these two screens give you most everything wanted when someone asks for "the basic descriptive statistics."

The mode is the piece of data with the highest frequency. It is **15** .

Exercise 2.59

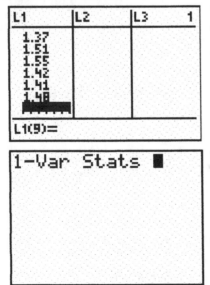

To begin this problem we enter the data into the List. I will use list 1.

Press: **STAT   ENTER**
This will take you to the Lists Menu.

Clear any old data from the list as it may interfere with this new problem.

After you have entered all the data *Press*:
    STAT ▶ ENTER
At this time you must enter the list designation where you placed the data. Since I put the data in List 1, I will press:
2nd 1

Next *Press*: **ENTER**

As can be seen from the display the mean daily ammonia level is **1.4713**

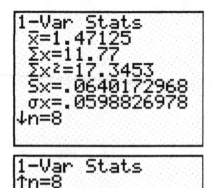

Press: ▼   a few times and you will see this new screen

The median, 50%-ile = **1.49**

Example 2.9

To calculate the Variance and Standard Deviation of this data set we will use the same procedures as finding the mean and median.

First enter the data into the calculator.

Press: **STAT ENTER**.   Enter the data. (this takes you to the list menu)
 After you have entered all the data

Press: **STAT ▶ ENTER** (this takes you to the calculations menu)

Once again you must enter the List where you placed the data. Again I used List 1, so I *press*: **2nd 1   ENTER**

 The Standard Deviation can be seen on the resulting screen.

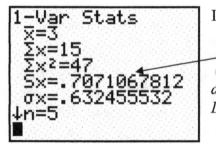

It is **.71** rounded to the hundredths place.

*(We choose "**Sx**" because this is a sample data set so we need the sample Standard Deviation.)*

The Variance can be calculated directly from the Standard Deviation as it is the square of the Standard Deviation. So

$0.7071067812^2 = 0.5$ Thus we calculate the **Variance = 0.5**

Exercise 2.83

The first step in solving this problem is to enter the data into the calculator.
Press: **STAT ENTER**
Enter all the data remembering to press ENTER after each entry.

After all the data has been entered *press*: **STAT   ▶   ENTER**
You should now be at the step where you enter the list on which the calculator
will perform the 1-Variable Statistics. I *press*: **2nd 1 ENTER** I used List 1.

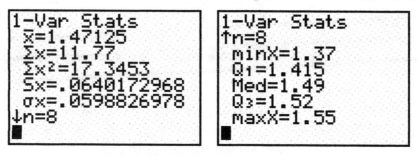

a.   The range is the maximum – the minimum. For our problem
       it is $1.55 - 1.37 = \mathbf{0.18}$
b.   Recall the variance is the standard deviation squared. For our problem
       it is $.0640172968^2 = \mathbf{.0041}$
c.   The Sample Standard Deviation is seen on the screen to be **.0640**
d.   The **morning** drive-times have the **greater** variable ammonia levels as
       $1.45 > .0641$

Example 2.16

Drawing a box-plot on the TI-84 is the nearly the same procedure as drawing
a histogram except for 1 step. First you must enter the data into the list.
Re-Consider the example problem : **Percentage of Student Loans in Default**

To enter the data turn on the calculator and *press*: **STAT ENTER**
you will see these two screens as you press each of the buttons:

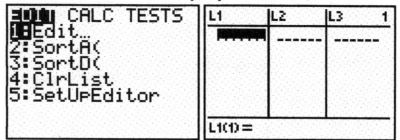

This is the LISTS Menu. You will be using this screen every time you enter data
into the calculator. (*The only exception is when entering data for a χ2 test. You will then use
the matrix menu*)

Type each number followed by pressing **ENTER**

Once you have entered all the data
press: **2ⁿᵈ Y=**

You will see this screen.

*This is the **Statistical Plots** Menu.*
Whenever you want to plot a graph you will need to adjust the parameters
within this menu. You may graph up to 3 things on the same screen by turning
on up to 3 Plots.

For now we only want 1 plot (a Boxplot) of the data we just entered so we will
only turn "on" Plot1. Enter the parameters section by pressing **ENTER**
After pressing the enter key you will enter the highlighted plot.

Press: **ENTER** *This will turn the Plot on.*
You will notice the highlighted switch was
moved to "On" as you pressed enter.

For our current problem we will highlight Boxplot with outliers,
by following these steps.

Press: ▼ ▶ ▶ ▶   *This will highlight the Boxplot.*
Press: **ENTER**
Press: ▼   *Beside the* **"Xlist :"** *you must enter the list where*
Press: **2ⁿᵈ 1**   *you put your data. (I used List1)*

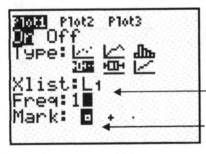

Press: **ALPHA 1** *This step is* **only**
needed if your screen does not show **1**.
In some cases the data comes to you
In a frequency table, in which case, you
Will need to enter the List which contains
the frequencies here.

For Mark, I choose the <u>*square*</u>*. Hey, It's Math*

By pressing **ZOOM 9** you will see this Boxplot.

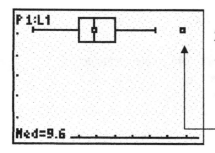

ZOOM 9 *is a built-in function of the TI-84&*
Plus. It will give you a graph, but that graph
may not be appropriate for your problem. For
any Boxplot, **Zoom 9** *is usually sufficient.*

This unconnected mark is the possible
outlier. In this case it is also the maximum.

By pressing **TRACE** *the calculator will display several characteristics of the*
graph. For a Boxplot it will display the minimum, Q1, Median, Q3, lower
fence(s), Maximum, one number at a time by flashing the curser on the
appropriate location as you arrow from left to right.

Note: the calculator does not display titles, units or numerical scale. If you are
printing any graph you must add those items yourself using your favorite word
processor. The calculator is a wonderful machine but it does not "know" what is
appropriate for your problem.☺

Exercise 2.126

For this problem we will use two lists and two stat plots.
First enter the data into the appropriate lists. I will put Sample **A** data into
List 1 and Sample **B** data into **list 2**.
Press: **STAT ENTER** (*Clear list 1 and list 2 by highlighting L1 and L2 at*
the very top of your screen & pressing: **CLEAR ENTER**)

L1	L2	L3	2
163	171		
145	185		
196	206		
172	169		
100	180		
171	188		

L2(22)=

It is good to glance at the two lists to see if
they are the same length.
If they are not the same length you probably
missed a piece of data in the shorter list.
Many mistakes occur at data entry.
Make sure you have entered the data correctly.
If you make a mistake here nothing else will be
correct!

Press: **2ⁿᵈ Y=**
You must turn on each plot one at a time. First plot 1.

Press: **ENTER** Make sure **On** is highlighted.
Press: **ENTER** Arrow over to highlight the boxplot and *press*: **ENTER**

We will use plot 1 for list 1 (A) and plot 2 for list 2 (B).

Again make sure the **Freq:** *is set to* **1**.
You may choose any Mark you wish.

Press: **2ⁿᵈ** **Y=**   This time arrow down to
2:Plot2…
Press: **ENTER** This will take you to plot 2.

Make sure On is highlighted.
 Press: **ENTER**
Arrow over to highlight the boxplot and press:
ENTER
Make sure **Xlist:** *is* **L2** (*press:* **2ⁿᵈ** **2**)
Freq: *must be* **1**
 Any Mark you wish.

Since these are boxplots we can
press: **ZOOM   9**

Pressing **TRACE** *will display all important
information associated with these boxplots.*

Arrow up/down to move from plot1/plot2.
Arrow right/left to move within a plot.

The similarity between these two box-plots is they both have possible outliers
on the low side. Plot 2 which is the Sample B data has a higher value in
every category. (e.g. minimum, Q1, Median, Q3, maximum) Sample A has
two possible outliers : **85, 100**. Sample B has 1 possible outlier: **140**.

Exercise 2.139

To construct a Scatter-plot using the TI-84 you must first enter the data into the
Lists. The Independent variable (**x**) usually goes in List 1 and the Dependent
Variable (**y**) usually goes in List 2. *It is not "carved in stone". It's just a general rule.*
For this problem enter Variable #1 into List 1 and Variable #2 into List2.

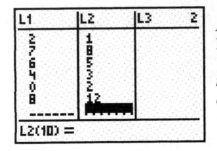

Press: **STAT ENTER**
Note: *For all data given in a frequency
 table or listed as two or more variables it is
 necessary to enter the data in the* **EXACT**
 order it is given.

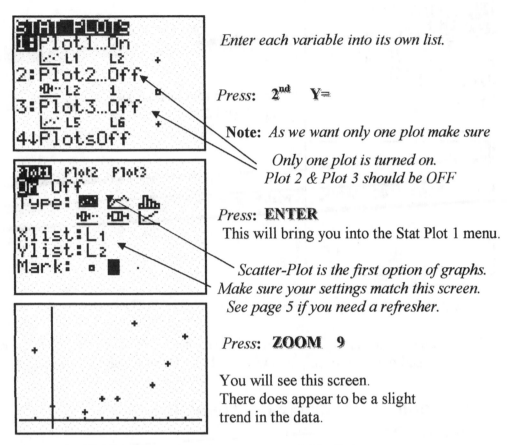

Enter each variable into its own list.

Press: **2ⁿᵈ Y=**

Note: *As we want only one plot make sure*

Only one plot is turned on.
Plot 2 & Plot 3 should be OFF

Press: **ENTER**
This will bring you into the Stat Plot 1 menu.

Scatter-Plot is the first option of graphs.
Make sure your settings match this screen.
See page 5 if you need a refresher.

Press: **ZOOM 9**

You will see this screen.
There does appear to be a slight
trend in the data.

It appears variable #2 tends to increase as variable #1 increases.

Exercise 2.142 *Feeding behavior of fish*

To construct a scatterplot for this data we must
enter the data into the list I will enter the age of
fish into list 1 and the number of strikes into
list 2.
*(I always enter the "x" variable into list 1 and the "y"
variable into List 2)*

After you have entered the data into the lists go to the STAT PLOTS Menu

Press: **2ⁿᵈ Y=**
Press: **ENTER**

Arrow down and select the first picture.

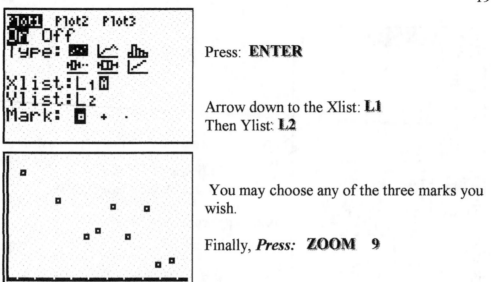

Press: **ENTER**

Arrow down to the Xlist: **L1**
Then Ylist: **L2**

You may choose any of the three marks you wish.

Finally, *Press:* **ZOOM 9**

You will see this scatter-gram. (Scatter-Plot)

Yes, there does appear to be a pattern, the data moves down as we move to the right. We call this relationship an inverse relationship or negative correlation.

The Counting Principle & Probability
Example 3.7

In a <u>combination</u> we are simply interested in the number of ways that objects can be selected. The calculator uses computer notation as follows: $\binom{N}{n}$ will be written as nCr

Where n ~ N, and r ~ n from the textbook's formula. So we want (20C5).
Start from a clear screen.
Press: **20**

Next *press:* **MATH**
Press: ◄ Highlight **"3 : nCr"**

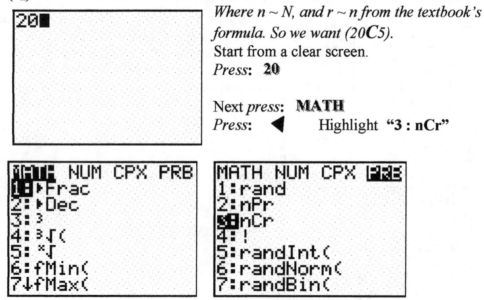

and *Press*: **ENTER**   Press: **5 ENTER**

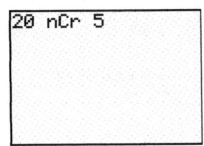

15,504 *is the number of combinations of* 5 *objects from* 20 *objects.*

Stated another way; it is the number of ways you can select 5 objects from a total of 20 objects.

Exercise 3.31   *Post Op Study*

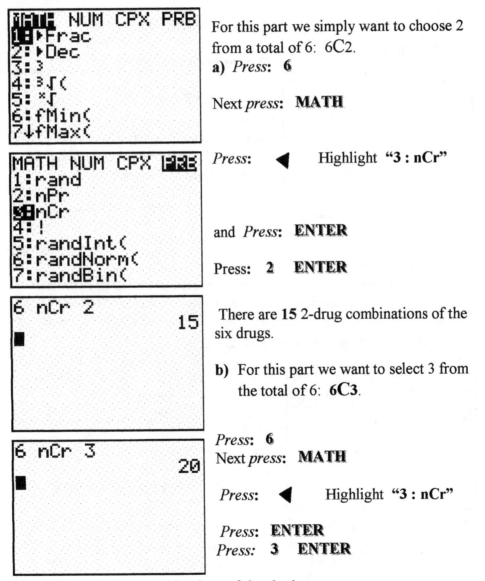

For this part we simply want to choose 2 from a total of 6: 6C2.
a) *Press*: **6**

Next *press*: **MATH**

Press: ◀   Highlight "**3 : nCr**"

and *Press*: **ENTER**

Press: **2 ENTER**

There are **15** 2-drug combinations of the six drugs.

b) For this part we want to select 3 from the total of 6: **6C3**.

Press: **6**
Next *press*: **MATH**

Press: ◀   Highlight "**3 : nCr**"

Press: **ENTER**
Press: **3 ENTER**

There are **20** 3-drug combinations of the six drugs.

Example 3.28

To permute 5 objects from a group of 5 objects we could actually use **!**
But for our problem let's use the new Permutations rule.

On the calculator it will be written as **(5P5)** , (nPr), *where n is the total
number of objects, and r is the desired number of objects.*

Start from a clear screen. First ***press*: 5** then *press*: **MATH**

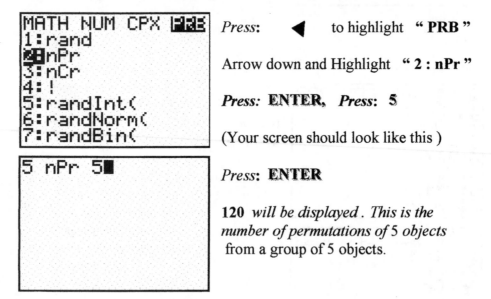

Press: ◀ to highlight " **PRB** "

Arrow down and Highlight " **2 : nPr** "

***Press*: ENTER, *Press*: 5**

(Your screen should look like this)

Press: **ENTER**

120 *will be displayed . This is the
number of permutations of 5 objects*
from a group of 5 objects.

Example 3.31

Recall: in a <u>combination</u> we are simply interested in the number of ways
that objects can be selected. The calculator uses computer notation as
follows: $\binom{N}{n}$ will be written as nCr. W*here n is the total number of objects,
and r is the desired number of objects. So we want (100C5).*
Start from a clear screen.

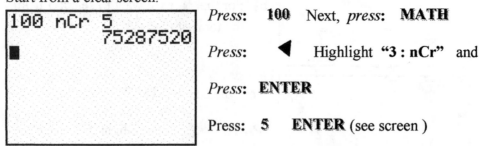

Press: **100**   Next, *press*: **MATH**

Press: ◀   Highlight **"3 : nCr"**   and

Press: **ENTER**

Press: **5   ENTER** (see screen)

75,287,520 *is the number of combinations of 5 objects from* 100 *objects.
Stated another way; it is the number of ways you can select 5 objects
from a total of* 100 *objects.*

Example 4.7   Skin Cancer Treatment

To begin this problem we first enter the data into the calculator. The (**x**)'s go into **list 1** and the p(**x**), [**y**'s] go into **list 2**. *We will use the same procedure as we used to calculate the mean and standard deviation from a frequency table. Press*: **STAT   ENTER**

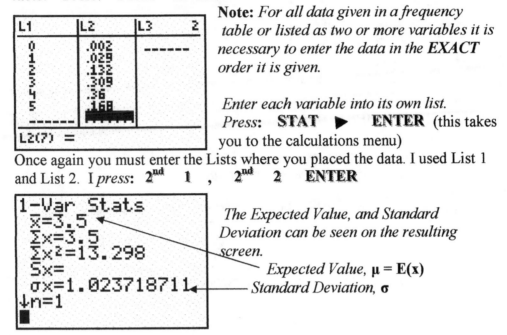

Note: *For all data given in a frequency table or listed as two or more variables it is necessary to enter the data in the EXACT order it is given.*

Enter each variable into its own list.
Press: **STAT ▶   ENTER** (this takes you to the calculations menu)

Once again you must enter the Lists where you placed the data. I used List 1 and List 2. I *press*: **2ⁿᵈ   1 ,   2ⁿᵈ   2   ENTER**

The Expected Value, and Standard Deviation can be seen on the resulting screen.

Expected Value, $\mu = E(x)$
Standard Deviation, σ

Exercise 4.35   Probability Distribution for Random Variable X

To begin this problem we first enter the data into the calculator. The (**x**)'s go into **list 1** and the p(**x**), [**y**'s] go into **list 2**. *We will use the same procedure as we used to calculate the mean and standard deviation from a frequency table.*

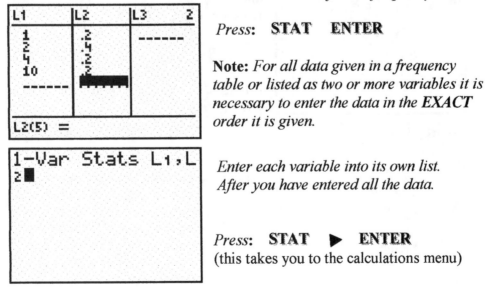

Press: **STAT   ENTER**

Note: *For all data given in a frequency table or listed as two or more variables it is necessary to enter the data in the EXACT order it is given.*

Enter each variable into its own list. After you have entered all the data.

Press: **STAT ▶   ENTER** (this takes you to the calculations menu)

You must enter the Lists where you placed the data.
I used List 1 and List 2 so ....

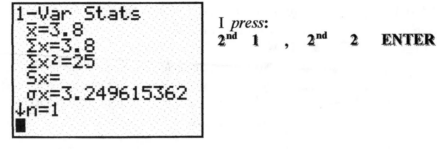

I *press:*

2nd **1** , 2nd **2** **ENTER**

a. $\mu = 3.8$
b. $\sigma^2 = 3.249615362^2 = 10.56$
c. $\sigma = 3.2496$
d. Over the long run the mean we would expect is **3.8**
e. *Not for this case, all x appear to be natural numbers.*

Consider the frequency distribution:

x	1	2	3
P(x)	.25	.50	.25

$\mu = 2$ is one of the values, so **yes** it can happen

Example 4.10   Physical Fitness Problem

We want to calculate the probability of **0, 1, 2, 3,** and **4** successes from a
total of **4** independent trials where the probability of success on
each trial is **0.10**   Here : **"n" = 4** ,**"x" = 0, 1, 2, 3, 4** and **"p"= 0.10**

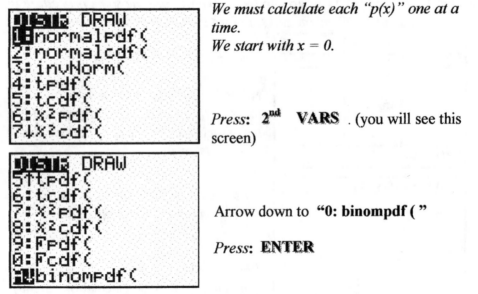

We must calculate each "p(x)" one at a
time.
We start with x = 0.

Press: 2nd **VARS** . (you will see this
screen)

Arrow down to **"0: binompdf ("**

Press: **ENTER**

You must enter **n , p , x** *in that order inside the parenthesis.*

For our problem Press: **4**   ,   **0 .10**   ,   **0**   **)**

Don't forget the **commas** *or the closed parenthesis!*

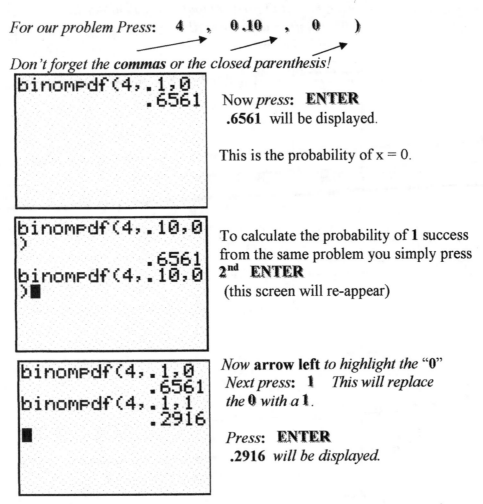

```
binompdf(4,.1,0
              .6561
```

Now *press:* **ENTER**
.6561 will be displayed.

This is the probability of x = 0.

```
binompdf(4,.10,0
)
              .6561
binompdf(4,.10,0
)█
```

To calculate the probability of **1** success from the same problem you simply press **2nd ENTER**
(this screen will re-appear)

```
binompdf(4,.1,0
              .6561
binompdf(4,.1,1
              .2916
█
```

Now **arrow left** *to highlight the* "**0**"
Next press: **1**   *This will replace the* **0** *with a* **1**.

Press: **ENTER**
.2916 *will be displayed.*

Each of the remaining values can be found following the same steps just outlined above.

** To graph a particular Binomial Distribution**

Recall this problem's data where **"n"= 4**, **"p"= 0.10**
Press: **STAT**   "**EDIT**" should be highlighted. *Press:* **ENTER**
Enter the numbers **{0,1,2,3,4}** into **List 1.** These are the **"x"** values.

```
L1      L2      L3      2
 0     ▓▓▓▓▓   ------
 1
 2
 3
 4
 -----
L2(1)=
```

Now we want to enter a **formula** into **List 2**

It is the binomial formula with n = 4, & P = .10

*If you have never entered a formula into a computer or calculator take
extreme care to have everything exactly correct!*

Even the parentheses, etc. must be exact.

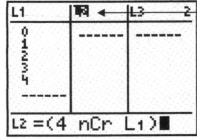

First highlight **L2** at the very top of the
column.

L2 =(4 nCr L1)

Press: **(4 MATH ◄ 3 2ⁿᵈ 1) (.10 ^ 2ⁿᵈ 1)**

You will see the text scroll across the bottom of the screen as you type.

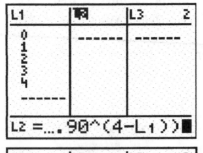

Press:

(.90 ^ (4 - 2nd 1))

*You will be able to see all the text by
pressing* **◄** *or* **►**

L2 =....90^(4−L1))

Press: **ENTER**
The probabilities will appear in **List2.**

*[The values in **List2** are the probabilities
associated with the "x" values from **List1**
for this Binomial problem]*

L2(1)=.6561

Recall: This is given in Scientific Notation

Now to see a plot we must go to the **STAT PLOT** so *press:* **2ⁿᵈ Y=**
"1:Plot1..." *should be highlighted. Press:* **ENTER,** *again*
Press: **ENTER** *(Note: the switch is moved to "On")*
Arrow down *and highlight the Histogram.*

Press: **ENTER**
Make sure **Xlist: L1 , Ylist: L2**

Now, as always, we need to adjust the **window** to fit our data.
Press: **WINDOW**

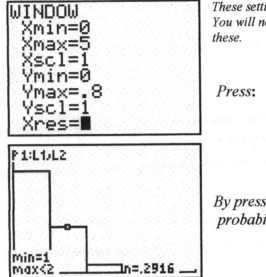

These settings are from our new data. You will need to change your settings to match these.

Press: **GRAPH**

By pressing **TRACE** *you can read the probability associated with each class.*

Exercise 4.63   Belief in an Afterlife

For this problem we see **p = .80, n = 10**   We will need to work 3 different problems: 1 for part a, 1 for part b, 1 for part c.

   a.   **P(x = 3)**

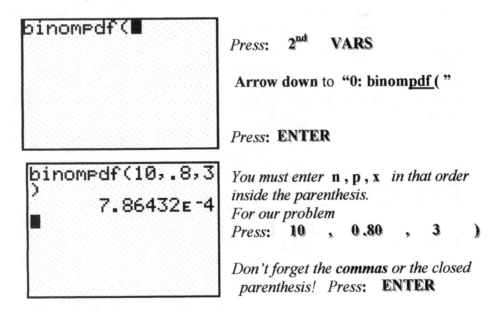

Press: **2ⁿᵈ   VARS**

 Arrow down to **"0: binompdf ("**

Press: **ENTER**

You must enter **n , p , x** *in that order inside the parenthesis.*
For our problem
Press: **10** , **0.80** , **3** **)**

*Don't forget the **commas** or the closed parenthesis! Press:* **ENTER**

Note: The answer is given in Scientific Notation.
In common notation it = **0.000786432**

b. P(x < 7)   For this part   *Press:* **2ⁿᵈ   VARS**

```
DISTR DRAW
9↑Fcdf(
0:binompdf(
A:binomcdf(
B:poissonpdf(
C:poissoncdf(
D:geometpdf(
E:geometcdf(
```

This is not the same as part a.
   !!! Look carefully !!!

Arrow down to **"A: binomcdf ("**

Press: **ENTER**

On the TI-84 the ending "cdf" will calculate a cumulative frequency, starting from 0 and continuing up to and including the number you enter.

Enter **n , p , x** in that order.
For our problem we enter **10 , .80 , 7)**

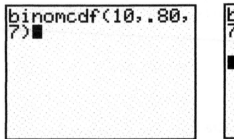

```
binomcdf(10,.80,
7)█
```

```
binomcdf(10,.80,
7)
            .322200474
█
```

Press: **ENTER**   (see screen above right)
As can be seen on the screen the probability is **.322200474**

c. P(x > 4)   Recall: this is the complement to P(x ≤ 4). We will work this complementary event and subtract the answer from 1.

```
binomcdf(10,.80,
7)
            .322200474
binomcdf(10,.80,
7)█
```

Press: **2ⁿᵈ   ENTER**

You will see this screen.

Next, **Arrow left** *and replace the 7 with* **4**

```
binomcdf(10,.80,
7)
            .322200474
binomcdf(10,.80,
4)
          .0063693824
█
```

Press: **ENTER**

We now need to subtract this complementary event from **1** *to get our answer:*

1 - .0063693824 = **.9936306176**

Example 4.13 *Poisson*   **Whale Sightings**

For this example part "a" is completely worked in the text and there is no "special way" to do it on the calculator. I will begin with part b.

b.   we want to find the probability of "fewer than 2" which means
      P(x < 2) and we have given information $\lambda = 2.6$

On the calculator we go to the distribution menu

Press: **2ⁿᵈ   VARS   Arrow down** to "C : poisson<u>cdf</u> ("

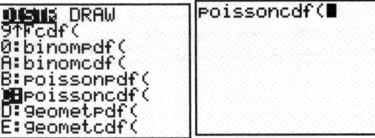

Press: **ENTER**   (you will see screen above right)
Now we must enter λ , **x**) *in that order.*

For our problem enter   **2.6** , **1**   **)**
Press: **ENTER**   *P(x < 2) means x = 0, or x = 1, So we must enter x = 1*

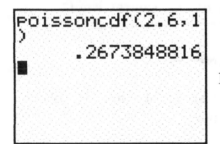

As can be seen from the screen
P(x < 2) is **0.2673848816**

c. Here we are looking for P(x > 5). λ still equals 2.6.
   Once again we must use the complement to find our answer.
   The complement to P(x > 5) is P(x ≤ 5).
   After we calculate the probability for the complementary event we will
   subtract that answer from 1 giving P(x > 5).
   Poisson<u>cdf</u>(2.6,5) = .9509628481 So.....
                    P(x > 5) = 1 - .9509628481 = **.0490371519**

d.   For this part we are looking for P(x = 5) so this will be
      "**B:Poisson<u>pdf</u>**"   Again λ = 2.6.
On the calculator we go to the distribution menu
Press: **2ⁿᵈ VARS**   Arrow down to "**B : poissonpdf ("**
Press: **ENTER**

You need only enter λ , **x)** for Poisson.

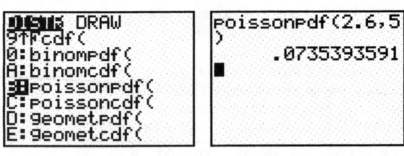

For our problem we *Press*:   **2.6** ,   **5**   **)**   **ENTER**

As can be seen from the screen $P(x = 5)$ is **.0735393591** or about **7.4%**

Example 5.2 $P(-z_0 < z < z_0)$

We want to calculate the area under the Standard Normal Curve between
-1.33 and positive 1.33. To begin this problem we will go to the
Distributions Menu.

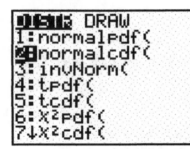

Press: **2nd   VARS**

Highlight option "**2:normalcdf (**"

Press: **ENTER**

To calculate the area under the curve from a **Standard** Normal Distribution
you need only enter 2 values: **lower**   ,   **upper**

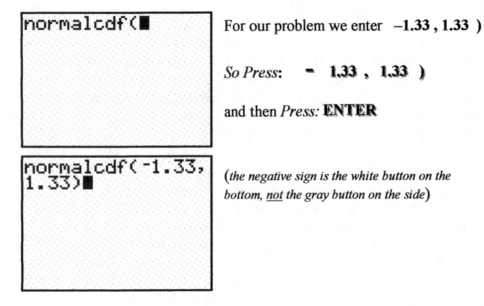

For our problem we enter   **–1.33 , 1.33)**

So Press: **–**   **1.33** ,   **1.33**   **)**

and then *Press:* **ENTER**

(*the negative sign is the white button on the
bottom, not the gray button on the side*)

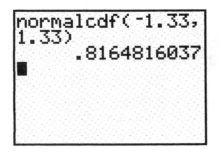

As can be seen from the screen the Probability is **.8164816037**

Example 5.3 $P(z > z_0)$

For this problem we are still trying to find the area under the curve so again we use **2:normalcdf**   *Press*:   **2nd**   **VARS**

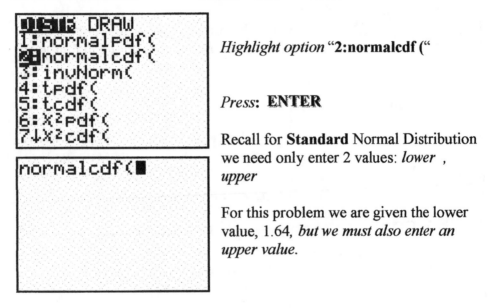

Highlight option "**2:normalcdf (**"

Press: **ENTER**

Recall for **Standard** Normal Distribution we need only enter 2 values: *lower* , *upper*

For this problem we are given the lower value, 1.64, *but we must also enter an upper value.*

Based on the theory of the Normal Distribution the upper limit should be + infinity. I choose an upper value of "**1E99**"   (*This will introduce an extremely small error. It is as close to $+\infty$ as we can get on the calculator*)

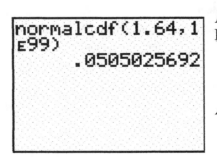

Press: **1.64** , **1 2ⁿᵈ** , **99**)
ENTER

As can be seen from the screen the probability is **.0505025692**

Example 5.4   P(z < z₀)

For this problem we are still trying to find the area under the curve so we
use **2:normalcdf** once again.   *Press*: **2nd** **VARS**

Highlight option "**2:normalcdf (**"

Press: **ENTER**

Recall, once again that for **Standard** Normal Distribution we need only
enter 2 values: *lower* , *upper*

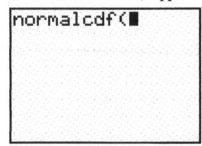

For this problem we are given the upper
value .67, *but we must first enter a lower
value.*

Based on the theory of the Normal Distribution the lower limit should be
negative infinity. I choose a lower value of -1 EE99 . This is as close to $-\infty$
as the calculator can get. *Certainly acceptable for our applications.*

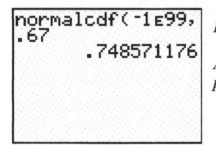

Press: $-$ **1** **2ⁿᵈ** , **99** , **.67**) **ENTER**

*As can be seen from the screen the
probability is* **.748571176**

Example 5.6   Cell Phone Application

For this problem we are still trying to find the area under the curve so again
we use **2:normalcdf**

To begin this problem we go to the
Distributions Menu.
Press: **2nd** **VARS**

Highlight option "**2:normalcdf (**"
Press: **ENTER**

Since we are using the <u>actual data</u> in this problem and <u>not</u> the z-scores we will need to enter 4 values. The first 2 will be the same *(i.e. lower, upper);* however, we now must add 2 more values ...*mean, standard deviation).*

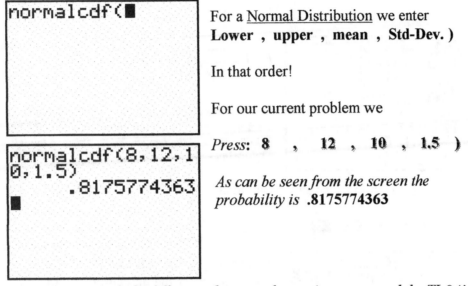

For a <u>Normal Distribution</u> we enter **Lower , upper , mean , Std-Dev.)**

In that order!

For our current problem we

Press: **8 , 12 , 10 , 1.5)**

As can be seen from the screen the probability is **.8175774363**

Note: *there is a slight difference between the text's answer and the TI-84's answer.* **Error is introduced when rounding** *the z-value for using the z-Tables.* The TI-84 displays the more accurate answer in this case.

Example 5.8   Using Normal tables in reverse

For this problem we will go to the Distributions Menu. But this time we will use **<u>3:invNorm</u> (** *because we are not looking for area, we are looking for z.*

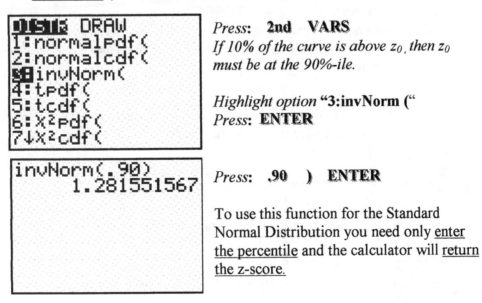

Press: **2nd VARS**
If 10% of the curve is above z_0, then z_0 must be at the 90%-ile.

Highlight option **"3:invNorm ("**
Press: **ENTER**

Press: **.90) ENTER**

To use this function for the Standard Normal Distribution you need only <u>enter the percentile</u> and the calculator will <u>return the z-score.</u>

As can be seen from the screen **z** = **1.281551567**

Example 5.9   Using Normal tables in reverse

For this problem we will go to the Distributions Menu. Again we will use
"3:invNorm (" *because we are looking for z, not the area.*

Press: **2nd   VARS**

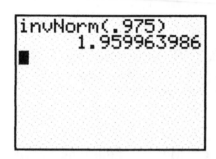

Highlight option **"3:invNorm ("**
Press: **ENTER**

*If 95% of the curve is between z and –z, and since
the normal distribution is symmetrical, then 47.5%
is to the right of the center and 47.5% is to the left
of the center, 50%-ile.*

Therefore the rightmost z-value must lie at $50\% + 47.5\% = $ **97.5%**-ile.

Press: **.975   )   ENTER**

```
invNorm(.975)
        1.959963986
■
```

To use this function for this problem you
need only enter the percentile and the
calculator will return the z-score.

As can be seen from the screen **z** = **1.959963986** and since the Normal
Distribution is symmetrical   **–z = -1.959963986**
*Recall: the z-score tells you how many Standard deviations you are away
from the mean, so it should not be a surprise that it is the same value no
matter which direction you move away from the mean* ☺

Example 5.10   College Entrance Exam Application

Again we are given area under the curve, but this time we are asked for the
actual value: not the z-score!
To begin this problem we return to the Distributions Menu.

```
DISTR DRAW
1:normalpdf(
2:normalcdf(
3:invNorm(
4:tpdf(
5:tcdf(
6:X²pdf(
7↓X²cdf(
```

Press: **2nd   VARS**

Highlight option "**3:invNorm ("**

To calculate the <u>value of the actual score</u> given the percentile, mean and standard deviation you need only enter 3 values: **%-ile , mean , std- dev**

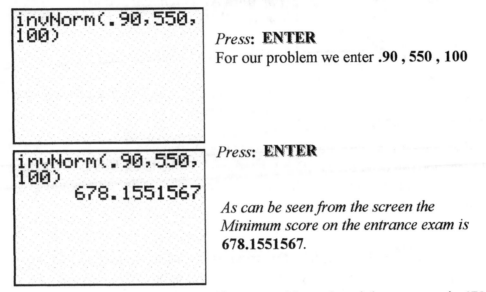

Press: **ENTER**
For our problem we enter **.90 , 550 , 100**

Press: **ENTER**

As can be seen from the screen the Minimum score on the entrance exam is **678.1551567**.

Since you probably can't score this we would say the minimum score is 678.

Exercise 5.58   Normal Probability Plot

Part d. To graph a Normal Probability plot for this problem we first enter the data into the calculator. I will use list 1.

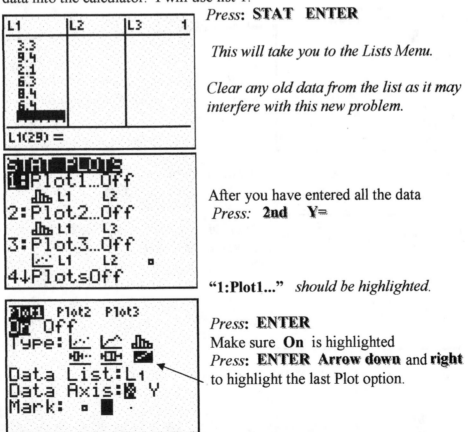

Press: **STAT ENTER**

This will take you to the Lists Menu.

Clear any old data from the list as it may interfere with this new problem.

After you have entered all the data
Press: **2nd   Y=**

"1:Plot1..." should be highlighted.

Press: **ENTER**
Make sure **On** is highlighted
Press: **ENTER Arrow down** and **right** to highlight the last Plot option.

For **Data List** *I entered* **L1** *as that is where I placed my data.*

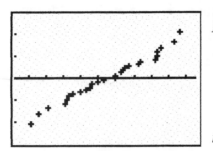

For **Data Axis** *highlight* **X** *and press:* **ENTER**

Press: **ZOOM   9**

As you can see from the screen there is no noticeable curve.

It does appear that this data is near Normally Distributed.
In fact you can enter the function y = 0.45x −2.5 (the line-of-best-fit for the given data) and you will see a line graphed through this data. The data follows the line very closely ☺

The Exponential Distribution Function is not a built-in function on the TI-84 so we will use a slightly different procedure for calculating the area under the curve. Since it is not a function that appears on the calculator we must enter the function into the Y= Function Line ourselves. Remember College Algebra?

Example 5.13   *Hospital Emergency Arrivals*
For this problem we know θ = 2 , and we want **x ≥ 5**

```
Plot1  Plot2  Plot3
\Y1=█
\Y2=
\Y3=
\Y4=
\Y5=
\Y6=
\Y7=
```

So our equation would be :
$$(1 / 2) e^{(- x / 2)}$$

*On the calculator you will go to the functions menu to enter the equation.
Press:* **Y=**

Now we will need to enter the exponential equation into Y=.
Press: **(1   ÷   2   )   2nd   LN   (-)   X,T,Θ,n   ÷   2   )**

```
Plot1  Plot2  Plot3
\Y1█(1/2)e^( -X/2
)█
\Y2=
\Y3=
\Y4=
\Y5=
\Y6=
```

(you will see this text appearing on screen as you type)

Please notice that all 3 STAT PLOTS are OFF. If any of these are highlighted you must turn them off. See p. 4,5 for a refresher.

Now we need to adjust the **WINDOW** to fit our data.

```
WINDOW
 Xmin=0
 Xmax=16
 Xscl=2
 Ymin=0
 Ymax=.5
 Yscl=1
 Xres=■
```

Press: **WINDOW**

Recall: $\theta = 2$, we want $P(x \geq 5)$

$Xmin = 0$, $Xmax = 8*\theta$, $Xscl = \theta$,
$Xmin = 0$, $Ymax = 1/\theta$, $Yscl < 1/\theta$

These settings will always give you a nice view of an Exponential graph.

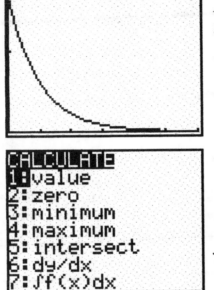

Press: **GRAPH** (you will see this graph)

To find the area under the curve past **x =5** we will need to go to the **Function's Calculations** menu.

```
CALCULATE
1:value
2:zero
3:minimum
4:maximum
5:intersect
6:dy/dx
7:∫f(x)dx
```

Press: **2nd TRACE**
(this screen will appear)

Arrow down to " **7 : { f (x) dx** "
This choice will integrate the function (calculate the Area) between two points

Press: **ENTER**
Now we need to enter the lower and upper limits for our problem.

[Since we wanted **x ≥ 5** , **5** *will be the lower limit and* **16** *will be our upper limit]* **16** *because our graph window only goes up to* **16**

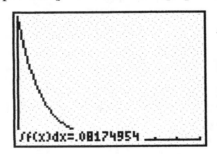

Press: **5 ENTER 16 ENTER**
This screen shows us that the area under the curve past 5 is **.08174954** So the probability that more than 5 hours will pass between emergency arrivals is $\approx$ **0.082**

[*To clear the screen after shading go to the* **DRAW** *menu.]*
Press: **2nd PRGM** "**1 : ClrDraw** " *will be highlighted,*
press: **ENTER** Wait a second then, *Press:* **CLEAR**

Example 6.8   Testing a Manufacturer's Claim

For Sampling Distribution problems we will use the Normal Distribution Functions found in the Distributions Menu. The only difference is: instead of entering the given standard deviation we will adjust the deviation to fit our sample size. ($\sigma_x = \sigma / \sqrt{n}$)

   a.   The distribution will be distributed normally with $\mu = 54$ and
            $\sigma_x = \sigma / \sqrt{n} = 6 / \sqrt{50} = .85$

   b.   Find $P(x \leq 52)$

                           We begin by going to the Distributions Menu.

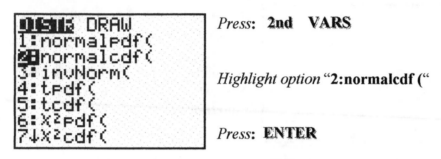

Press: **2nd VARS**

Highlight option "**2:normalcdf (**"

Press: **ENTER**

Since we are using the actual data in this problem and not the z-scores we will need to enter 4 values. The first 3 will be the same (i.e. lower, upper, mean); however, we now must adjust the standard deviation to fit our sample size.

For a Sampling Distribution we must enter
Lower , upper , mean , adjusted standard deviation) *In that order!*

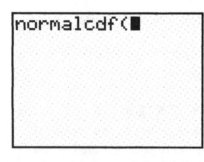

From part a we know σ_x = .85

For our problem we enter:
0 , 52 , 54 , .85)
After you have entered all four values
Press: **ENTER**

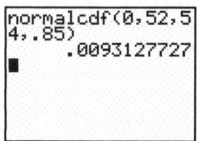

As can be seen from the screen the Probability a light bulb last 52 hours or less is **.0093**

The value in the textbook is calculated using the z-tables and therefore has a small rounding error.

Exercise 6.31

This is a Sampling Distribution problem. We will use the Normal Distribution Functions found in the Distributions Menu. The only difference is: instead of entering the given standard deviation we will adjust the deviation, as found in part a, to fit our sample size.

We have given information: n = 100, μ = 30, σ = 16.

    a. $\mu_x = 30$, $\sigma_x = 16/\sqrt{100} = 1.6$

    b.   The shape will be approximately normal.

    c.   Find $P(x \geq 28)$

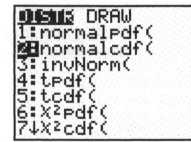

Press: **2nd   VARS**

Highlight option "**2:normalcdf (**"

Press: **ENTER**

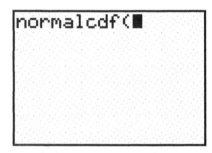

Since we are using the actual data in this problem and not the z-scores we will need to enter 4 values. We now must adjust the standard deviation to fit our sample size. $\sigma_x = 16 / \sqrt{100} = 1.6$

For a Sampling Distribution we must enter
Lower , upper , mean , adjusted standard deviation) *In that order!*
Don't forget the comma's!!
For our problem we enter: **28 , 1000 , 30 , 1.6)**

I picked 1000 because it is a very large value for this problem. (> 600 standard deviations from the mean) It will introduce very little error.
After you have entered all 4 values

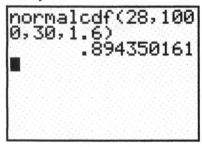

Press: **ENTER**

As can be seen from the screen the probability is **.894350161**

d. Find $P(22.1 \leq x \leq 26.8)$ *Press:*   **2nd   VARS**

Highlight option "2:normalcdf ("

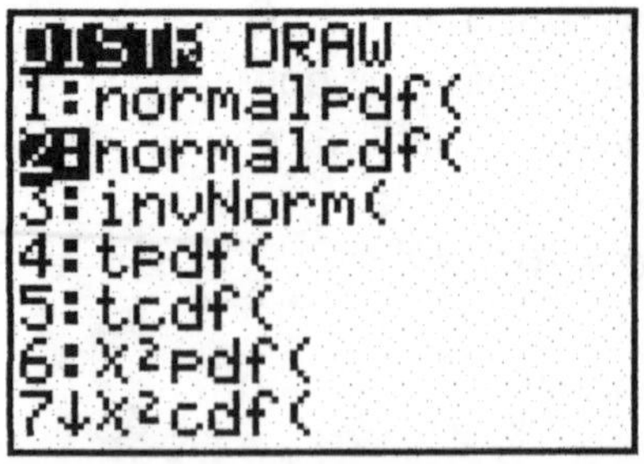

```
DISTR DRAW
1:normalpdf(
2:normalcdf(
3:invNorm(
4:tpdf(
5:tcdf(
6:X²pdf(
7↓X²cdf(
```

Press: **ENTER**

For a Sampling Distribution we must enter
**Lower , upper , mean , adjusted
standard deviation)** *In that order!*

```
normalcdf(22.1,2
6.8,30,1.6)
          .0227496658
■
```

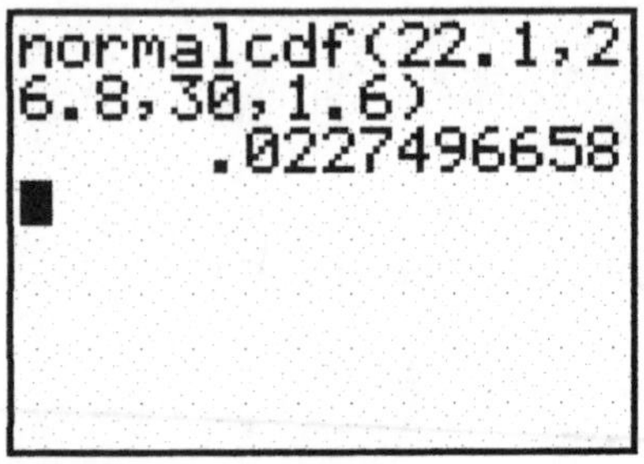

So we enter: **22.1 , 26.8 , 30 , 1.6)**
Don't forget the comma's!!

After you have entered all four values
press: **ENTER**

*As can be seen from the screen the
probability is* **.0227496658**

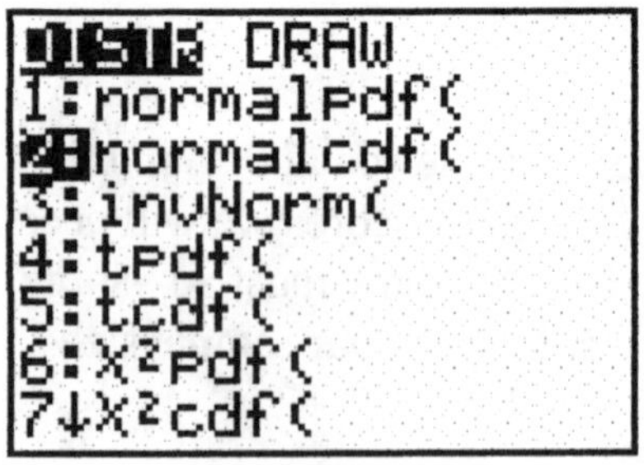

```
DISTR DRAW
1:normalpdf(
2:normalcdf(
3:invNorm(
4:tpdf(
5:tcdf(
6:X²pdf(
7↓X²cdf(
```

e.   Find P(x ≤ 28.2)

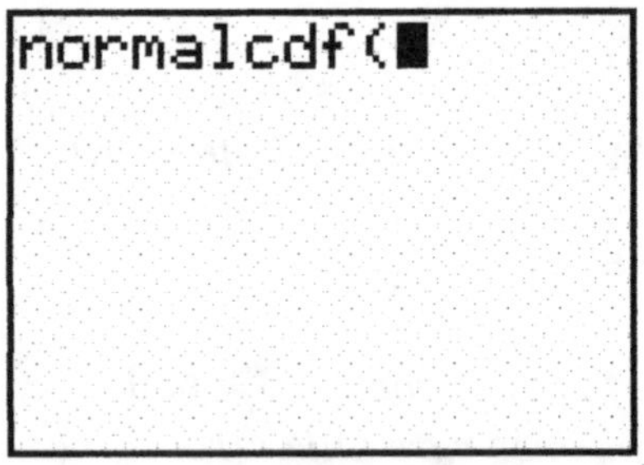

```
normalcdf(■
```

Press: **2nd VARS**
Highlight option "2:normalcdf ("
Press: **ENTER**

*For a Sampling Distribution we must
enter:*
**Lower , upper , mean , adjusted
standard deviation)** *In that order!*

```
normalcdf(0,28.2
,30,1.6)
          .1302945643
■
```

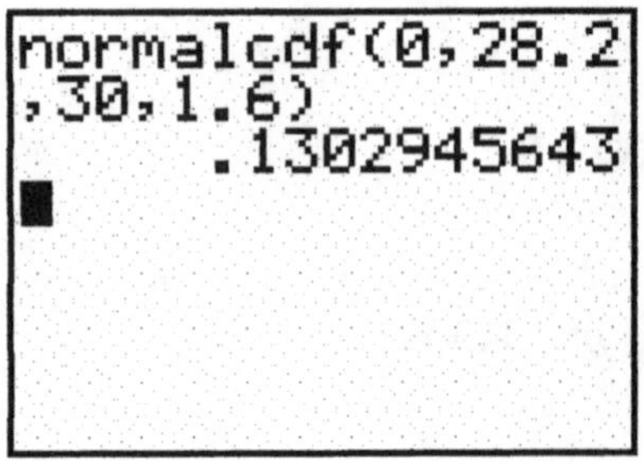

So we enter: **0 , 28.2 , 30 , 1.6)**
Don't forget the comma's!!

After you have entered all four values
press: **ENTER**

As can be seen from the screen the probability is **.1302945643**

f.   Find   P(x ≥ 27.0)   To start we *Press*:   **2nd   VARS**

Highlight option "**2:normalcdf (**"   *Press*: **ENTER**

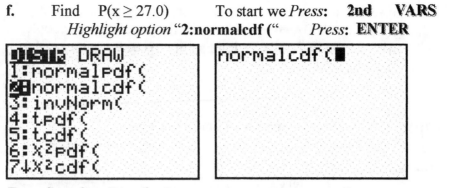

For a Sampling Distribution we must enter:

Lower , upper , mean , adjusted standard deviation) *In that order!*

So for our problem we enter **27.0 , 1EE99 , 30 , 1.6)**

Recall our upper limit should be as close to infinity as possible so we use 1EE99. ☺.

After you have entered all 4 values *press*: **ENTER**

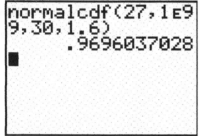

As can be seen from the screen the probability is **.9696037028**

Exercise 6.43   *Exposure to Teflon Cookware*

For this problem we will once again use the Normal Distribution Function found in the Distributions Menu. The following information is given in the problem: $\mu = 6$, $\sigma = 10$, n = 326.   We know $\sigma_x = 10/\sqrt{326} = .5538$

a.   We are looking for P(x ≥ 7.5)

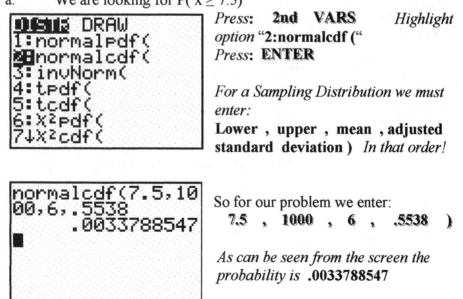

Press: **2nd   VARS**   *Highlight option* "**2:normalcdf (**"
Press: **ENTER**

For a Sampling Distribution we must enter:

Lower , upper , mean , adjusted standard deviation) *In that order!*

So for our problem we enter:

7.5 , 1000 , 6 , .5538   )

As can be seen from the screen the probability is **.0033788547**

b. We expect a sample mean 6 ppb. It is very highly probable that the true mean μ PFOA concentration for the population of people who live near the DuPont's Teflon Facility exceeds the mean of 6 ppb for the general population.

Example 7.1   **Mean Number of Seats Per Flight**

When calculating a confidence interval there are three one-sample choices on the calculator: 1. **Z-Interval**
                                        2. **T-Interval**
                                        3. **Proportion Interval**
All Intervals are found under the **STAT / TESTS** Menu. For this problem we want a **Z-Interval**. We have given information: $\overline{x} = 11.6$, $s = 4.1$, and n = 225 To find the 90% Confidence Interval on the calculator
Press: **STAT**   ▶   ▶

Arrow down *to* "**7 : Z Interval**"
Press: **ENTER**

See screen

For **Inpt** you have 2 choices:

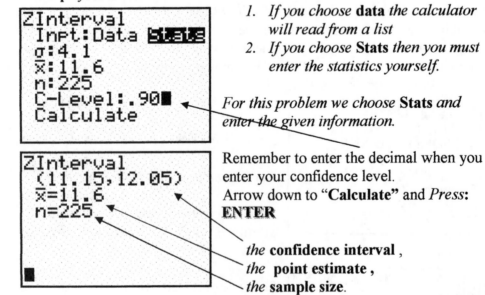

1. If you choose **data** *the calculator will read from a list*
2. If you choose **Stats** *then you must enter the statistics yourself.*

For this problem we choose **Stats** *and enter the given information.*

Remember to enter the decimal when you enter your confidence level.
Arrow down to "**Calculate**" and *Press*: **ENTER**

the **confidence interval** ,
the **point estimate** ,
the **sample size**.

As can be seen from the screen the 90% Confidence is: **(11.15 , 12.05)**;
Interval Notation for **11.6 ± .45** ☺

Example 7.2   *Small Sample Destructive Sampling*

For this problem we want a **T-Interval.**

a.   To find the 99% confidence on the calculator

First enter the data into List 1 in the calculator. *Press*: **STAT   ENTER**

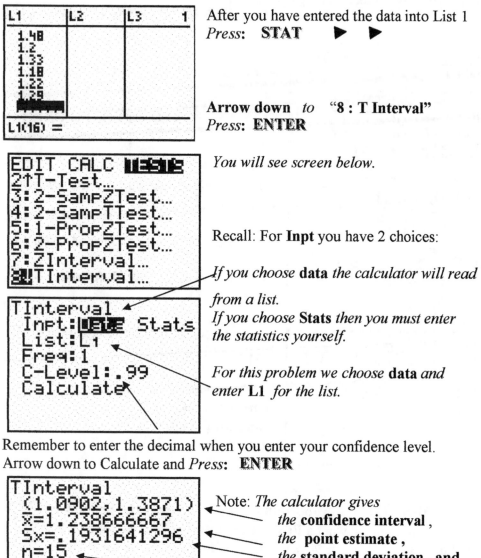

After you have entered the data into List 1
Press: **STAT   ▶   ▶**

Arrow down *to*   "**8 : T Interval**"
Press: **ENTER**

You will see screen below.

Recall: For **Inpt** you have 2 choices:

If you choose **data** *the calculator will read*
from a list.
If you choose **Stats** *then you must enter*
the statistics yourself.

For this problem we choose **data** *and*
enter **L1** *for the list.*

Remember to enter the decimal when you enter your confidence level.
Arrow down to Calculate and *Press*: **ENTER**

Note: *The calculator gives*
          the **confidence interval** ,
          the **point estimate** ,
          the **standard deviation** , **and**
          the **sample size**.

As can be seen from the screen the 99% Confidence Interval is **(1.09 , 1.39)**
Of course for part b. we are assuming that the sample is random and comes
from a nearly Normal Distribution.

Exercise 7.38   *Treating Alzheimer's Disease*

To find the 99% Confidence Interval on the calculator we first enter the data into the List.

Press: **STAT**   **ENTER** Type each piece of data into your chosen List. After you have entered all the data into the List

Press: **STAT** ▶ ▶   This will bring you to the TESTS Menu.

```
EDIT CALC TESTS
2↑T-Test…
3:2-SampZTest…
4:2-SampTTest…
5:1-PropZTest…
6:2-PropZTest…
7:ZInterval…
8↓TInterval…
```

Arrow down *to* **"8 : TInterval…"**

Press: **ENTER**

```
TInterval
 Inpt:Data Stats
 List:L1
 Freq:1
 C-Level:.99
 Calculate
```

For this problem we choose **data** *and enter* **L1** *for the list.*

Set the C-Level to the requested value.

Arrow down *to* "**Calculate**" *and*
Press: **ENTER**

```
TInterval
 (17.137,20.863)
 x̄=19
 Sx=2.108484326
 n=13
```

As can be seen from the screen the 99% CI is **(17.137 , 20.863)**

Also given is:
point estimate and
standard deviation
sample size.

Example 7.4   **Proportion Optimistic About the Economy**

For this problem we want a **Proportion-Interval**. We are given x = 257, and n = 484. To find the 90% Confidence Interval on the calculator

```
EDIT CALC TESTS
5↑1-PropZTest…
6:2-PropZTest…
7:ZInterval…
8:TInterval…
9:2-SampZInt…
0:2-SampTInt…
A↓1-PropZInt…
```

Press: **STAT**   ▶   ▶

Arrow down *to* "**A : 1-PropZInt…**"

Press: **ENTER**

Recall: **x** *is the number of successes*, **n** *is the sample size and*
        C-level *is confidence level.*

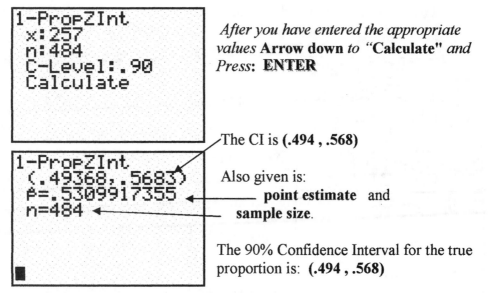

After you have entered the appropriate values **Arrow down** *to* "**Calculate**" *and Press*: **ENTER**

The CI is **(.494 , .568)**

Also given is:
— **point estimate** and
— **sample size**.

The 90% Confidence Interval for the true proportion is: **(.494 , .568)**

Exercise 7.99   *Brown Bag Lunches*

For this problem we want 1 Proportion-Interval for part a, and 1 for part b.

a.   We are given x = 665, and n = 1007. To find the 95% Confidence Interval on the calculator

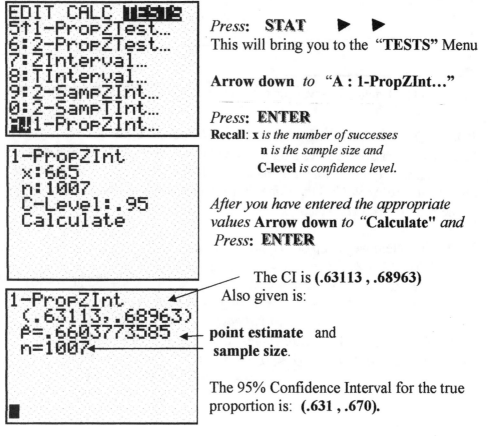

Press: **STAT** ▶ ▶
This will bring you to the "**TESTS**" Menu

Arrow down *to* "**A : 1-PropZInt...**"

Press: **ENTER**
Recall: **x** *is the number of successes*
        n *is the sample size and*
        C-level *is confidence level.*

After you have entered the appropriate values **Arrow down** *to* "**Calculate**" *and Press*: **ENTER**

The CI is **(.63113 , .68963)**
Also given is:

point estimate   and
sample size.

The 95% Confidence Interval for the true proportion is: **(.631 , .670).**

b. We are given x = 200, and n = 665.
To find the 95% Confidence Interval on the calculator

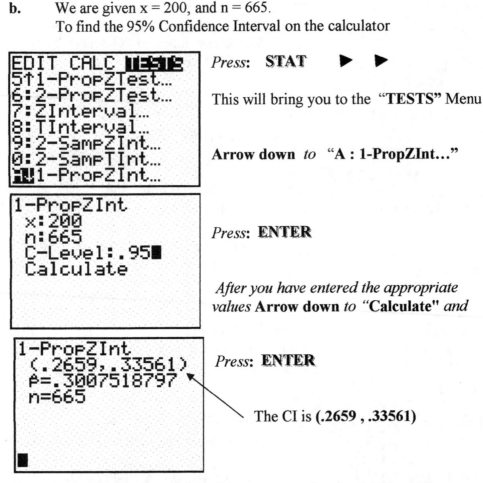

Press: **STAT** ▶ ▶

This will bring you to the "**TESTS**" Menu

Arrow down *to* "**A : 1-PropZInt...**"

Press: **ENTER**

After you have entered the appropriate
values **Arrow down** *to* "**Calculate**" *and*

Press: **ENTER**

The CI is **(.2659 , .33561)**

The 95% Confidence Interval for the true proportion is: **(.266 , .336)**

Examples 8.1 & 8.2 Mean Drug Response Time

Working a Test of Hypothesis on the calculator is very different and much
less complicated than doing it by hand. While the methods are different the
end results are ALWAYS the same. For **all** Tests of Hypothesis you will go
to the **STAT TESTS** Menu.

For this problem we set up our Hypothesis as:

Ho: $\mu \geq 1.2$

Ha: $\mu < 1.2$ we have given: n = 100, x = 1.05, s = .5 and α = .01

Remember .... the equal sign *always* ***goes with*** **Ho**

And you *always*   ***test*** **Ha**

Following the test if the **p-value** is less than the level of significance, α
we reject **Ho**.   In short: **p < α >>>>>>>>>>>>>> reject Ho,**

p $\geq$ α >>>>>>>> do not reject Ho.

Press: **STAT**

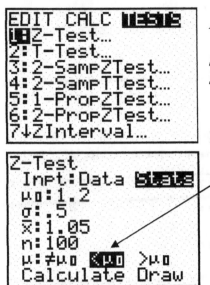

Arrow right *to highlight* **TESTS**

Since **Z-Test** *is already highlighted we simply press*: **ENTER**

Notice: once again you choose between **Data & Stats**.

There is one new line; the **"Test Line"***. You MUST* **match** *the inequality in this line to the inequality in* **your Ha**.

Notice: I highlighted the "$<\mu_0$" because that is the symbol in Ha.

The value beside μ_o: under **Inpt** is <u>always</u> the value associated with the Null Hypothesis. *For this problem it is* **1.2**

After you enter the appropriate values **Arrow Down** to **Calculate** and *Press*: **ENTER**

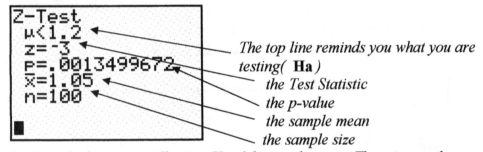

The top line reminds you what you are testing(**Ha** *)*
    the Test Statistic
    the p-value
    the sample mean
    the sample size

On the calculator you will reject Ho if the p-value $< \alpha$. <u>*There is no other Rule!*</u> *It will always work!* Since **.0013 < .01** we will **reject Ho**.

Exercise 8.43

Everything we need to work this problem is given in the problem. To begin

Press: **STAT**
Arrow right to highlight **TESTS**

Since **Z-Test** *is already highlighted we simply press*: **ENTER**

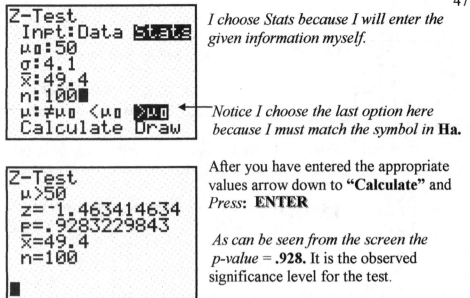

```
Z-Test
 Inpt:Data ████
 μo:50
 σ:4.1
 x̄:49.4
 n:100█
 μ:≠μo <μo ████
 Calculate Draw
```

I choose Stats because I will enter the given information myself.

Notice I choose the last option here because I must match the symbol in **Ha.**

```
Z-Test
 μ>50
 z=-1.463414634
 p=.9283229843
 x̄=49.4
 n=100
 █
```

After you have entered the appropriate values arrow down to **"Calculate"** and *Press:* **ENTER**

As can be seen from the screen the p-value = .928. It is the observed significance level for the test.

If we choose to reject Ho there is a .928 = 92.8% chance we are committing an error.

Example 8.5   Does New Engine Meet Air-Pollution Standards

For this problem we have Ho: $\mu \geq 20$ and Ha: $\mu < 20$.

```
L1      L2     L3    1
20.5
16.4
19.4
19.6
17.9
13.7
████
L1(10) =14.9
```

We also have given information: $\alpha = .01$ and we test **Ha: $\mu < 20$.**

After you have entered all the data into the list *Press:* **STAT**

```
EDIT CALC ████
1:Z-Test…
▓T-Test…
3:2-SampZTest…
4:2-SampTTest…
5:1-PropZTest…
6:2-PropZTest…
7↓ZInterval…
```

Arrow right *to*   "TESTS"

Arrow down *to*   "2 : T-Test…"

Press: **ENTER**

```
T-Test
 Inpt:████ Stats
 μo:20
 List:L1
 Freq:1
 μ:≠μo ████ >μo
 Calculate Draw
```

For this problem we choose **Data** *and enter* **L1** *for List*

Notice we still match this line to the inequality in **Ha.**

After you have adjusted your settings to match **arrow down** to
"**Calculate**" and *Press*: **ENTER**

```
T-Test
 μ<20
 t=-2.602890915
 p=.0143011456
 x̄=17.57
 Sx=2.952230795
 n=10
```

*As can be seen from the screen the p-value
is .0143 which is not less than .01
therefore we can not reject Ho.
The test is inconclusive.*

So............. no, there is not sufficient evidence at $\alpha = .01$ to conclude this
new type of engine meets the standard that emissions be less than 20 ppm.

Exercise 8.59

a. To begin we note the given information: $\bar{x} = 4.8$, $s = 1.3$, $\alpha = .05$, $n = 5$

      Ho: $\mu \geq 6$

      Ha: $\mu < 6$

         So we will reject Ho if $p < .05$

Press: **STAT**

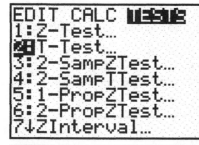

```
EDIT CALC TESTS
1:Z-Test...
2:T-Test...
3:2-SampZTest...
4:2-SampTTest...
5:1-PropZTest...
6:2-PropZTest...
7↓ZInterval...
```

Arrow left to "**TESTS**"

Arrow down to "**2 : T-Test...**"

Press: **ENTER**

```
T-Test
 Inpt:Data Stats
 μ₀:6
 x̄:4.8
 Sx:1.3
 n:5
 μ:≠μ₀ <μ₀ >μ₀
 Calculate Draw
```

Make sure you choose: **Stats**

*Notice this inequality must match the
symbol in* **Ha**.

After you have entered all the values from
your problem arrow down to "**Calculate**"

```
T-Test
 μ<6
 t=-2.064062748
 p=.0539770368
 x̄=4.8
 Sx=1.3
 n=5
```

Press: **ENTER**

As can be seen from the screen **p = .054**
which is not less than .05 therefore we
do not reject Ho.
The test is inconclusive.

b. To begin, note the given information: x = 4.8, s = 1.3, α = .05, n = 5.

Ho: μ = 6

Ha: μ ≠ 6   So we will reject Ho if p < .05

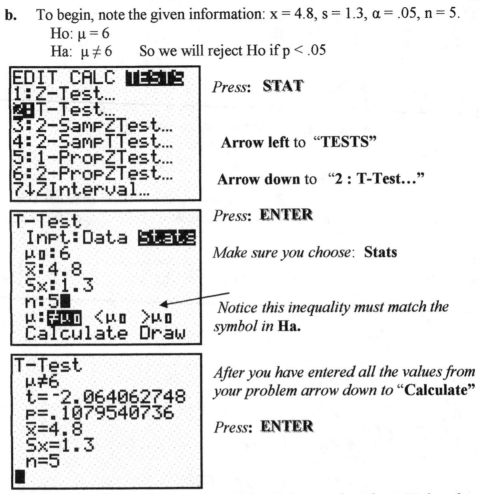

Press: **STAT**

Arrow left to **"TESTS"**

Arrow down to **"2 : T-Test..."**

Press: **ENTER**

Make sure you choose: **Stats**

Notice this inequality must match the symbol in **Ha.**

After you have entered all the values from your problem arrow down to "**Calculate**"

Press: **ENTER**

As can be seen from the screen **p = .108** *which is not less than .05 therefore we* **do not reject Ho.** The Test is inconclusive.

b. Since on the calculator we use p-values not Test-Statistics we simply give the p-values as displayed on the screens:

For part (a) **p = .054**, for part (b) **p = .108**

Example 8.7 **Proportion of Defective Batteries**

Again, on the calculator this problem will be worked quite differently than it is by hand yet the results will be the same. Start by noting the given information. N = 300, x = 10, α = .01,

Ho: p ≥ 5

Ha: p < 5   So we will reject

Ho if our p-value < .01

Press: **STAT**

Arrow left to **"TESTS"**

Press: **STAT**
Arrow left to **"TESTS"**

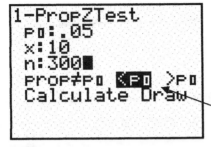

Arrow Down to **"5 : 1-PropZTest…"**

Press: **ENTER**
You must enter all appropriate values
Once again, this line must match the
inequality in your **Ha**

After you have entered the correct values arrow down to **"Calculate"** and
Press: **ENTER**

As can be seen from the screen **p = .093**

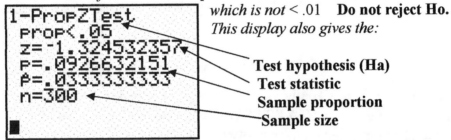

which is not < .01   **Do not reject Ho.**
This display also gives the:

Test hypothesis (Ha)
Test statistic
Sample proportion
Sample size

Example 9.1   Comparing Mean Weight Loss for Two Diets

On the calculator we will go to the **STAT TESTS** menu once again.
There are 3 different 2-Sample **Intervals** and 3 different 2-Sample **Tests**
under the **TESTS** Menu. Because of the large sample sizes found in this
problem we will use the **2-Sample Z-Interval**. Begin by entering all data
into List1 and List2.(low-fat diet into list1 & regular diet into list2)

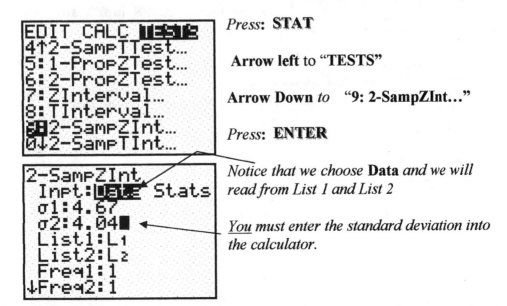

Press: **STAT**

Arrow left to **"TESTS"**

Arrow Down *to* **"9: 2-SampZInt…"**

Press: **ENTER**

Notice that we choose **Data** *and we will*
read from List 1 and List 2

You must enter the standard deviation into
the calculator.

If the standard deviation is not given in the problem, as is our case you must first run "1-var stats" on each list to obtain the standard deviations.

```
2-SampZInt
↑σ2:4.04
 List1:L₁
 List2:L₂
 Freq1:1
 Freq2:1
 C-Level:.95
```

After you have entered all the values for your problem **Arrow Down** *to* "Calculate" *and press:* **ENTER**

```
2-SampZInt
 (.69072,3.1293)
 x̄1=9.31
 x̄2=7.4
 n1=100
 n2=100

■
```

As can be seen from the screen the 95% CI of the difference between the means is: **(.68 , 3.12)**

Example 9.2   Comparing Mean Weight Loss for Two Diets

Once again we go to the **STAT TESTS** Menu, and for this problem we will choose a **2-Sample Z-Test** as requested by the problem. We will test the Alternative hypothesis: the two means are not equal, and use alpha = .05 as given in the problem.

Press: **STAT**

Arrow left to **TESTS**

Arrow Down *to* **3: 2-SampZTest...**

Press: **ENTER**

Once again you must choose **Data** *or* **Stats** *. For our problem we choose* **Stats** *because we will enter the statistics ourselves.*

Don't forget to arrow down and enter all values associated with your problem.

Recall: *you must match this inequality to the one found in your* **Ha**.

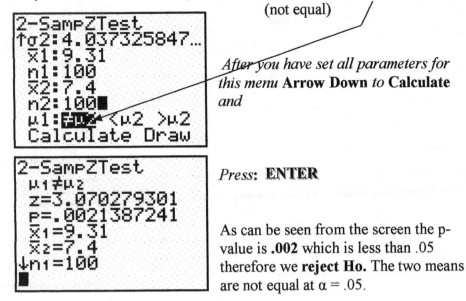

(not equal)

After you have set all parameters for this menu **Arrow Down** *to* **Calculate** *and*

Press: **ENTER**

As can be seen from the screen the p-value is **.002** which is less than .05 therefore we **reject Ho.** The two means are not equal at $\alpha = .05$.

Example 9.4 *Reading test scores for slow learners*

For this problem we will go back to the **STAT TESTS** Menu, but this time we will choose a **2-Sample T-Interval** as the sample size dictates.

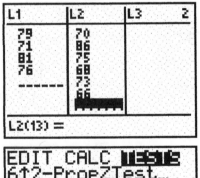

First enter the data.
Each Method into its own List.

I will enter "**New Method**" *into* **List 1** *and* "**Standard Method**" *into* **List 2**.

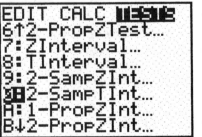

Press: **ENTER**
Arrow left to **TESTS** and **Arrow Down** to "**0: 2-SampTInt...**"

Press: **ENTER**

This time we choose **Data** as we entered the data into **List 1** and **List 2**.

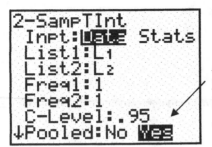

Notice we choose **Yes,** *since based on the F-test we cannot reject the claim the two sample variances are equal.*** *See note that follows*

The F Test : $H_o : \text{var(pop1)} = \text{var(pop2)} >>> s_1^2 = s_2^2$ **(equal)**

$H_a : \text{var(pop1)} \neq \text{var(pop2)} >> s_1^2 \neq s_2^2$ **(not equal)**

This test can be performed on the calculator. Go to the tests menu by pressing **STAT** and choose option "D: 2-SampFTest...". You will have the same choices as with any of the other tests. We will highlight **Stats** for this problem since we are entering the statistics ourselves. Adjust your screen to this problem's statistics and at the test line highlight $\sigma 1$: "$\neq \sigma 2$" and press **ENTER**. Arrow down to calculate and press **ENTER** We would reject H_o only if $p < \alpha$ If you reject H_o then choose **NO**, else choose **YES**.

After you adjust all the fields in this menu for your problem **Arrow Down** to **Calculate** and *Press*: **ENTER**

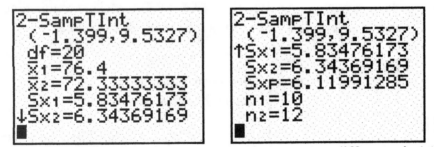

As seen from the screen the 95% CI for the true mean difference is **(-1.4 , 9.5)**

Exercise 9.12

For this problem we will use a 2-Sample T Test and consider the p-value to determine if these two populations produce the same mean.
Press: **STAT**

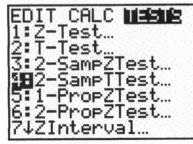

Arrow left to "**TESTS**"

Arrow Down *to* "**4: 2-SampTTest...**"

Press: **ENTER** For this problem we choose **Stats**.

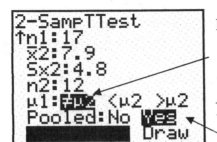

For the test line we choose $\neq$ as dictated by the problem.
  Recall it must match **Ha**.

Again, based on the results of the F-Test we poole the variances.

After you have adjusted all the fields
Arrow Down *to* **Calculate** *and Press:* **ENTER**

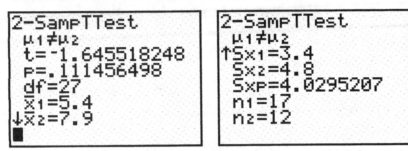

As can be seen from the screen the p-value is .11 which is **not less** than .05 therefore we **do not reject Ho**.

b.   Go to **TESTS Arrow Down** to **0 2-SampTInt….**

```
2-SampTInt
 (-5.617,.6173)
 df=27
 x̄₁=5.4
 x̄₂=7.9
 Sx₁=3.4
↓Sx₂=4.8
■
```

Press: **ENTER**
Arrow Down *to* **Calculate** *and again*
Press: **ENTER** you will see this screen.

As can be seen from the screen the 95% CI is **(-5.617 , .6173)** *which does include zero.*

Example 9.5   Mean Salaries for Males vs. Females

To perform a Paired Difference Confidence Interval on the calculator first enter the data into List 1 and List 2. I will enter Male into List 1 and Female into List 2. The next step is to create List 3 which will be (List 1 - List 2). Enter your data into List 1 and List 2.

```
L1      L2      L3      3
29300   28800   ------
41500   41600
40400   39800
38500   38500
43500   42600
37800   38000
69500   69200
L3 =
```

*For matched pair calculations you **MUST** enter the data in the **EXACT** order it is given.*

*After you have entered all the data Arrow up and highlight the **L3** itself (not the top line of the data)*

Press: **(** **2ⁿᵈ** **1** **-** **2ⁿᵈ** **2** **)** **ENTER**

You will see the expression scroll across the bottom of your screen.
List 3 will be **created** as you hit ENTER

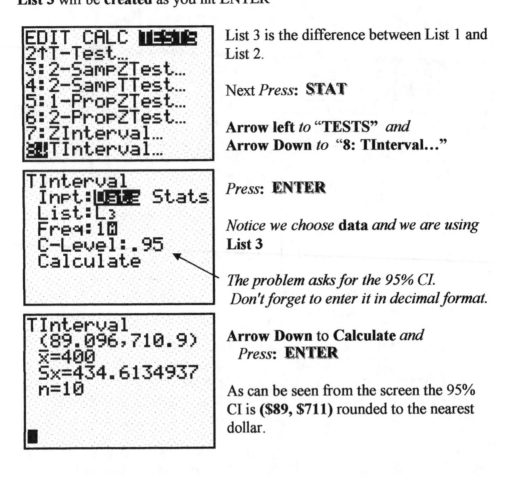

List 3 is the difference between List 1 and List 2.

Next *Press:* **STAT**

Arrow left *to* **"TESTS"** *and*
Arrow Down *to* "8: TInterval..."

Press: **ENTER**

Notice we choose **data** *and we are using* **List 3**

The problem asks for the 95% CI.
Don't forget to enter it in decimal format.

Arrow Down to **Calculate** *and*
Press: **ENTER**

As can be seen from the screen the 95% CI is **($89, $711)** rounded to the nearest dollar.

Example 9.6   Comparing Fractions of Smokers for Two Years

For the 2-Sample Proportion test we will again use the **STAT TESTS** menu
The given information is: $n_1 = 1500$, $x_1 = 555$, $n_2 = 1750$, $x_2 = 578$
Ho: $(p_1 - p_2) \leq 0$: *the proportion has not decreased*
Ha: $(p_1 - p_2) > 0$: *the proportion has decreased*, and we are given $\alpha = .05$
So on the calculator we will reject Ho if the p-value < .05

Press: **STAT**
 Arrow Left to **TESTS** and **Arrow Down** to **6 : 2-PropZTest...**

Press: **ENTER**
 Fill in the appropriate information for your problem

*The inequality in this line must match your **Ha**.*

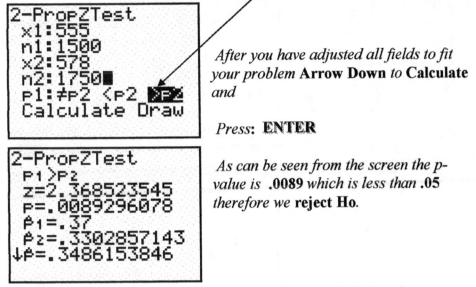

*After you have adjusted all fields to fit your problem **Arrow Down** to **Calculate** and*

Press: **ENTER**

*As can be seen from the screen the p-value is **.0089** which is less than **.05** therefore we **reject Ho**.*

There is sufficient evidence at the $\alpha = .05$ level to conclude that the proportion of adult smokers has decreased.

Example 9.10   Comparing Weight Variations in Mice

To perform an F-Test on the calculator there is one thing you must remember. Data from Supplier 1 goes into list 1 and supplier 2 into list 2.

> No matter what order the text puts the data in, on the calculator the data set that has the **largest variance is called sample 1!**
> <u>This only applies to F-Test on the calculator</u>.

The Hypothesis for the F-Test are

$$Ho: \sigma_1^2 = \sigma_2^2$$
$$Ha: \sigma_1^2 \neq \sigma_2^2 , \quad \text{and we will reject Ho if our p-value} < .10$$

We will once again use the **STAT TESTS** Menu.

Press: **STAT**

Arrow Left to **TESTS** and **Arrow up** to "E: 2-SampFTest…"

Press: **ENTER**

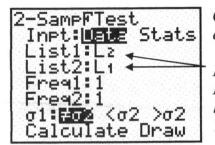

Once again we choose Data as we will enter the statistics ourselves.

Note the order. .The calculator always puts List 2 on top(it must be the sample with the largest variance)
The test line still must match Ha.

After you adjust all fields for your problem **Arrow Down** to **Calculate** and
Press: **ENTER**

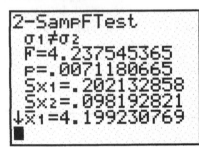

As can be seen from the screen the p-value = .007 < .10, therefore we **reject Ho**.

At the $\alpha = .10$ level the two variances are not equal.

Example 9.12   *Reading Tests Scores for Slow Learners*

To perform an F-Test on the calculator there is one thing you must remember:

No matter what order the text puts the data in, on the calculator, the data set that has the **largest variance is called sample 1 !**
This only applies to F-Test on the calculator.

The Hypothesis for the F-Test are
$$Ho: \sigma_1^2 = \sigma_2^2$$
$$Ha: \sigma_1^2 \neq \sigma_2^2, \text{ and we will reject Ho if our p-value} < .10$$

We will once again use the **STAT TESTS** Menu.
Press: **STAT**

Arrow Left to **TESTS** and **Arrow up** to "**E: 2-SampFTest...**"

Press: **ENTER**

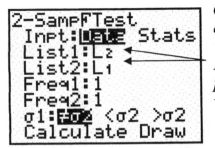

Once again we choose Data as we will enter the statistics ourselves.

Note the order. .The calculator always puts List 2 on top(it must be the sample with the largest variance)
The test line still must match Ha.

After you adjust all fields for your problem **Arrow Down** to **Calculate** and

```
2-SampFTest
 σ1≠σ2
F=.8459839357
p=.8147505689
Sx1=5.83476173
Sx2=6.34369169
↓x̄1=76.4
■
```

Press: **ENTER**

As can be seen from the screen the p-value = **.8148** *which is not less than* **.10,** *therefore we do not* **reject Ho**.

At the $\alpha = .10$ we can not conclude there are differences in the variances.

Exercise 9.96

For this problem we begin by entering the data into the calculator.

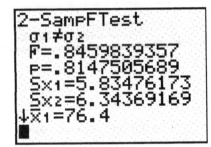

I entered the Sample1 data into List 1 and Sample2 data into List 2.

After you have entered all the data

Press: **STAT**

```
EDIT CALC TESTS
0↑2-SampTInt...
A:1-PropZInt...
B:2-PropZInt...
C:χ²-Test...
D:2-SampFTest...
E:LinRegTTest...
F:ANOVA(
```

Arrow Left to **TESTS** *and* **Arrow up** *to* **D: 2-SampFTest...**

Press: **ENTER**

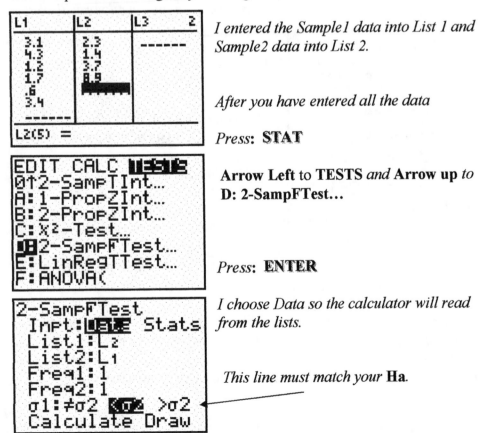

I choose Data so the calculator will read from the lists.

This line must match your **Ha**.

After you change all fields for your problem **Arrow Down** to **Calculate** and

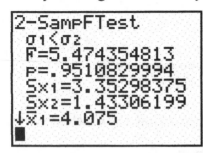

Press: **ENTER**

As can be seen from the screen the p-value is **.951** which is not less than .01 therefore we **do not reject Ho**.

b. For this part every step will be the same except one. We will change the test line since the Alternative Hypothesis for part b is $\neq$

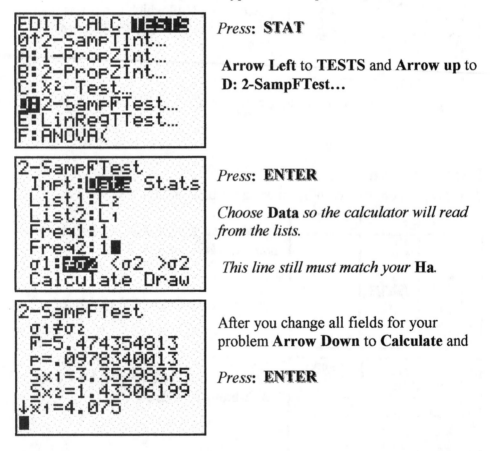

Press: **STAT**

Arrow Left to **TESTS** and **Arrow up** to **D: 2-SampFTest…**

Press: **ENTER**

Choose **Data** *so the calculator will read from the lists.*

This line still must match your **Ha**.

After you change all fields for your problem **Arrow Down** to **Calculate** and

Press: **ENTER**

As can be seen from the screen the p-value is **.0978** *which is slightly less than .10, therefore* **we reject Ho**.

If we were to round our p-value to 2 decimal places it would be **.10** *which is not less than itself. Care must be taken to round according to the instructions given.*
As you can see from this problem it could affect your answer!

Supplemental Exercise 9.116   Tapeworms in Brill Fish

Here we are looking for a 2-Sample Proportion Interval. We will once again go to the **STAT TESTS** Menu.

Press: **STAT**

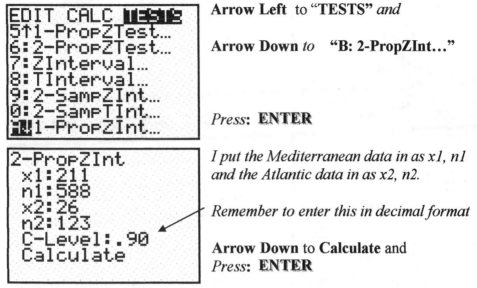

Arrow Left to "TESTS" *and*

Arrow Down *to*   "B: 2-PropZInt..."

Press: **ENTER**

I put the Mediterranean data in as x1, n1 and the Atlantic data in as x2, n2.

Remember to enter this in decimal format

Arrow Down to **Calculate** and
Press: **ENTER**

As can be seen from the screen the 90% CI is **(.079 , .216)**

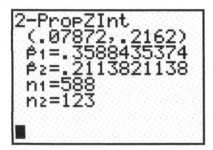

There is from a 7.9% increase to a 21.6% increase in the number of bacteria found within the brill captured from the Mediterranean Sea.

Exercise 9.137   Isolation Timeouts for Students

For this problem we will use a **2-Sample Z-Test** and consider the p-value to determine if "...Option III classrooms had significantly more timeout incidents..." I'll put data for Option II into List 1, and Option III into List 2.

Press: **STAT**

Arrow left to "TESTS"

Arrow Down *to* "**3: 2-SampZTest…**"

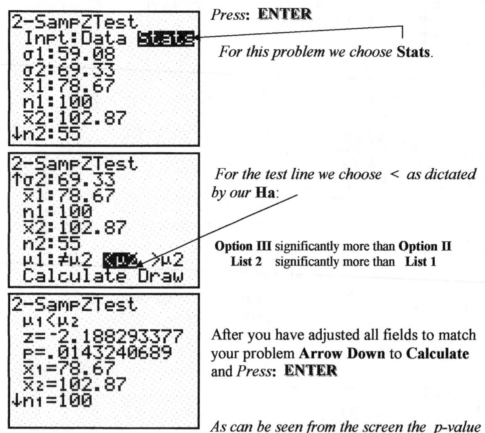

Press: **ENTER**

For this problem we choose **Stats**.

For the test line we choose < *as dictated by our* **Ha**:

Option III significantly more than **Option II**
List 2   significantly more than   **List 1**

After you have adjusted all fields to match your problem **Arrow Down** to **Calculate** and *Press:* **ENTER**

As can be seen from the screen the p-value is .014 which means if we choose to reject Ho there is a 1.4% chance we are committing a Type-I error.

So at alpha = .014, or anything greater, **yes Option III has significantly more Timeout incidents.**

Example 10.4   **Comparing Golf Ball Brands**

We begin by entering all data into the calculator. Enter each brand into its own list. As with all test of Hypothesis we will test Ha and reject Ho only if our p-value < α

L2	L3	L4	4
264.3	270.5	246.5	
257	265.5	251.3	
262.8	270.7	261.8	
264.4	272.9	249	
260.6	275.6	247.1	
255.9	266.5	245.9	
------	------	------	

L4(11) =

For all ANOVA problems the Hypothesis will be Ho: all means are equal
Ha: at least one mean is different

After you have entered all the data

```
EDIT CALC TESTS
B↑2-PropZInt…
C:X²-Test…
D:X²GOF-Test…
E:2-SampFTest…
F:LinRegTTest…
G:LinRegTInt…
H:ANOVA(
```

Press: **STAT**

Arrow Left to **TESTS** and **Arrow up** to "**H : ANOVA(**"

```
ANOVA(■
```

Press: **ENTER**

For ANOVA you simply enter the list where you placed the data.

For this problem I will enter L1 , L2 , L3 , L4 as I used said lists.

```
ANOVA(L₁,L₂,L₃,L
4)■
```

Press: **2ⁿᵈ 1 , 2ⁿᵈ 2 , 2ⁿᵈ 3 , 2ⁿᵈ 4)**

Press: **ENTER**
To see all the results you must arrow down.

All values requested can be read directly from the screens.
As can be seen the p-value is given in Scientific Notation.
In common Notation it is **0.000000000003973108**
So we feel very confident that all means are not equal!

```
One-way ANOVA
 F=43.98874592
 P=3.973108E⁻12
 Factor
  df=3
  SS=2794.38875
↓ MS=931.462917
■
```

```
One-way ANOVA
↑ MS=931.462917
 Error
  df=36
  SS=762.301
  MS=21.1750278
 Sxp=4.60163316
■
```

Exercise 10.24

We begin by entering all data into the appropriate lists.

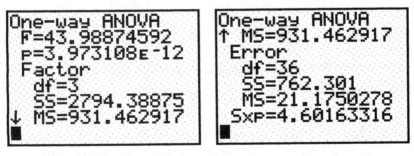

Note: In ANOVA it is not necessary that the list be of equal length.

Press: **STAT**

```
EDIT CALC TESTS
B↑2-PropZInt...
C:X²-Test...
D:X²GOF-Test...
E:2-SampFTest...
F:LinRegTTest...
G:LinRegTInt...
:HANOVA(
```

Arrow Left to "TESTS" and **Arrow up** to "H : ANOVA("

```
ANOVA(L₁,L₂,L₃)█
```

Press: **ENTER**

For ANOVA you simply enter each list where you have the data.

For this problem I will enter L1 , L2 , L3 as I used these three lists.

Press: **2ⁿᵈ 1 , 2ⁿᵈ 2 , 2ⁿᵈ 3)**

Press: **ENTER**   *To see all the results you must arrow down.*

```
One-way ANOVA
 F=3.387603014
 p=.0752767198
 Factor
  df=2
  SS=12.4203077
↓ MS=6.21015385
█
```

```
One-way ANOVA
↑ MS=6.21015385
 Error
  df=10
  SS=18.332
  MS=1.8332
 Sxp=1.35395716
█
```

All values requested can be read directly from the screens as the calculator uses standard notation.

b.   We see from the first screen the p-value is **.075** which is not less than .01 therefore we **do not reject Ho**.

Exercise 11.28   *Sweetness of Orange Juice*

We begin by entering the data into the calculator. I will put the x's into List 1 and the y's into List 2.

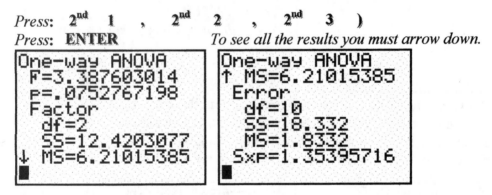

After you have entered all the data you should be looking at this screen.

Notice I put the pectin level(independent) into list 1 and the sweetness index(dependent) into list 2, since the sweetness index should be based upon the

pectin level and not vice/versa.
Press: **STAT**

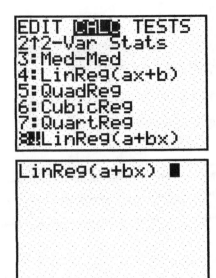

Arrow Right *to* **CALC** *and* **Arrow Down**
to **8: LinReg(a+bx)**

Notice option 4 can also be used but it will
give the equation inform (ax + b). Like
algebra (mx+b) form ☺

Press: ENTER

For regression you need only tell the
calculator where you put your data and
which place to store the equation.

It must however be in this order:
x's , y's , where

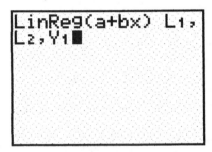

I used List 1 and List 2 for the x's and y's
respectively and I will place the results in
Y1.

Press: **2ⁿᵈ 1 , 2ⁿᵈ 2 , VARS ▶**
ENTER ENTER ENTER

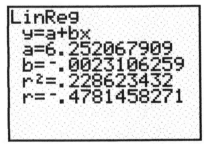

a) As can be seen from the screen the
Least Squares Line is **y = 6.25 - .0023x**
Press: **ZOOM 9**

Once you have the graph displayed you can use the **CALC** *feature of the*
calculator to predict specific y-values.

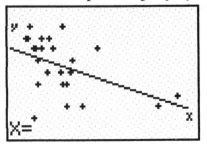

To **predict a y-value** from a **given x-value**
follow these steps:

Press: **2nd   TRACE   ENTER**

Simply enter the given x-value and *press:*
ENTER

c)   *As can be seen from the screen if x = 300 we predict* **y = 5.56**

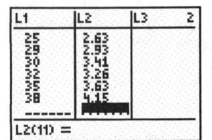

Example 11.4   *Relating Crime Rate & Casino Employment*

Begin by entering the data into the calculator.

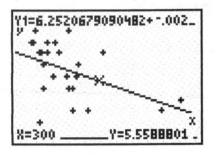

Be careful to enter the data in the exact order it is given.

Notice I place the Crime Rate(dependent) into List 2 and the Number of Casinos (Independent) into List 1.

After you have entered all the data into the calculator

Press: **STAT**
Arrow Right *to* **"CALC"** *and*
Arrow Down *to* **"8: LinReg(a+bx)"**

Notice option 4 can also be used but it will give the equation in (ax + b).
Like algebra (mx+b) form ☺
Press: **ENTER**

For regression you need only tell the calculator where you put your data and which place to store the equation.

It must however be in this order: x's , y's , where

I used List 1 and List 2 for the x's and y's respectively, placed results in Y1.

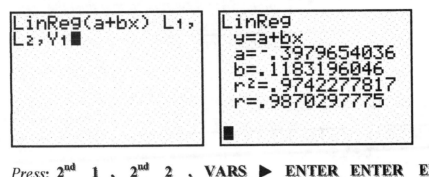

Press: **2ⁿᵈ 1 , 2ⁿᵈ 2 , VARS ▶ ENTER ENTER ENTER**

As can be seen from the screen **r = .987** and **r² = .974**

Example 11.5   *Drug Reaction Regression*

To perform a Linear Regression on the calculator first enter the data into the calculator. Enter the Independent variables(x's) into List 1 and the dependent variables(y's) into List 2.

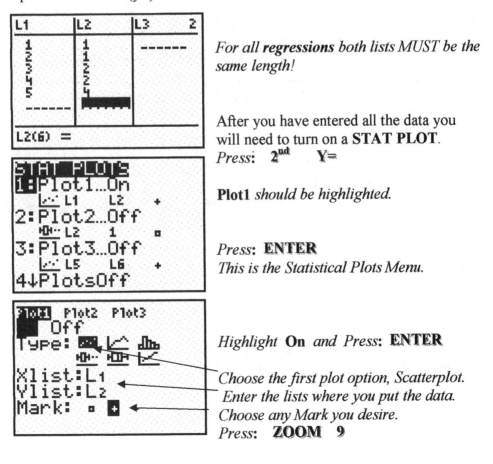

*For all **regressions** both lists MUST be the same length!*

After you have entered all the data you will need to turn on a **STAT PLOT**.
Press: **2ⁿᵈ   Y=**

Plot1 *should be highlighted.*

Press: **ENTER**
This is the Statistical Plots Menu.

Highlight **On** *and Press:* **ENTER**

Choose the first plot option, Scatterplot.
Enter the lists where you put the data.
Choose any Mark you desire.
Press: **ZOOM 9**

This will give you a scatter plot of the data.

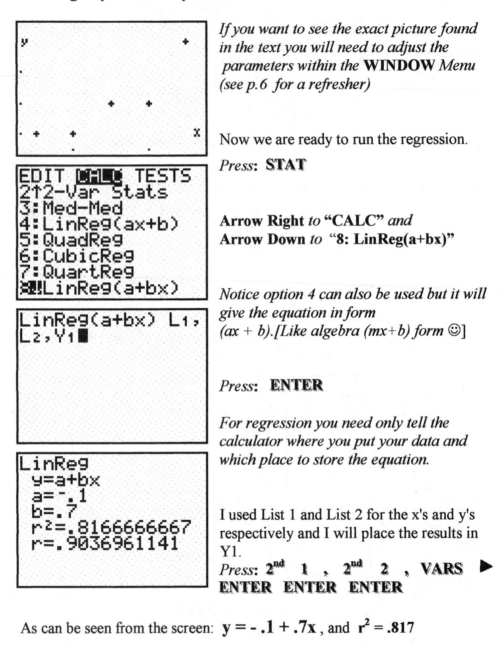

If you want to see the exact picture found in the text you will need to adjust the parameters within the WINDOW Menu (see p.6 for a refresher)

Now we are ready to run the regression.

Press: **STAT**

Arrow Right *to* **"CALC"** *and*
Arrow Down *to* **"8: LinReg(a+bx)"**

Notice option 4 can also be used but it will give the equation in form
(ax + b).[Like algebra (mx+b) form ☺]

Press: **ENTER**

For regression you need only tell the calculator where you put your data and which place to store the equation.

I used List 1 and List 2 for the x's and y's respectively and I will place the results in Y1.
Press: 2nd **1** , 2nd **2** , **VARS** ▶
ENTER ENTER ENTER

As can be seen from the screen: $y = -.1 + .7x$, and $r^2 = .817$

If you do not see the r and r^2 you will need to follow these steps:
Press: **2nd** **0** x^{-1}
Next **Arrow Down** to **"DiagnosticsOn"**
Press: **ENTER ENTER** You only have to do this 1 time. ☺

Exercise 11.115   *Feeding Habits of Fish*

Begin this problem by entering the data into the calculator. I will put the x's into list 1 and the y's into list 2.

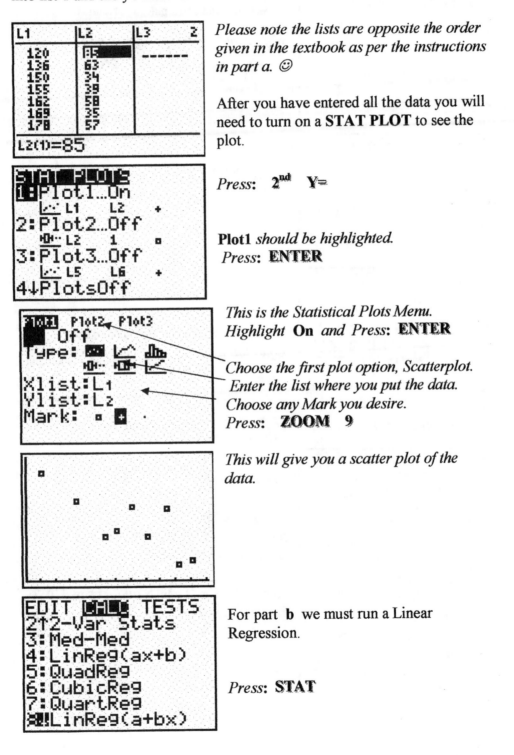

Please note the lists are opposite the order given in the textbook as per the instructions in part a. ☺

After you have entered all the data you will need to turn on a **STAT PLOT** to see the plot.

Press: **2ⁿᵈ Y=**

Plot1 *should be highlighted.*
Press: **ENTER**

This is the Statistical Plots Menu.
Highlight **On** *and Press:* **ENTER**

Choose the first plot option, Scatterplot.
Enter the list where you put the data.
Choose any Mark you desire.
Press: **ZOOM 9**

This will give you a scatter plot of the data.

For part **b** we must run a Linear Regression.

Press: **STAT**

Arrow Right *to* **"CALC"** *and* **Arrow Down** *to* **"8: LinReg(a+bx)"**
Press: ENTER

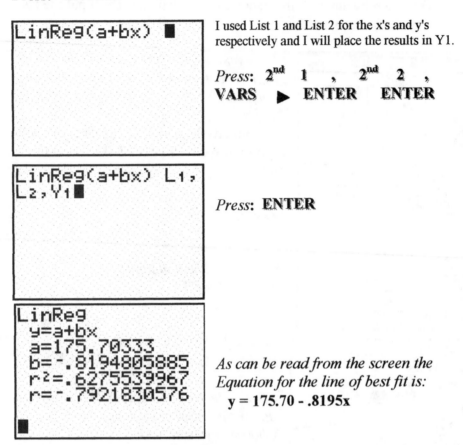

I used List 1 and List 2 for the x's and y's respectively and I will place the results in Y1.

Press: **2ⁿᵈ 1 , 2ⁿᵈ 2 ,**
VARS ▶ ENTER ENTER

Press: **ENTER**

As can be read from the screen the Equation for the line of best fit is:
y = 175.70 - .8195x

To conduct a test we go back to the **STAT TESTS** Menu.

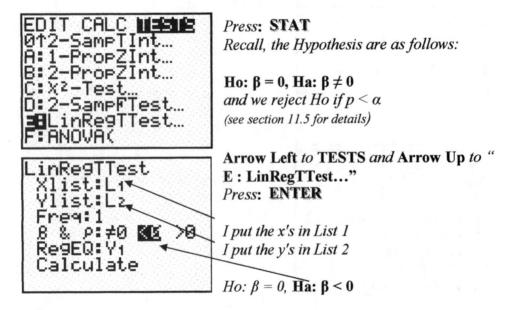

Press: **STAT**
Recall, the Hypothesis are as follows:

Ho: $\beta = 0$, Ha: $\beta \neq 0$
and we reject Ho if $p < \alpha$
(see section 11.5 for details)

Arrow Left *to* **TESTS** *and* **Arrow Up** *to* **"E : LinRegTTest..."**
Press: **ENTER**

I put the x's in List 1
I put the y's in List 2

Ho: $\beta = 0$, **Ha: $\beta < 0$**

After you have adjusted all fields to match your problem **Arrow Down** to "**Calculate**" and *Press*: **ENTER**

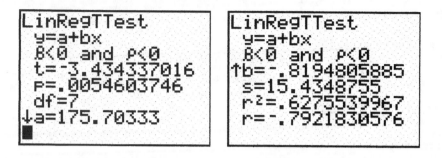

As can be seen from the screen the p-value is **.005** which is less than **.10** Therefore we **reject Ho**.

Example 12.7   Quadratic Model Predicting Electrical Usage

We begin this problem by entering the data into the calculator. After which we will run the Quadratic regression.

Enter size of home into List 1 and Monthly-usage into List 2.

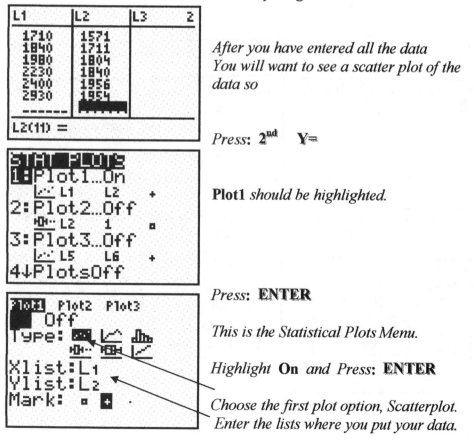

After you have entered all the data
You will want to see a scatter plot of the data so

Press: **2ⁿᵈ   Y=**

Plot1 *should be highlighted.*

Press: **ENTER**

This is the Statistical Plots Menu.

Highlight **On** *and Press:* **ENTER**

Choose the first plot option, Scatterplot.
Enter the lists where you put your data.

Choose any Mark you desire.

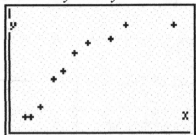

Press: **ZOOM** **9**

As can be seen from the screen there is a noticeable curve. A linear regression would not be appropriate.

To run the Quadratic regression

Press: **STAT** **Arrow Right** to "**CALC**" and **Arrow Down** to "**5:QuadReg**"

At the flashing curser you must enter the lists where you placed the data in this order: x's , y's .

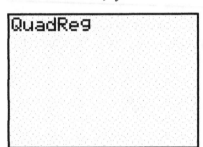

Also, if you want to see the graph you will need to tell the calculator where to send the regression equation.

Make sure your Y1 is clear of any old equations

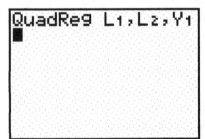

Press: **2ⁿᵈ** **1** , **2ⁿᵈ** **2** , **VARS**
▶ **ENTER** **ENTER**
Don't forget the commas ☺

Press: **ENTER**

As can be seen from the screen the Quadratic equation that best describes the data is: $Y = -.00045x^2 + 2.3989x - 1216.143887$

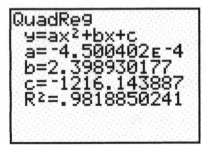

To fit the variables given in the text
β_1 = -.00045
β_2 = 2.3989
β_3 = -1216.1439

To see the graph of the regression equation along with the data press:
GRAPH

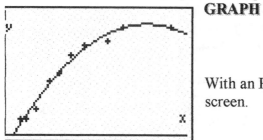

With an $R^2 = .9819$ we expect the close fit we see on the screen.

Example 12.17 Using residuals to check for constant error variance

For this problem we will begin by looking at a plot of the residuals and then proceed to the data transformation, and finally we will determine the Line of best fit for the transformed data.

L1	L2	L3	2
20	36530		
19	52745		
27	67282		
25	80931		
12	32303		
11	38371		
------	------		

L2(51) =

Enter all data into its appropriate list.

After you have entered all the data

Press: **STAT**
Arrow right to **CALC** and **Arrow Down** to "**8 : LinReg(a+bx)**"
Press: **ENTER**
Enter the location for the X's and Y's

Press: **2nd 1 , 2nd 2**

```
EDIT CALC TESTS
2↑2-Var Stats
3:Med-Med
4:LinReg(ax+b)
5:QuadReg
6:CubicReg
7:QuartReg
8▮LinReg(a+bx)
```

```
LinReg(a+bx) L1,
L2
```

You must run the regression first in order to have the residuals to plot.

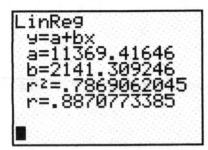

Note: The R^2 vaule = 79%

Now we will plot the residuals

Now we will go to the Stat Plot menu and plot the residuals and the original x values using the scatter plot option.

Press: 2nd $Y =$

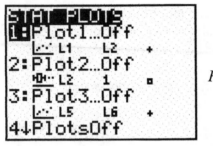

Press: **ENTER**

Make sure **On** is highlighted and *Press:* **ENTER**

Highlight the scatter plot option and
Press: **ENTER**

Arrow Down to **Ylist:** and *Press:* **CLEAR**

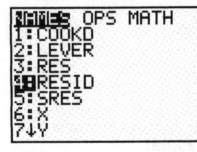

This is where we will place the residuals.

Press: 2nd **STAT**

Arrow Down to "**7: RESID**"
and *Press:* **ENTER**

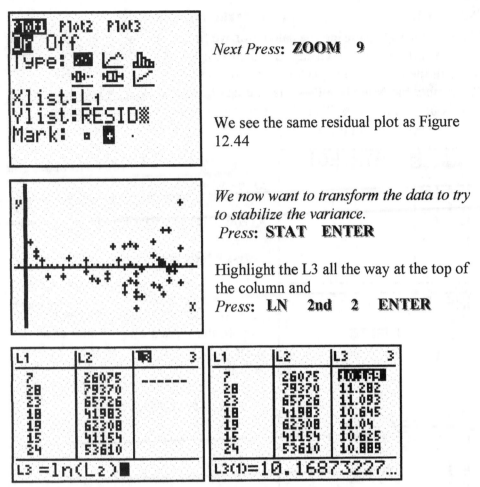

Next Press: **ZOOM 9**

We see the same residual plot as Figure 12.44

We now want to transform the data to try to stabilize the variance.
Press: **STAT ENTER**

Highlight the L3 all the way at the top of the column and
Press: **LN 2nd 2 ENTER**

The transformed data will appear in the List when you press ENTER.

Now we want to re-plot the residuals using the newly created list 3.
Return to the **STAT CALC** Menu and run the regression using L1 and L3.
This will replace the previous residuals with the residuals formed by using the newly transformed data.

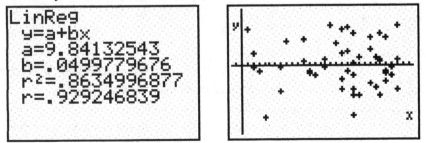

Note: *The R^2 value has also increased.*
To see the new plot of the residuals simply Press: **ZOOM 9**
As you can see from the screen the transformation stabilized the error variances. Our plot resembles fig. 12.46

Section: 13.3   Journal of Marketing Example

To perform a χ^2 test using the calculator we first enter the **observed** data into a Matrix (usually A) using the calculator's Matrix Menu.
We will then enter the **expected** data into another Matrix (usually B).
Remember the Null Hypothesis is: there is Independence, and we will reject Ho if our p-value is less than the significance level, α for our problem.
Now we can perform the χ^2 test.

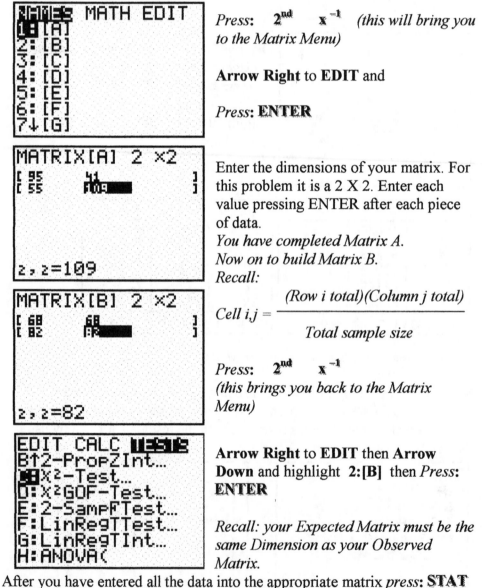

Press: 2nd x^{-1} *(this will bring you to the Matrix Menu)*

Arrow Right to **EDIT** and

Press: **ENTER**

Enter the dimensions of your matrix. For this problem it is a 2 X 2. Enter each value pressing ENTER after each piece of data.
You have completed Matrix A.
Now on to build Matrix B.
Recall:

$$Cell\ i,j = \frac{(Row\ i\ total)(Column\ j\ total)}{Total\ sample\ size}$$

Press: 2nd x^{-1}
(this brings you back to the Matrix Menu)

Arrow Right to **EDIT** then **Arrow Down** and highlight **2:[B]** then *Press:* **ENTER**

Recall: your Expected Matrix must be the same Dimension as your Observed Matrix.

After you have entered all the data into the appropriate matrix *press:* **STAT**
Arrow Left to **TESTS** and **Arrow Up** to C: χ^2–Test….
Press: **ENTER**   This is the Chi square test.

```
X²-Test
 Observed: [A]
 Expected: [B]
 Calculate Draw
```

Recall we placed the Observed values into Matrix A and the Expected values into Matrix B.

If you do not automatically see **[A]** & **[B]**
Arrow Down *and beside* **Observed:**
Press: **2ⁿᵈ   x⁻¹ [A]   ENTER**
Arrow Down *and beside* **Expected:**
Press: **2ⁿᵈ   x⁻¹ [B]   ENTER**

```
X²-Test
 X²=39.22166428
 P=3.783191ε⁻10
 df=1
```

Next, **Arrow Down** to **Calculate** and
Press: **ENTER**

As can be seen from the screen the χ^2 value is **39.222.**

The associated p-value is **.000000000378**; which is far below any reasonable α -value. We therefore conclude that Gender and Brand Awareness are **not** Independent events. We say event A, viewer gender and event B, brand awareness are **dependent** events

EXAMPLE 13.3 Marital Status and Religion

To perform a χ^2 test using the calculator we first enter the **observed** data into a Matrix (usually A), and then enter the **expected** data into another Matrix (usually B). Remember the Null Hypothesis is: there is Independence, and we will reject Ho if our p-value is less than the significance level, α for our problem. Now we can perform the χ^2 test.

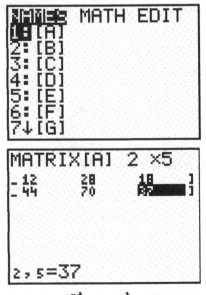

Press: **2ⁿᵈ   x⁻¹** *(this will bring you to the Matrix Menu)*

Arrow Right to **EDIT** and

Press: **ENTER**

Enter the dimensions of your matrix. For this problem it is a 2 X 5. Enter each value pressing ENTER after each piece of data.
You have completed Matrix A.
Now, ... on to build Matrix B.

Press: **2ⁿᵈ   x⁻¹** *(this brings you back to the Matrix Menu)*

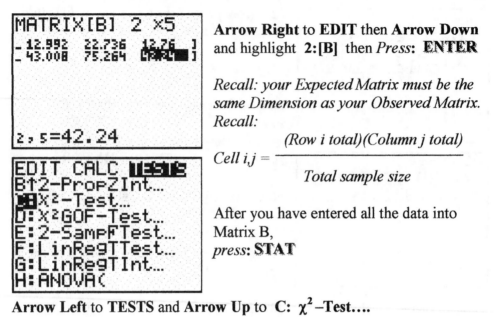

Arrow Right to **EDIT** then **Arrow Down** and highlight **2:[B]** then *Press*: **ENTER**

Recall: your Expected Matrix must be the same Dimension as your Observed Matrix.
Recall:

$$Cell\ i,j = \frac{(Row\ i\ total)(Column\ j\ total)}{Total\ sample\ size}$$

After you have entered all the data into Matrix B,
press: **STAT**

Arrow Left to **TESTS** and **Arrow Up** to **C:** χ^2 **–Test….**

Press: **ENTER**

Recall we placed the Observed values into Matrix A and the Expected values into Matrix B.

If you do not automatically see **[A]** & **[B]**
Arrow Down *and beside* **Observed:**
Press: **2**nd **x**−1 **[A]** **ENTER**
Arrow Down and beside **Expected:**
Press: **2**nd **x**−1 **[B]** **ENTER**

Next, **Arrow Down** to **Calculate** and *Press*: **ENTER**

As can be seen from the screen the χ^2 value is **7.1355.**

The associated p-value is **.1289**; which is **not** below our α -value of .01.
We therefore conclude that Marital Status and Religious affiliation are **not** dependent events.

b.   To graph a histogram of the data we will first enter the data into the **LISTS.** In List 1 we will enter the numbers $1 - 10$. $1 - 5$ for the five different religious affiliations - divorced, and $6 - 10$ for the five different religious affiliations - never married.
In List2 we will enter the percentages for each of the 5 classes – divorced, and in List3 we will enter the percentages of the 5 classes – never married.

The percentages are calculated as follows: (the number within that class / the total number)   Class A divorced = 39/211 = .184834

Class B divorced = 19/80 = .2375

Class C divorced = 12/56 = .21429, etc

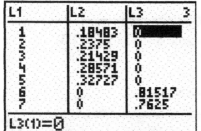

L3(1)=0

Class A Never Married = 172/211 = .815166

Class B Never Married = 61/80 = .7625, etc

Note the zeros I used as place holders to adjust the position of the "bars" in the desired graph.

L3(10) = .672727272...

After you have entered all the percentages and Place holders into the Lists you will need to go to the **STAT PLOTS.**

Press: **2nd Y=**

We will want to enter each of the first 2 stat plots and adjust the parameters.

Press: **ENTER**

Press: **ENTER** again to turn the Plot ON.

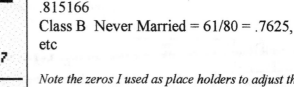

You must highlight the **Histogram** and make sure your **Xlist:** and **Freq:** are as shown.

After you have adjusted your settings
Press: **2nd Y=**

This brings you back to the STAT PLOT Menu.

Arrow Down to **2 : Plot2…Off**
And *Press:* **ENTER**

Press: **ENTER** again to turn the Plot **ON.**

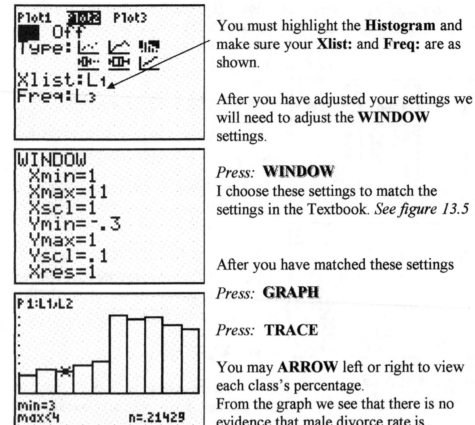

You must highlight the **Histogram** and make sure your **Xlist:** and **Freq:** are as shown.

After you have adjusted your settings we will need to adjust the **WINDOW** settings.

Press: **WINDOW**
I choose these settings to match the settings in the Textbook. *See figure 13.5*

After you have matched these settings

Press: **GRAPH**

Press: **TRACE**

You may **ARROW** left or right to view each class's percentage.
From the graph we see that there is no evidence that male divorce rate is dependent on religious affiliation.

Exercise 13.25

To perform the χ^2 test we first enter the observed data into Matrix A and the expected data into Matrix B. Next we perform the χ^2 test using $\alpha = .05$

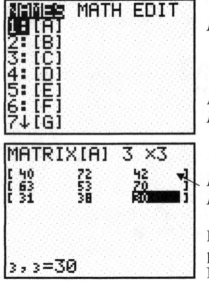

Press: **2ⁿᵈ**   **x⁻¹**   **ENTER**

Arrow Right to **EDIT** and
Press: **ENTER**

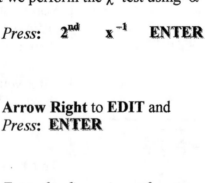

Enter the dimensions of your matrix.
For this problem it is a 3 X 3

Enter each value into the Observed Matrix pressing **ENTER** after each piece of data.
Now on to build Matrix B.

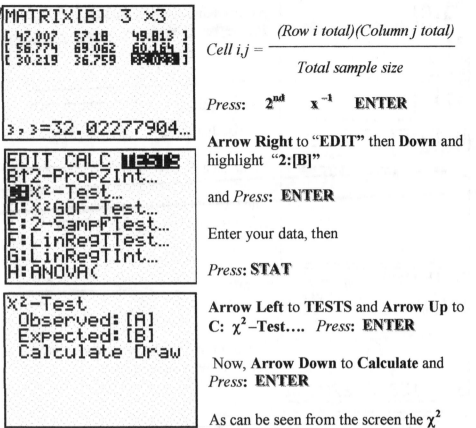

$$Cell\ i,j = \frac{(Row\ i\ total)(Column\ j\ total)}{Total\ sample\ size}$$

Press: **2ⁿᵈ x⁻¹ ENTER**

Arrow Right to "**EDIT**" then **Down** and highlight "**2:[B]**"

and *Press:* **ENTER**

Enter your data, then

Press: **STAT**

Arrow Left to **TESTS** and **Arrow Up** to C: χ^2–**Test....** *Press:* **ENTER**

Now, **Arrow Down** to **Calculate** and *Press:* **ENTER**

As can be seen from the screen the χ^2 value is **12.327.** *The associated p-value is* **.015;** *which is not less than* **.01** *therefore we cannot reject Ho: the events are Independent.*
We therefore conclude that **A** and **B** are **Independent events.** ☺

Exercise 13.31 IQ and Mental Retardation

To conduct the Chi-Squared test for this problem I will place the Observed data into Matrix A and the Expected values into Matrix B.
 For our problem we are trying to determine if the types of method of test/retest are independent of MR diagnosis so there will be 3 columns and 2 rows. We perform the χ^2 test using $\alpha = .01$ so we reject Ho, (Independence) if our p-value exceeds 0.01. The Observed Matrix is shown below:

Row Totals		WISC-R/WISC-R	WISC-R/WISC-III/	WISC-III/WISC-III	
115	Diagnosed with MR	25	54	36	
411	Above MR cutoff	167	103	141	Grand Total
	Column Totals =	**192**	**157**	**177**	**526**

Press: **2ⁿᵈ x⁻¹ ENTER**

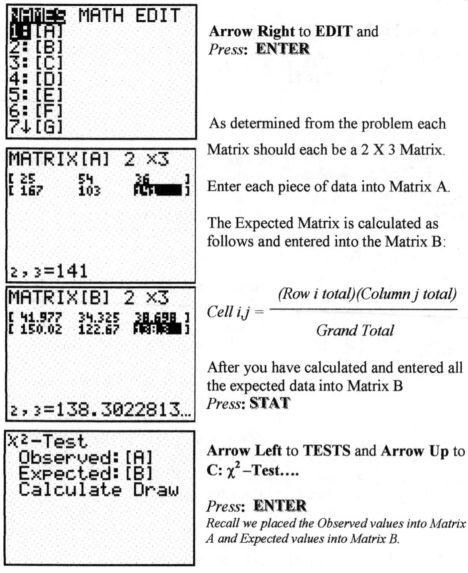

Arrow Right to **EDIT** and
Press: **ENTER**

As determined from the problem each Matrix should each be a 2 X 3 Matrix.

Enter each piece of data into Matrix A.

The Expected Matrix is calculated as follows and entered into the Matrix B:

$$Cell\ i,j = \frac{(Row\ i\ total)(Column\ j\ total)}{Grand\ Total}$$

After you have calculated and entered all the expected data into Matrix B
Press: **STAT**

Arrow Left to **TESTS** and **Arrow Up** to
C: χ^2 –Test….

Press: **ENTER**
Recall we placed the Observed values into Matrix A and Expected values into Matrix B.

Now, Arrow Down to Calculate and *Press*: **ENTER**

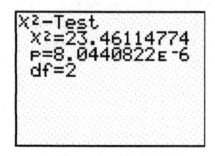

As can be seen from the screen the p-value is **0.0000080440822** which is less than 0.01 so we **reject** Ho.

We conclude that Test/retest method and Test score are **not** independent events. They are Dependent.

----------ERROR Messages------------

DIM MISMATCH is by far the most common **Error** message.

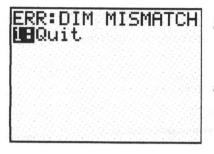

To clear this message
press **CLEAR** several times

and/or

2nd **MODE** then **CLEAR**

The steps above will return you to the home screen, but they will **not** remedy the reason for the error! This message is displayed indicating an error in one of two main categories:

A   the Lists are different lengths (dimensions)

B   the designated List may be empty

Error Category A. for DIM MISMATCH

Look at this screen below carefully.

The error at the top of this page was given when I tried **ZOOM 9** *using* **List 1 , List 2** to graph a Histogram.

Look carefully at List 1 and List 2. They are different lengths.

I thought I was careful in entering the data, yet I received the ERROR message.

First I went back to the Stat Plot.

After checking my Stat Plot and seeing that everything looked correct I decided to check the actual data.

That is when I found the error! Lists must be the same length!

Error Category B for DIM MISMATCH

I was trying to graph a Box & Whisker plot when I received this message:

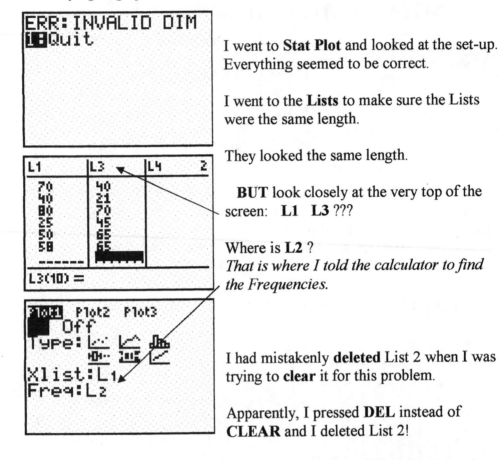

I went to **Stat Plot** and looked at the set-up. Everything seemed to be correct.

I went to the **Lists** to make sure the Lists were the same length.

They looked the same length.

BUT look closely at the very top of the screen: **L1   L3** ???

Where is **L2** ?
That is where I told the calculator to find the Frequencies.

I had mistakenly **deleted** List 2 when I was trying to **clear** it for this problem.

Apparently, I pressed **DEL** instead of **CLEAR** and I deleted List 2!

At this point I could just go back to Stat Plot and replace **L2** with **L3** and graph my Box & Whisker plot, but that will not return my List 2.

To return any or all the lists to the editor follow these steps:
From the home screen *Press:* **STAT Arrow down** to **"5:SetUpEditor"**
Press: **ENTER   ENTER**

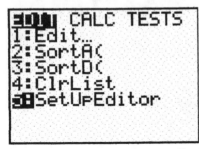

The second screen above tells me that I have corrected my deletion and that all Lists are back to where they should be ☺
To verify I *Press:* **STAT ENTER**

I see that List 2 is back between List 1 and List 3 where it belongs.

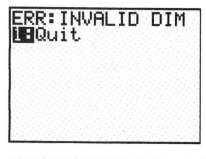

Now, to get the Box & Whisker plot I simply go back to Stat Plot and change the frequency list from **L2** to **L3**.

Or

Copy the frequencies from list 3 into List 2.

You can clear the error message by following the same steps as described on the top right of page 81. ☺

INVALID DIM is another **ERROR** message:

This error occurs when there is an entry that doesn't "make sense" to the calculator. It most often occurs when you mistakenly enter a stray keystroke.

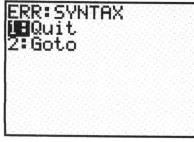

I was trying to graph a Histogram when I mistakenly entered this **"Y"** when I thought I was entering a **"1"** for **Freq:** To enter a **"1"** you must *Press*: **ALPHA 1**

When you arrow down to **Xlist:** & **Freq:** *the calculator is in* **ALPHA MODE** by default *so by pressing* **ALPHA** *again you return the calculator to* **STANDARD MODE**. Yes, I know it seems backwards but it works☺

SYNTAX is another **ERROR** message that occurs with some regularity.

This error occurs when the calculator is expecting a particular type or number of response(s) and you have neglected to give it what it wants.

This particular screen appeared when I entered a "**-**" sign instead of a "**(-)**" sign when I was adjusting the window.

```
WINDOW
 Xmin=-5█
 Xmax=10
 Xscl=1
 Ymin=-10
 Ymax=10
 Yscl=1
 Xres=1
```

Please note the position and size of the **subtraction** sign, $-$.

It is larger and placed lower than the **negative** sign, (-).

```
normalcdf(-1,1)█
```

You must also be careful when using **any** of the functions from the **Distributions Menu.**

Here I am trying to find the area under the standard normal distribution curve between **negative 1 and positive 1**, but what I've entered is **subtract 1 comma 1.**

```
ERR:SYNTAX
1█Quit
2:Goto
```

Subtract 1 comma 1 makes no sense and the calculator returns the Syntax error message.

You must take care to use the correct syntax when entering values into your calculator.

DOMAIN is another common **error** message.

This error occurs when you have entered a value that is outside the expected *domain* (remember algebra).

```
ERR:DOMAIN
1█Quit
```

I was trying to perform a 1-Proportion Z-Test when I received the error message.

The error occurred on the very first line. I entered the whole number **30,** even though

I meant 30% for the p_0 :

```
1-PropZTest
 p0:30
 x:80
 n:300
 prop≠p0 <p0 >p0
 Calculate Draw
```

The calculator is expecting a decimal *(number between 0 and 1)*

The correct entry would be **.30** This type of error can occur when you are using any of the Statistical Tests or Distributions.

Here is another example of a value outside the expected domain that resulted in this error message.

```
ERR:DOMAIN
1█Quit
2:Goto
```

I was trying to calculate the probability for a problem using the binomial distribution function when I received this **ERROR** message.

```
binomcdf(10,25,4
█
```

As you can see the second entry is **25.** It is the percent that was listed in my book, but recall the proper notation for **25%** is **.25** in decimal notation.

I should have entered **.25** instead of **25**.

While it is impossible to explain every error for every student I have tried to give examples of the most common error types my students encounter. If you are stuck or confounded you may email me skelly@pjc.edu. ☺

GETTING STARTED
WITH
MICROSOFT EXCEL

Pamela Lancaster
University of Kentucky

Statistics ELEVENTH EDITION

McClave | Sincich

PEARSON
Prentice Hall

Upper Saddle River, NJ 07458

1.1 Introduction and Overview

The purpose of this manual is to provide students with a step-by-step process for solving statistical problems introduced in the textbook. By designing a "user-friendly" manual that replicates solutions to specific examples and homework problems within the text, it is hoped that the study of statistical applications may be made more palatable and enjoyable. It is important to remember that statistical software packages such as Excel often allow the user to bypass tedious and rigorous calculations with a few clicks of the mouse and arrive at a solution to a specific problem within a few seconds. However, the derivation of the manual calculations is only half the battle. It is secondary to understanding the philosophy and ideology of the statistical concepts and theorems presented in the textbook. Spending less time on computation and more time on developing the analytical skills necessary to interpret output is the primary goal of this manual. Excel is a good tool for a beginning statistical student to use to avoid performing calculations; other statistical software should be used for any professional needs.

By design, each chapter in this manual follows a three-step process. First, a brief introduction of the statistical concepts that need to be reviewed prior to using Excel software is presented. Remember that it will be difficult to perform calculations in Excel unless the student has a working knowledge of the concepts in the textbook. Second, a step-by-step process on how to find the relevant statistical functions is provided using pictures of where each function is located on the Excel menu. Third, examples and/or homework problems from the text are used in order to illustrate how to use Excel to derive solutions

For those that have worked with spreadsheets in Excel, Minitab, SPSS, SAS, Lotus 1-2-3, or Quattro, most of this introductory chapter may be skipped. It is important to note that all spreadsheets differ slightly in terms of presentation of menus and terminology in describing statistical functions. If new to Excel or new to spreadsheets, this chapter will be beneficial.

1.2 The Basics of Spreadsheets: Using the Mouse

Examine the mouse that accompanies the laptop or desktop computer. Each mouse should have two buttons that may be used to click once, twice, or hold in order to perform various operations. If using a laptop, these buttons may not be contingent to the mouse but located at the front end of the keyboard. There are some primary functions for the buttons on the mouse: point, click, double-click, and hold to drag.

1. **Changing the Mouse Buttons**: The following discussion about the functions of the mouse assumes that the settings for the mouse are set at default. This usually implies that the mouse is designed to be used by the right hand for a right-handed individual. If left-handed, the description of the functions still applies; however, the movements of the mouse may be slightly uncomfortable.

2. **Right-Click**: The term "click" means to press and then release the appropriate mouse button in order to execute a command or highlight a cell in Excel. When the user performs a right-click (press and then release the right mouse button on the mouse), a shortcut menu appears in most programs.

3. **Left-Click**: If the user points to a cell within the Excel spreadsheet and left-click (press and release the left-hand button on the mouse), the cell will be activated in Excel. Visually, the cell will be highlighted on all four sides in bold and implies that the cell is ready to receive qualitative or

1

quantitative information. When the command "click" is used within this manual, it is assumed to be a left-click.

4. **Double-Left-Click**: The double-click feature is used to rapidly open a program or to execute a command quickly in order to bypass the OK option after a single click. To double click to activate program, press and release the left-hand mouse button twice quickly. Practice by placing the mouse over any icon on the desktop such as the Recycle Bin icon or the My Computer icon. Notice that a single left-click highlights the icon. A double left-click activates the icon and allows the user to enter into the program.

5. **Left-Click and Drag**: By holding down the left-click button in one cell of a spreadsheet and moving the mouse at the same time, the user may highlight data. If an options box is open, then it is possible to use this maneuver to select a data location that is to be entered into the menu. If done properly, the data will be surrounded by a dashed border.

1.3 Excel Setup

Although most examples in this manual are presented using Excel 2007, Excel 2003 may also be used. Most applications are very similar in both versions, but the layout has changed from Excel 2003 to Excel 2007.

Excel 2003 uses the **Main Menu bar** and the **Toolbar** at the top of the page to organize commands and tools. The **Main Menu bar** contains options: **File**, **Edit**, **View**, **Insert**, **Format**, **Tools**, **Data**, **Window**, and **Help**. Once PHStat is installed, a **PHStat** option will also appear. There are several icons within the **Toolbar** that allow the user to access tools. Not all of these are used in this manual.

2

Excel 2007 uses the **Ribbon** with different tabs to help organize tools. The **Ribbon** consists of several tabs: **Home**, **Insert**, **Page Layout**, **Formulas**, **Data**, **Review**, and **View**. Once PHStat is installed, a **PHStat** option will also appear within a new tab, **Add-Ins**. The **Microsoft Button**, located in the top left corner in Excel 2007, has replaced the **File** option in the Main Menu of Excel 2003 for accessing some basic commands such as: opening a new blank workbook, opening an existing data file, saving the current workbook, and printing the current workbook.

The **Formula Bar** is contained in both Excel versions. It is the row directly above the spreadsheet grid. It is comprised of two primary windows. The first window shows the cell that is currently active within the worksheet. When opening a new worksheet for the first time, the active cell by default is A1. In the second window, the user may begin to input data or a formula. All version of Excel also have the **Worksheet Area**. It is comprised of the grid of columns labeled with letters and rows labeled with numbers.

1.4   Access to Analysis ToolPak

For the most part, once Microsoft Office has been installed on a computer, Excel has the necessary functions to complete the tasks presented in this manual. However, it is necessary to make sure that the Analysis ToolPak is available. Follow these steps to ensure access to the Analysis ToolPak.

For Excel 2007:

Step 1: Click on the **Microsoft Button** and then click on **Excel Options**.

Step 2: A window for **Excel Options** will appear. Click on **Add-Ins** on the left side of this window.

Step 3: Click on the **Analysis ToolPak** in the middle of the screen. Click on the **OK** button.

For Excel 2003:

Step 1: Click on **Tools** within the **Main Menu Bar** and then click on **Add-Ins**.

Step 2: A window for **Add-Ins** will appear. Click on the box in front of **Analysis ToolPak** so that there is a check in it.

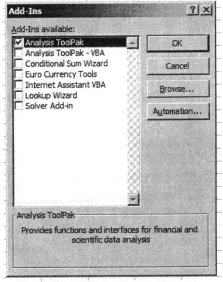

Step 3: Click on the **OK** button.

Note: If the **Add-Ins** are not visible, then they may not be installed on the computer, and the Microsoft Office installation CD will be required. The user will need administrative access in order to install the **Add-Ins.**

1.5   Using PHStat

PHStat is a statistical add-in for Excel. The user should follow the instructions accompanying the PHStat CD to install the product. Excel should be installed prior to installing PHStat and the security level for Excel should be set to **Medium** or lower so that PHStat may function properly. Once PHStat is installed on the computer, it may be accessed either via the start up menu or by a shortcut icon on the desktop. When performing examples from this manual, Excel should always be accessed via PHStat. Notice that when PHStat is opened, the user must click on the **Enable Macros** button in order to have this add in work properly. When Excel is accessed via PHStat, it will not automatically open a new worksheet, so the user will need to open a new workbook before accessing other commands. In Excel 2003, access the PHStat commands via clicking on **PHStat** in the menu bar. In Excel 2007, access PHStat commands via clicking on the **Add-Ins** tab and then **PHStat**.

1.6   Using Excel 2007 versus Excel 2003

As stated before, most examples presented in the manual are done using PHStat and Excel 2007; however it is possible to use PHStat and Excel 2003. For those using Excel 2003, take note of the following:

1. If the steps instruct the user to access **PHStat** under the **Add-Ins** tab, then simply click on the **PHStat** button at the top of the toolbar. Subsequent steps will be exactly the same.

2. If the steps instruct the user to access **More Functions** and **Statistical** under the **Formulas** tab, then the user will click on the **Insert** button at the top of the toolbar, and drop down to **Functions**. After clicking on **Functions**, a box with options will appear. Use the drop-down menu to choose **Statistical** and then the statistical functions will become available. Once the user clicks on the desired function, subsequent steps will be exactly the same.

3. If the steps instruct the user to access **Data Analysis** under the **Data** tab, then simply click on **Tools** at the top of the toolbar. Drop down to **Data Analysis** at the bottom of the list of options and click on it. Subsequent steps will be exactly the same.

4. If the steps instruct the user to access a graphing option under the **Insert** tab, then separate instructions will be given for creating a graph using Excel 2003.

1.7 Accessing Help

In order to quickly find out where to locate a function or command, the user may access the Help function in Excel and PHStat.

In Excel 2003, click on **Help** within the Toolbar and then click on **Microsoft Excel Help**. A help menu will appear in the right hand side of the screen. There will be several options for how to access help, including typing a key word into the blank box under **Assistance** and clicking on the green arrow.

In Excel 2007, click on the circular question mark icon at the top left of the page. A separate box for **Excel Help** will appear. This contains a table of contents that the user may access. It also contains a blank box that allows the user to search on a key word. If the computer being used has an active internet connection, then **Excel Help** will also do an internet search on the Excel topic of interest.

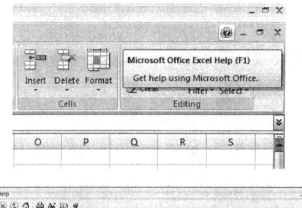

For help in PHStat, click on **PHStat**, drop down to the last option, **Help for PHStat**. A window labeled **PHStat2 Help** will appear. Click on **Help Topics** at the top of the window and a new box, **Help Topics: PHStat2 Help**, will appear. Using this, it is possible for the user to search on a key word or scroll through a list of topics.

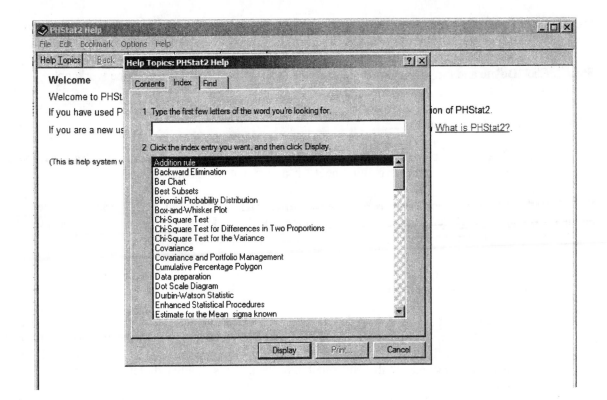

1.8   Status Bar

At the very bottom of the Excel screen, the **Status Bar** indicates what function is currently being undertaken by the user. Notice that **Ready** is displayed at the left of the status bar when a cell is idle or inactive. When editing a cell, the display changes from **Ready** to **Edit**. To indicate the difference, highlight cell A1 and type in "123456". Click the green check button to accept the number sequence. Notice that once the green check button is clicked, the number sequence "123456" is displayed in A1 and in the Formula Bar. Notice how the three icons, including the green check mark, disappear. Now, edit the number sequence by changing the number "4" to "9". Move the pointer to the numbers in the formula bar and notice that the pointer changes to an **I-beam**. Place the pointer to the right of the number "4" and then click in that location. Now the three icons, including the green check mark, reappear. Press the backspace key and the number "4" disappears. Manually enter the number "9" in its place and click on the green check mark to indicate acceptance.

1.9   Changing the Name of a Worksheet and Inserting/Deleting Worksheets

If Excel is started by clicking on an Excel icon, Excel will automatically open to a new blank worksheet. If Excel is accessed via PHStat, then the user will need to open a new worksheet. By default, the title at the top is called **Book 1** and the three worksheets accompanying it are titled **Sheet 1, Sheet 2,** and **Sheet 3**. All three sheets are stored in a unit called a **workbook**, which is simply a collection of worksheets.

Step 1: By default, each worksheet is titled Sheet1, Sheet 2, etc. Right click over the title Sheet 1 and notice that a series of subcommands (**Insert, Delete, Rename**, **Move or Copy**, etc.) appear within the current worksheet.

Step 2: Click **Rename** and title it with a descriptive name representative of the data set being analyzed. Click **Insert** and the user is prompted (among other things) to add a new worksheet to the workbook.

Step 3: Click **Delete** and the user is prompted to delete the existing worksheet from the workbook.

1.10 Moving A Worksheet

Step 1: There is also the option to change the position of the worksheet in its array. Suppose that a researcher prefers that Sheet 2 appear before Sheet 1 in an array of worksheets.

Step 2: Right click over Sheet 1 and click **Move or Copy**. The user may move the worksheet, Sheet 1, to another workbook altogether or move Sheet 1 **before sheet** 3. This would place sheet 1 in position behind Sheet 2 as preferred by the user.

1.11 Activating a Cell or a Range of Cells

Each **cell** is defined by a combination of letters and numbers located at the top and left side of the spreadsheet. The letters refer to the column location and the numbers refer to the row location. For example, the cell indicating the very first row and column of the spreadsheet is referred to as cell A1. The label A1 is considered the **address** of that cell, much like the address of any individual residence.

Notice that when a new worksheet is opened, cell A1 is automatically the active cell. A cell is "active" when its address (in this case, A1) appears displayed in the Name Box, next to the formula bar. It is also active when a cell is highlighted, in bold, on all four sides as indicated. For practice, click on cell B5. Notice that B5 appears in the Name Box and is highlighted in bold on all four sides.

A **range** of cells is a group of cells. Now that B5 is active, click on B5, hold the mouse button down and drag to B8. When the mouse button is released, cell B6 through B8 is highlighted. The address of this range is now called B5:B8 where a colon separates the top cell from the bottom cell. Notice how the address within the name box does not change.

To practice, activate cell B5. Once it is active, click on B5, hold the mouse button down and drag to C8. When the mouse button is released, cell B6 through C8 is highlighted in bold. The address of this range is now called B5:C8 where B5 refers to the upper left cell and C8 refers to the lower right cell.

Continue to practice activating many cells by placing the cursor in the upper left hand cell and dragging to the lower right. Or place the cursor in the lower right hand cell and drag to the upper left.

1.12   Cutting and Copying Information into Cells

In order to remove information from one cell and insert that information into a new cell or to copy information into a new cell, use the **cut-copy** commands located within Excel. Notice that cutting a cell means to remove the original cell contents from that cell with the intent of copying it elsewhere. This implies that the cell is empty after the data is "cut" from it. Copying a cell implies that the data will exist in two locations after the command is executed.

Use the same number sequence, 13579, to illustrate the technique of copying. Enter the number sequence into cell A1 on the worksheet.

Step 1: To copy the number sequence into cell B1, first activate A1 by clicking on it. Notice that the cell is framed.

Step 2: For Excel 2003 users, click **Edit** from the menu option and select **Copy** from the list of subcommands. For users of Excel 2007, **Copy** may be found directly below the **Home** tab. Notice how the cell is no longer highlighted but is illustrated by a dotted line surrounding the cell that blinks.

Step 3: Click on the cell where the information is to be copied, B1. This cell should now be activated.

Step 4: For Excel 2003 users, click **Edit** from the menu option and select **Paste**. For users of Excel 2007, **Paste** may be found directly below the **Home** tab. Now the number sequence, 13579, appears in two places.

Step 5: Activate cell C1, click **Edit** or the **Home** tab as appropriate, and then **Paste**. Notice the number sequence is copied again.

11

1.13 Formatting Numbers that Exceed Seven or Eight Digits

For some raw data in a spreadsheet or possibly for some output, the numerical display may be too large to be shown in its entirety. Follow these steps to format large numbers in Excel 2003 and Excel 2007.

Step 1: In an empty cell (A1), type in the number sequence 123456789123456789. Notice that when enter is pressed, the number displayed in cell A1 is 1.23457E+17. This display may be different given the default settings for Excel.

Step 2: Highlight cell A1.

Step 3: For Excel 2003, go to the main menu bar and click **Format-Cell**. For Excel 2007, go to the **Home** tab and click on **Format**, then **Format Cells…**.

Step 4: Open the number tab from the list of options (this should be the default) and select **NUMBER**. Click **OK**. (At this point Excel 2007 will have format the number. Follow the rest of the steps for Excel 2003.)

Step 5: Notice how cell A1 displays #######. This symbol indicates that the number is too large to be placed in cell A1 given its existing width. Place the cursor between "A" and "B" at the top of the column.

Step 6: Double left-click between "A" and "B" and the entire number sequence will be displayed.

1.14 Inserting or Deleting Rows and Columns

Step 1: Use a simple two column spreadsheet. Suppose that a column needs to be inserted between columns A and B. To do this, click on the letter "B" at the top of the column to highlight it.

Step 2: For Excel 2003, on the main menu bar, select **Insert-Columns.** For Excel 2007, go to the **Home** tab and click on **Insert**, then **Insert Sheet Columns…**.

Step 3: Suppose instead that two columns need to be inserted between A and B. To do this, highlight columns B and C simultaneously and perform **Step 2**.

Step 4: To delete a column(s), highlight it/them and then click **Edit-Delete** if using Excel 2003. For Excel 2007, highlight the column(s), go to the **Home** tab and click on **Delete**, then **Delete Sheet Columns…**.

The same procedure is followed for inserting or deleting rows. Suppose that a row needs to be inserted between rows 1 and 2. To do this:

Step 1: Click on the number "2" (designating row 2) at the left of the sheet to highlight it.

Step 2: For Excel 2003, on the main menu bar, select, **Insert-Rows.** For Excel 2007, go to the **Home** tab and click on **Insert**, then **Insert Sheet Rows…**.

1.15   Importing Data

There are many data sets included on the CD that accompanies the text. Many of these contain data for text examples and exercises. It is possible to import these data sets into Excel by clicking on the **Open** command. This command may be found under the **Microsoft Button** in Excel 2007, or under **File** Excel 2003. After **Open** is accessed, a dialog box will open that allows the user to dictate which file to open. Depending on the configuration of the machine being used, if the text CD has been placed into the machine, the data should be found in the CD/DVD drive.

Although many data sets are included on the CD, most examples in this manual include a step that requests the user enter the data manually. If the data is accessed via the CD instead, make sure that the format of the data is exactly as required by the example.

1.16   Saving Information

It is always recommended that work be saved regularly in case of accidents or power outages.

Step 1: For Excel 2003, within a new worksheet, click on **File** within the Main Menu and click on **Save**. For Excel 2007, click on the **Microsoft Button** and click on **Save**.

Step 2: The program will prompt the user to provide a descriptive file name to the collection of worksheets.

Step 3: Clicking **OK** confirms that the file is saved however it will be saved in the last location used. This could be on the hard drive or on the network for example. Some Excel programs will save work into a folder called **My Documents**.

It is recommended that once a workbook has been saved, the user save subsequent changes:

Step 1: For Excel 2003, click on **File** then **Save As**. For Excel 2007, click on the **Microsoft Button** and click on **Save As**. A dialog box will appear. At the top of the dialog box, the program prompts the user to save work in a file or folder.

Step 2: Click on the arrow in the **Save in** display at the top of the dialog box and select where the information should be saved. Be sure to provide the file with a specific name that will be remembered.

Step 3: Save as type should be labeled as **Excel Workbook** or a specific Excel file indicated by default within Excel. Click on the arrow key and scroll through the options listed if another Excel version than the default is desired.

1.17   Algebraic Symbols in Excel

When creating formulas in Excel, it is necessary to remember the basic symbols that are used within Excel. Formulas may be created in the formula bar within Excel. It is important to remember that the equal sign, =, must be clicked prior to starting a calculation. Before creating formulas in Excel, become acquainted with the basic symbols needed for formulas.

1. **Addition**: The symbol for addition is the plus sign, +. To write a formula to find the sum of 2 and 3, simply write 2+3.

2. **Multiplication**: The symbol for addition is the asterisk, *. To write a formula to find 2 multiplied by 3, write 2*3.
3. **Subtraction**: The symbol for subtraction is the minus sign, -. To write a formula subtracting 2 from 3, write 3-2.
4. **Division**: The symbol for division is the backward slash, /. To divide 3 by 2, write 3/2 in Excel
5. **Exponentiation**: The symbol for raising a number to a power, or exponent, is the carat sign, ^. To raise 2 to the third power, write 2^3 in Excel.
6. **Percent**: The symbol for expressing a number in percentage terms is %. To express 2 as a percent, write 2% in Excel and the number will be converted to 0.02.
7. **Negation**: The symbol for negation is the minus sign, -. To indicate 2 as a negative number, write -2 in Excel.

1.18   Order of Operations in Excel

By convention, Excel will calculate expressions in parentheses first and then use results to continue on with the rest of the formula. Without parentheses, the order of operations is:
1. Negation
2. Percent
3. Exponentiation
4. Multiplication/Division
5. Addition/Subtraction

Using an example, calculate the following expressions. Go to the formula bar in Excel and click on the equal sign. Now that the formula bar is activated, proceed with the following calculations.
1. =3 + 4*5   produces 23 because multiplication occurs before addition
2. =(3+4)*5   produces 35 because parenthetical operations occur first
3. =3 + 4*5^2   produces 103 because exponentiation occurs first, then multiplication, then addition

2.1   Introduction

Chapter 1 of the text was designed to introduce many of the basic definitions and concepts needed to interpret various types of datasets. Within that chapter, two primary areas of statistics are introduced: descriptive statistics and inferential statistics. Chapter 2 of the text sets the foundation for understanding and interpreting descriptive statistics by employing techniques to summarize and categorize data numerically as well as pictorially through various types of graphs.

Within Excel 2007, there are a variety of options that can be utilized to graph qualitative as well as quantitative data. For qualitative data (nonnumeric in nature), pie charts and bar graphs can be customized to the user's preference in terms of presentation. For quantitative data, box plots, histograms, dot plots, stem-and-leaf displays, and scattergrams can be used to create a visible picture of numeric data for data investigation presentation purposes.

In addition to various graphing options presented in Excel, the primary measures of central tendency (mean, median, and mode) and dispersion (range, variance, and standard deviation) can all be calculated.

The following examples will be used to illustrate techniques in PHStat and Excel.

> Text Exercise 2.14, page 12 (Sociology field work methods)
> Table 2.2, page 14 (EPA Mileage Rating on 100 Cars)
> Text Exercise 2.142, page 68 (Feeding behavior of fish)
> Sample Exercise #1: Problem 2.168, page 102(Competition for bird next holes)
> Sample Exercise #2: Problem 2.59, page 59 (Ammonia in car exhaust)
> Sample Exercise #3: Problem 2.83, page 67 (Ammonia in car exhaust)

2.2   Graphical Methods for Describing Qualitative Data: Pie Charts

The following example from the text will be used to illustrate how to construct a pie chart. Please refer to page 12 of the text, Exercise 2.14 which contains summary data on the sociology field work methods.

University of New Mexico Professor Jane Hood investigated the fieldwork methods used by qualitative sociologists. (*Teaching Sociology*, June 2006) Searching for all published journal articles, dissertations, and conference proceedings over the past seven years in the Sociological Abstracts database, she discovered that fieldwork methods could be categorized as follows: Interview, Observation plus Participation, Observation Only, and Grounded Theory. The accompanying table shows the number of papers in each category. Use an appropriate graph to portray the results of the study.

FIELDWORK	
Fieldwork Method	Number of Papers
Interview	5,079
Observation+Participation	1,042
Observation Only	848
Grounded Theory	537

Steps for constructing a pie chart in Excel 2007:

Step 1: Open a new workbook and place the cursor in the upper left-hand corner of the spreadsheet.

Step 2: Input the data manually. Enter the qualitative variable in the first column and the count (or percent may also be used) for that category into the second column.

Step 3: Highlight both columns of data.

Step 4: Select the **Insert** tab by clicking on it.

Step 5: Click on **Pie** from the choices available and pictures of options for creating a pie chart will be displayed. There are many options for types of pie chart styles; choose the option in the upper left corner for a simple, straightforward, easy to read 2-D pie chart.

Step 6: Excel will now have created a pie chart within the same sheet as the data. The default chart style (Style 2) has multicolor pieces of pie. The Ribbon at the top of the page will now display **Chart Tools**. To the left of **Chart Styles** are the **Chart Layouts**. Click on the downward pointing arrow with a line over it to view all pie chart layouts. Click on chart layout 6; it creates a titled pie chart with a legend.

Step 7: The graph will now be labeled as "Chart Title". Click on these words to activate the title and then click on them again so that they may be highlighted. Type in a more descriptive title for the pie chart data. In this case the graph may be labeled "Fieldwork Methods".

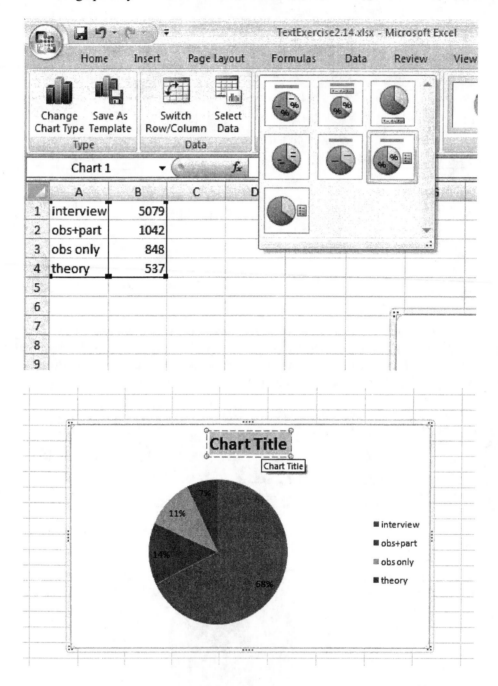

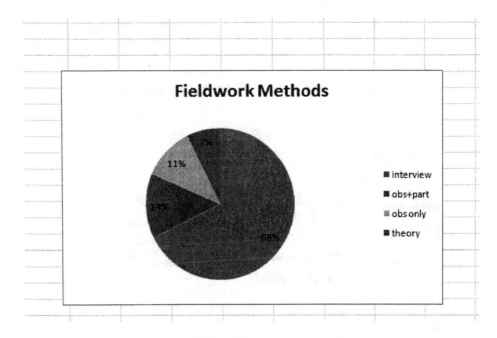

Steps for constructing a pie chart in Excel 2003:

Step 1: Open a new workbook and place the cursor in the upper left-hand corner of the spreadsheet.

Step 2: Input the data manually. Enter the qualitative variable in the first column and the count (or percent may also be used) for that category into the second column.

Step 3: Highlight both columns of data.

Step 4: Click on the **Chart Wizard** icon in the tool bar.

Step 5: Click on **Pie** from the choices available and pictures of options for creating a pie chart will be displayed. There are many options for types of pie chart styles; keep the default option for a simple, straightforward, easy to read 2-D pie chart. Click on the **Next** button.

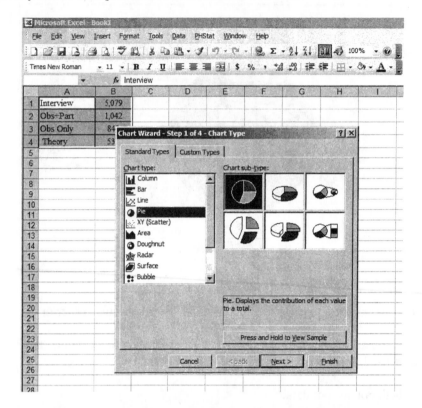

Step 6: Excel will now display a pie chart within the Chart Wizard. Click on the **Next** button.

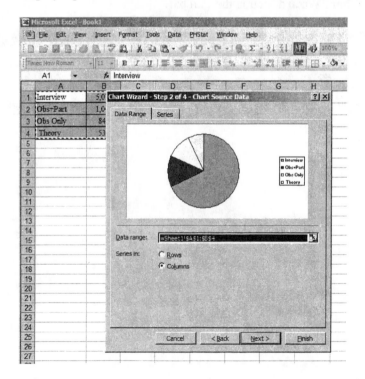

Step 7: The Chart Wizard will now display a box for entering a title for the graph. In this case the graph may be labeled "Fieldwork Methods". Click on the **Next** button.

Step 8: The Chart Wizard will now display an option for where to place the graph. The default is for it to be placed within the current sheet. Click on the **Finish** button and the graph will be displayed within the current sheet. A box for changing the graph options will appear over the graph. Click on the "x" in the right hand corner to remove this item.

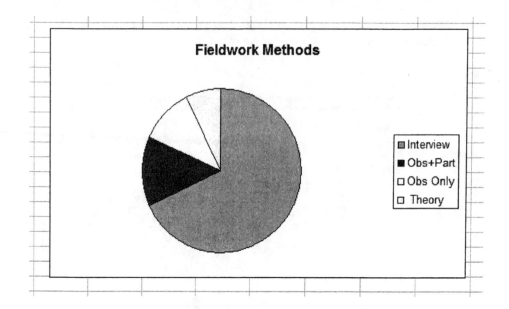

Fieldwork Methods

- ▣ Interview
- ■ Obs+Part
- ▢ Obs Only
- ▢ Theory

2.3 Graphical Methods for Describing Qualitative Data: Bar Graphs

The following example will be used from the text to illustrate how to construct a bar graph. Again, please refer page 12 of the text, Exercise 2.14 which contains summary data on the sociology field work methods.

Steps for constructing a bar graph in Excel 2007:

Step 1: Open a new workbook and place the cursor in the upper left-hand corner of the spreadsheet.

Step 2: Input the data manually. Enter the qualitative variable in the first column and the count (or percent may also be used) for that category into the second column.

Step 3: Highlight both columns of data.

Step 4: Select the **Insert** tab by clicking on it.

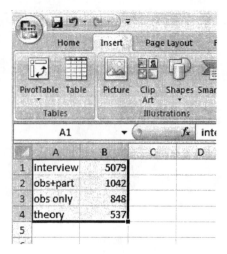

Step 5: Select **Column** from the choices available and pictures of options for creating a bar graph will be displayed. There are many options for types of bar graph styles; choose the option in the upper left corner for a simple, straightforward, easy to read 2-D bar graph. (There is a **Bar** option under the **Insert** tab, but **Column** creates the more standard bar graph with vertical bars.)

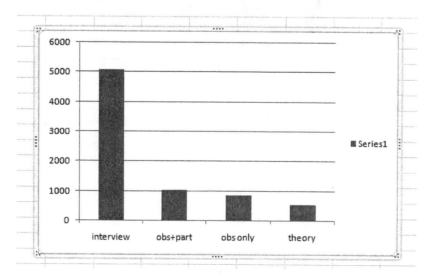

Step 6: Excel will now have created a bar graph within the same sheet as the data. The default chart style (Style 2) has blue bars. The Ribbon at the top of the page will now display **Chart Tools**. To the left of **Chart Styles** are the **Chart Layouts**. Click on the downward pointing arrow with a line over it to view all bar graph layouts. Click on chart layout 1, the first image; it creates a titled bar graph with a legend. If one were to have data broken down by an additional variable, then one of the other layouts would be useful.

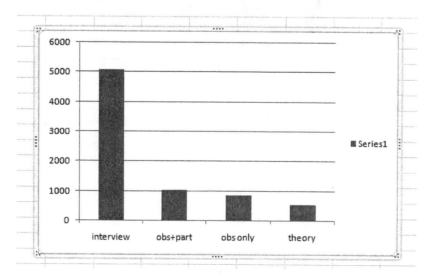

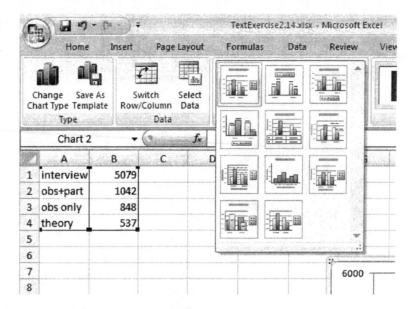

Step 7: The graph will now be labeled as "Chart Title". Click on these words to activate the title and then click on them again so that they may be highlighted. Type in a more descriptive title for the bar graph data. In this case the graph may be labeled "Fieldwork Methods".

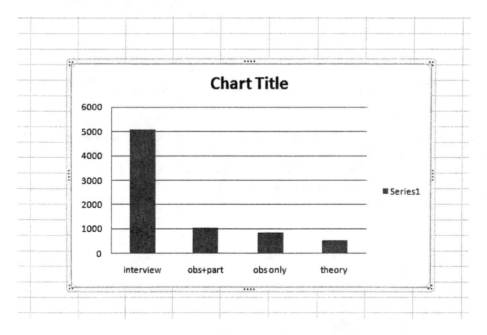

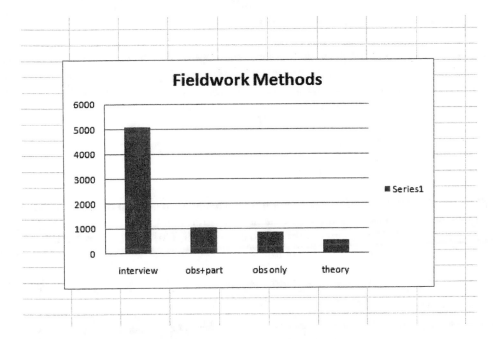

Step 8: Click on the graph to activate it. Now click on the **Add-Ins** tab in the top ribbon of the page. Click on **PHStat** and drop down to **Utilities** and then click on **Fix Up Chart**. This will result in a cleaner looking chart.

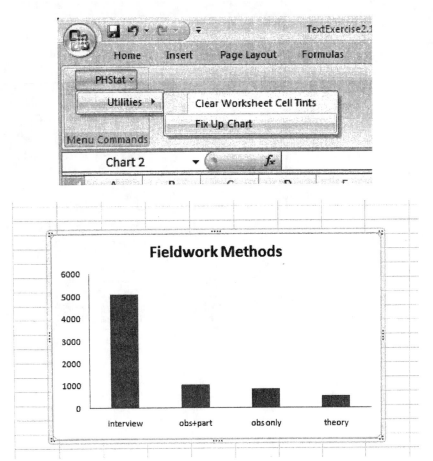

Steps for constructing a bar graph in Excel 2003:

Step 1: Open a new workbook and place the cursor in the upper left-hand corner of the spreadsheet.

Step 2: Input the data manually. Enter the qualitative variable in the first column and the count (or percent may also be used) for that category into the second column.

Step 3: Highlight both columns of data.

Step 4: Click on the **Chart Wizard** icon in the tool bar.

Step 5: Column is the default choice from the graph choices available; keep this option. There are many options for types of bar graph styles; keep the default option for a simple, straightforward, easy to read 2-D bar graph. (There is a **Bar** option within the graph choices, but **Column** creates the more standard bar graph with vertical bars.) Click on the **Next** button.

Step 6: Excel will now display a bar graph within the Chart Wizard. Click on the **Next** button.

Step 7: The Chart Wizard will now display a box for entering a title for the graph. In this case the graph may be labeled "Fieldwork Methods". There are also options for labeling the X axis and Y axis. Label these "method" and "number of papers" respectively. Click on the **Next** button.

Step 8: The Chart Wizard will now display an option for where to place the graph. The default is for it to be placed within the current sheet. Click on the **Finish** button and the graph will be displayed within the current sheet.

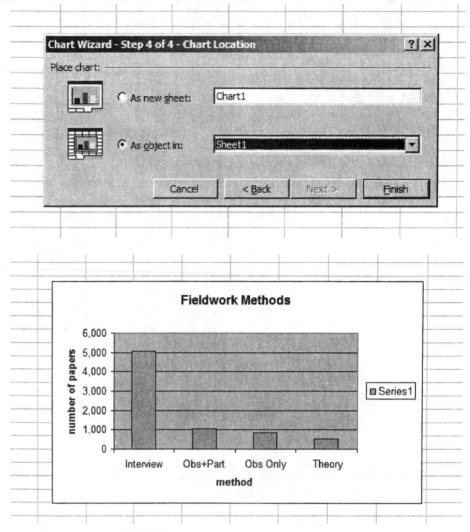

2.4   Graphical Methods for Describing Quantitative Data: Histograms

The following example will be used from the text to illustrate how to construct a histogram. Please refer page 14 of the text, Table 2.2 which contains raw data for mileage ratings for 100 cars.

The Environmental Protection Agency performs extensive tests on all new car models to determine their mileage ratings. Suppose that the 100 measurements in Table 2.2 represent the results of such tests on a certain new car model. How can we summarize the information in this rather large sample?

Steps for constructing a histogram:

Step 1: Open a clean workbook and enter the raw data into column A. A title may be included in the first row. The upper class boundaries should be entered into column B. For the histogram function in PHStat the upper boundaries for each class are defined to be the largest value contained in that class. When creating histograms for other data sets, use the guidelines from page 17 of the textbook to help determine equally spaced class boundaries.

	A	B
1	MPG	bounds
2	36.3	31.4
3	41	32.9
4	36.9	34.4
5	37.1	35.9
6	44.9	37.4
7	36.8	38.9
8	30	40.4
9	37.2	41.9
10	42.1	43.4
11	36.7	44.9
12	32.7	
13	37.3	
14	41.2	
15	36.6	
16	32.9	
17	36.5	
18	33.2	
19	37.4	
20	37.5	
21	33.6	
22	40.5	
23	36.5	
24	37.6	
25	33.9	
26	40.2	
27	36.4	
28	37.7	
29	37.7	
30	40	

⁄⁄ ◂ ▸ ▸⁄ **Sheet1** / Sheet2

Step 2: Click on the **Add-Ins** tab.

	A	B	C	D	E	F	G	H	I
1	MPG	bounds							
2	36.3	31.4							
3	41	32.9							
4	36.9	34.4							
5	37.1	35.9							
6	44.9	37.4							
7	36.8	38.9							
8	30	40.4							
9	37.2	41.9							
10	42.1	43.4							
11	36.7	44.9							
12	32.7								

Step 3: Click on **PHStat**.

Step 4: Drop down to **Descriptive Statistics** and click on **Histogram & Polygons....**

Step 5: A box for the histogram options will appear.

- Click on the entry box by **Variable Cell Range**. Enter the rows and column(s) where the data are located. This may be done by typing the range in manually or by clicking and dragging the appropriate range of data cells. In this example the data are located in Sheet1!$A$1:$A$101.
- Click on the entry box by **Bins Cell Range**. Enter the rows and column(s) where the data are located. This may be done by typing the range in manually or by clicking and dragging the appropriate range of data cells. In this example the data are located in Sheet1!$B$1:$B$11.
- Within the options box, the default setting is **First cell in each range contains label** (one would click on this box to unselect this option if there were no labels for the data set).
- Click the **OK** button and the histogram, ogive, and frequency table will appear in a new sheet labeled **Frequencies**.

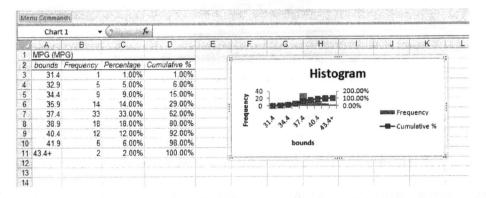

Step 6: Click and drag the lower right corner of the histogram box to increase the size of the image. A larger image will give the viewer a better view of the distribution. Note that the frequencies for the histogram are labeled to the left of the graphs and the percentages to the right of the graphs pertain only to the ogive.

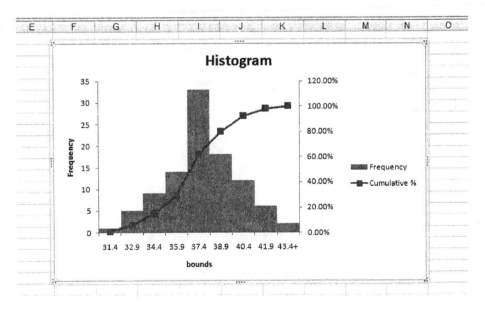

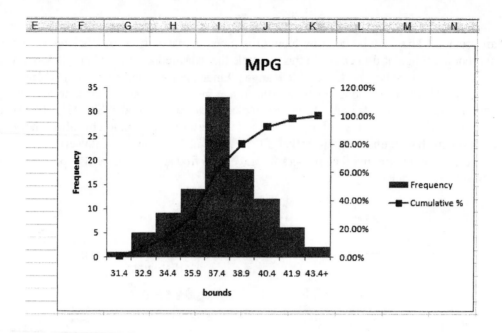

2.5 Graphical Methods for Describing Quantitative Data: Stem-and-Leaf Displays

The following example will be used from the text to illustrate how to construct a stem-and-leaf display. Again, please refer page 14 of the text, Table 2.2 which contains raw data for mileage ratings for 100 cars.

Steps for constructing a stem-and-leaf display:

Step 1: Open a clean workbook and enter the raw data into column A. A title may be included in the first row.

Step 2: Click on the **Add-Ins** tab.

Step 3: Click on **PHStat**.

Step 4: Drop down to **Descriptive Statistics** and click on **Stem-and-Leaf Display...**.

Step 5: A box for the **Stem-and-Leaf Display** options will appear.

- Click on the entry box by **Variable Cell Range**. Enter the rows and column(s) where the data are located. This may be done by typing the range in manually or by clicking and dragging the appropriate range of data cells. In this example the data are located in Sheet1!$A$1:$A$101.
- Within the options box, the default setting is **First cell contains label** (one would click on this box to unselect this option if there were no label for the data set).
- There is a box available for entering a title for the stem-and-leaf plot if a title for the output is desired.

- Click the **OK** button and the stem-and-leaf plot and summary statistics will appear in a new sheet labeled **StemLeafPlot.**

	A	B	C	D	E	F
1				MPG		
2						
3				Stem unit 1		
4						
5	Statistics			30	0	
6	Sample Size	100		31	8	
7	Mean	36.994		32	5 7 9 9	
8	Median	37		33	1 2 6 8 9 9	
9	Std. Deviation	2.417897		34	0 2 4 5 8 8	
10	Minimum	30		35	0 1 2 3 5 6 6 7 8 9 9	
11	Maximum	44.9		36	0 1 2 3 3 4 4 5 5 6 6 7 7 7 8 8 8 9 9 9	
12				37	0 0 0 0 1 1 1 2 2 3 3 4 4 5 6 6 7 7 8 9 9	
13				38	0 1 2 2 3 4 5 6 7 8	
14				39	0 0 3 4 5 7 8 9	
15				40	0 1 2 3 5 5 7	
16				41	0 0 2	
17				42	1	
18				43		
19				44	9	
20						

2.6   Graphical Methods for Describing Quantitative Data: Dotplot

The following example will be used from the text to illustrate how to construct a dotplot. Again, please refer page 14 of the text, Table 2.2 which contains raw data for mileage ratings for 100 cars.

Steps for constructing a dotplot:

Step 1: Open a clean workbook and enter the raw data into column A. A title may be included in the first row.

Step 2: Click on the **Add-Ins** tab.

Step 3: Click on **PHStat.**

Step 4: Drop down to **Descriptive Statistics** and click on **Dot Scale Diagram....**

Step 5: A box for the **Dot Scale Diagram** options will appear.

- Click on the entry box by **Variable Cell Range**. Enter the rows and column(s) where the data are located. This may be done by typing the range in manually or by clicking and dragging the appropriate range of data cells. In this example the data are located in Sheet1!$A$1:$A$101.
- Within the options box, the default setting is **First cell contains label** (one would click on this box to unselect this option if there were no label for the data set).
- There is a box available for entering a title for the stem-and-leaf plot if a title for the output is desired. For this example use "MPG".
- Click the **OK** button and the stem-and-leaf plot and summary statistics will appear in a new sheet labeled **Dot Scale**.

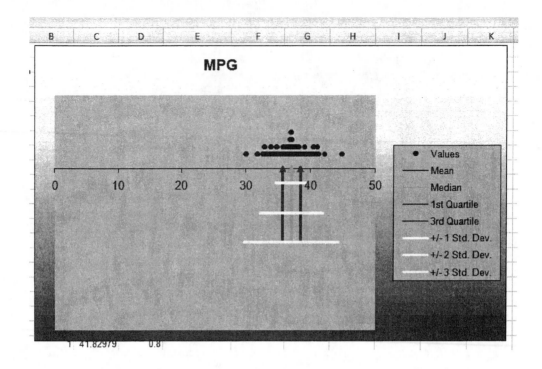

| | 41.82979 | 0.8 |

2.7 Graphical Methods for Describing Quantitative Data: Box Pots

The following example will be used from the text to illustrate how to construct a box plot. Again, please refer page 14 of the text, Table 2.2 which contains raw data for mileage ratings for 100 cars.

Steps for constructing a box plot:

Step 1: Open a clean workbook and enter the raw data into column A. A title may be included in the first row.

Step 2: Click on the **Add-Ins** tab.

Step 3: Click on **PHStat**.

Step 4: Drop down to **Descriptive Statistics** and click on **Box-and-Whisker Plot...**

Step 5: A box for the **Box-and-Whisker Plot** options will appear.

- Click on the entry box by **Variable Cell Range**. Enter the rows and column(s) where the data are located. This may be done by typing the range in manually or by clicking and dragging the appropriate range of data cells. In this example the data are located in Sheet1!$A$1:$A$101.
- Within the options box, the default setting is **First cell contains label** (one would click on this box to unselect this option if there were no label for the data set).
- There is a box available for entering a title for the box plot.
- Click on the box next to **Five-Number Summary** in order to obtain this output.

	A	B
1	MPG	
2		
3	Five-number Summary	
4	Minimum	30
5	First Quartile	35.6
6	Median	37
7	Third Quartile	38.4
8	Maximum	44.9
9		
10		

- Click the **OK** button and the box plot will appear in a new sheet labeled **BoxWhiskerPlot** and summary statistics will appear in a new sheet labeled **FiveNumbers**.

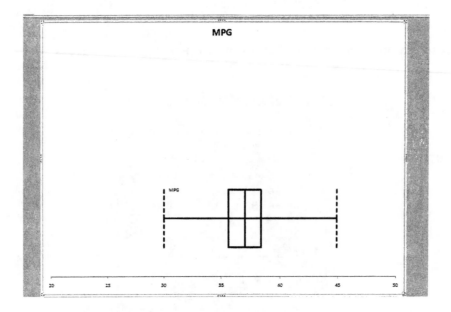

2.8 Measures of Central Tendency and Measures of Dispersion

The following example will be used from the text to illustrate how to obtain measures of central tendency and measures of dispersion. Again, please refer page 14 of the text, Table 2.2 which contains raw data for mileage ratings for 100 cars.

Steps for obtaining measures of central tendency and measures of dispersion:

Step 1: Open a clean workbook and enter the raw data into column A.

	A	B
1	MPG	
2	36.3	
3	41	
4	36.9	
5	37.1	
6	44.9	
7	36.8	
8	30	
9	37.2	
10	42.1	
11	36.7	
12	32.7	
13	37.3	
14	41.2	
15	36.6	
16	32.9	

Step 2: Click on the **Data** tab.

Step 3: Click on **Data Analysis**.

Step 4: A box for **Data Analysis** options will appear. Click on **Descriptive Statistics** and then click the **OK** button.

Step 5: A box for the **Descriptive Statistics** options will appear.

- Click on the entry box by **Input Range**. Enter the rows and column(s) where the data are located. This may be done by typing the range in manually or by clicking and dragging the appropriate range of data cells. In this example the data are located in Sheet1!$A$1:$A$101.
- Since there is a label in the first row, click on the box in front of **Labels in First Row** so that there is a check mark beside it.
- Click on the box in front of **Summary statistics** so that there is a check mark beside it.
- Click the **OK** button and the summary statistics will appear in a new sheet. The width of column A may need to be adjusted in order to read all of the output. The mean, median, mode, sample standard deviation, sample variance, and range are included in the output.

	A	B	C	D	E
		MPG			
1		MPG			
2					
3	Mean	36.994			
4	Standard Error	0.24179			
5	Median	37			
6	Mode	37			
7	Standard Deviation	2.417897			
8	Sample Variance	5.846226			
9	Kurtosis	0.769923			
10	Skewness	0.050909			
11	Range	14.9			
12	Minimum	30			
13	Maximum	44.9			
14	Sum	3699.4			
15	Count	100			
16					

2.9   Graphical Methods for Describing Bivariate Data: Scattergram

The following example will be used from the text to illustrate how to construct a scattergram. Please refer to page 68 of the text, Exercise 2.142 which contains data on the feeding behavior of fish.

Zoologists at the University of Western Australia conducted a study of the feeding behavior of black bream (a type of fish) spawned in aquariums. (*Brain and Behavior Evolution*, April 2000) In one experiment, the zoologists recorded the number of aggressive strikes of two black bream feeding at the bottom of the aquarium in the 10-minute period following the addition of food. The number of strikes and the age of fish (in days) were recorded approximately each week for nine weeks, as shown in the table.

Steps for constructing a scattergram in Excel 2007:

Step 1: Open a new workbook and place the cursor in the upper left-hand corner of the spreadsheet.

Step 2: Input the data manually. Enter the data for the variable measured along the horizontal axis (typically labeled x) in the first column and the variable measured along the vertical axis (typically labeled y) into the second column. Include variable names in the first row.

	A	B	C
1	fish	strikes	
2	120	85	
3	136	63	
4	150	34	
5	155	39	
6	162	58	
7	169	35	
8	178	57	
9	184	12	
10	190	15	
11			

Step 3: Highlight both columns of data.

Step 4: Select the **Insert** tab by clicking on it.

Step 5: Select **Scatter** from the choices available and pictures of options for creating a scattergram will be displayed. There are many options for types of graph styles; choose the option in the upper left corner for a simple, straightforward, easy to read scattergram.

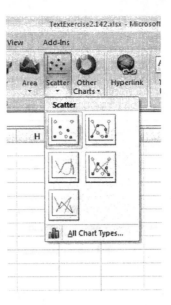

Step 6: Excel will now have created a scattergram within the same sheet as the data. The default chart style (Style 2) has blue dots. The Ribbon at the top of the page will now display **Chart Tools**. To the left of **Chart Styles** are the **Chart Layouts**. Click on the downward pointing arrow with a line over it to view all bar graph layouts. Click on chart layout 1, the first image; it creates a titled scattergram with a legend.

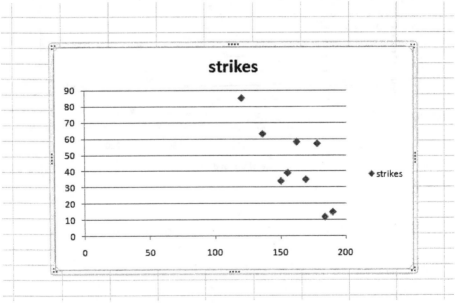

Step 7: The graph will now be labeled with the label from the second column of data. In this case that is "strikes". Click on these words to activate the title and then click on them again so that they may be highlighted. Now type in a more descriptive title for the scattergram data. In this case the graph may be labeled "Feeding behavior of fish". The same may be done for the axis labels.

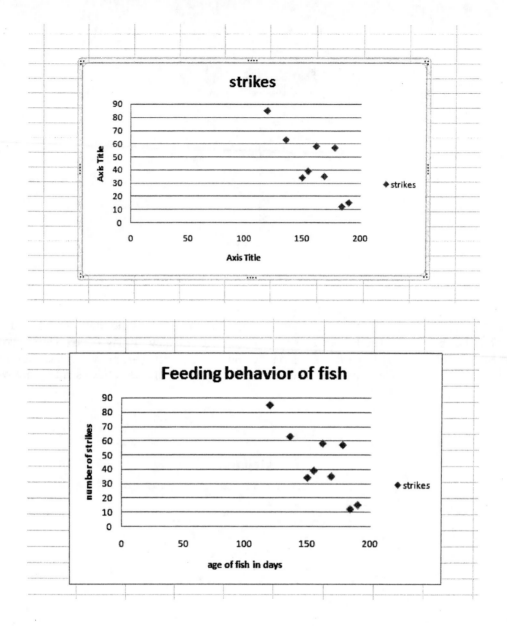

Step 8: Click on the graph to activate it. Now click on the **Add-Ins** tab in the top ribbon of the page. Click on **PHStat** and drop down to **Utilities** and then click on **Fix Up Chart**. This will result in a cleaner looking chart.

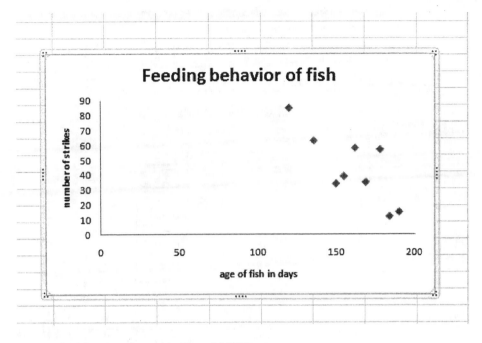

Steps for constructing a scattergram in Excel 2003:

Step 1: Open a new workbook and place the cursor in the upper left-hand corner of the spreadsheet.

Step 2: Input the data manually. Enter the data for the variable measured along the horizontal axis (typically labeled x) in the first column and the variable measured along the vertical axis (typically labeled y) into the second column. Include variable names in the first row.

Step 3: Highlight both columns of data.

Step 4: Click on the **Chart Wizard** icon in the tool bar.

Step 5: Select **Scatter** from the choices available and pictures of options for creating a scattergram will be displayed. There are many options for types of graph styles; keep the default option for a simple, straightforward, easy to read scattergram. Click on the **Next** button.

Step 6: Excel will now have created a scattergram within the Chart Wizard. Click on the **Next** button.

Step 7: The graph will now be labeled with the label from the second column of data. In this case that is "strikes". The Chart Wizard will now display a box for entering a title for the graph. Change the title from the default to a more descriptive title for the scattergram data. In this case the graph may be labeled "Feeding behavior of fish". The same may be done for the axis labels. Click on the **Next** button.

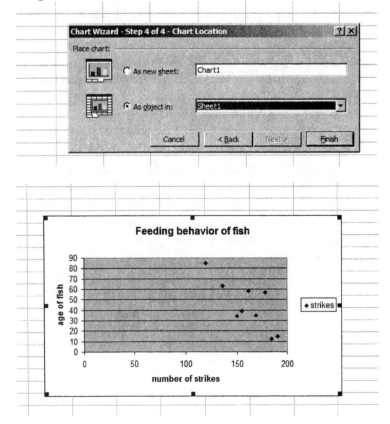

Step 8: The Chart Wizard will now display an option for where to place the graph. The default is for it to be placed within the current sheet. Click on the **Finish** button and the graph will be displayed within the current sheet. Click on **PHStat** and drop down to **Utilities** and then click on **Fix Up Chart**. This will result in a cleaner looking chart.

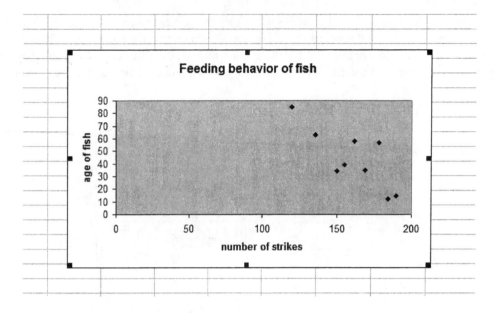

2.10   Sample Exercises

The following exercises were selected from the textbook to test the reader's understanding of Excel and its usage within the chapter. Try to replicate the output for the following problems.

Problem 2.168 (Competition for bird nest holes), page 102

The Condor (May 1995) published a study of competition for nest holes among collared flycatchers, a bird species. The authors collected the data for the study by periodically inspecting nest boxes located on the island of Gotland in Sweden. The nest boxes were grouped into 14 discrete locations (called plots). The table presents the number of flycatchers killed and the number of flycatchers breeding at each plot. Calculate the mean, median, and mode for the number of flycatchers killed at the 14 study plots. Graphically

Plot Number	Number Killed	Number of Breeders
1	5	30
2	4	28
3	3	38
4	2	34
5	2	26
6	1	124
7	1	68
8	1	86
9	1	32
10	0	30
11	0	46
12	0	132
13	0	100
14	0	6

Solution to Problem 2.168

Number Killed	
Mean	1.428571429
Standard Error	0.428571429
Median	1
Mode	0
Standard Deviation	1.603567451
Sample Variance	2.571428571
Kurtosis	0.484071484
Skewness	1.129959211
Range	5
Minimum	0
Maximum	5
Sum	20
Count	14

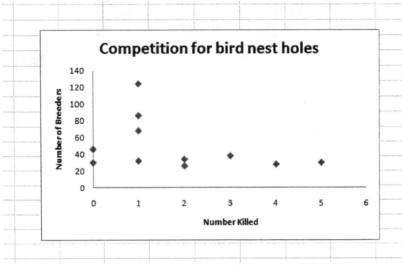

Problem 2.59 & 2.83 (Ammonia in car exhaust), pages 59 & 67

Refer to the *Environmental Science & Technology* (Sep. 1, 2000) study on the ammonia levels near the exit ramp of a San Francisco highway tunnel. The data (in parts per million) for 8 days during afternoon drive-time are reproduced in the table. Find the mean, median, range, variance, and standard deviation of the ammonia levels.

1.53	1.50	1.37	1.51	1.55	1.42	1.41	1.48

Row1	
Mean	1.47125
Standard Error	0.02263353 2
Median	1.49
Mode	#N/A
Standard Deviation	0.06401729 7
Sample Variance	0.00409821 4
Kurtosis	-1.23809065
Skewness	-0.45637176
Range	0.18
Minimum	1.37
Maximum	1.55
Sum	11.77
Count	8

3.1   Introduction

Chapter 3 in the text presents the basics of probability theory and the concept of random sampling. PHStat is a supplementary addition to Excel that provides the user with an alternative technique to calculating probabilities and selecting samples from a population. PHStat allows the user to compute probabilities associated with contingency tables. It may also be utilized to select random samples from a given population.

The following examples will be used to demonstrate techniques in PHStat and Excel.

> Example 3.15, page 139 (Probabilities of Smoking & Developing Cancer)
> Exercise 3.99, page 157 (Choosing n elements from N)
> Table 2.4, page 42 (Random Sampling from a given population)
> Sample Exercise #1: Problem 3.162, page 174 (Study of aggressiveness and birth order)
> Sample Exercise #2: Problem 3.104, page 157 (Auditing an accounting system)

3.2   Probabilities using PHStat: 2x2 Contingency Tables

PHStat will only calculate the probabilities of unions and intersections if the data are presented in a 2x2 contingency table. Refer to Example 3.15, page 139 of the text to illustrate how to compute these probabilities.

Many medical researchers have conducted experiments to examine the relationship between cigarette smoking and cancer. Consider an individual randomly selected from the adult male population. Let A represent the event that the individual smokes, and let A^c denote the complement of A (the event that the individual does not smoke). Similarly, let B represent the event that the individual develops cancer and let B^c be the complement of that event. Use the data presented in the table to find the following.

a. Find P(A), the probability that an adult male smokes
b. Find P(B|A), the probability that an adult male smoker develops cancer.
c. Find $P(B^c|A)$, the probability that an adult male smoker does not develop cancer.
d. Find $P(B|A^c)$, the probability that an adult male non-smoker develops cancer.
e. Find $P(B^c|A^c)$, the probability that an adult male nonsmoker does not develop cancer.

Probabilities of Smoking and Developing Cancer		
	Develops Cancer	
	Yes, B (Cancer)	No, B^c (Non cancer)
Yes, A (Smoker)	0.05	0.20
No, A^c (Nonsmoker)	0.03	0.72

Steps for calculating probabilities:

Step 1: Open a clean workbook.

Step 2: Click on the **Add-Ins** tab.

Step 3: Click on **PHStat**.

Step 4: Drop down to **Probability & Prob. Distributions** and click on **Simple & Joint Probabilities**.

A new sheet labeled **Basic Probabilities** will appear.

	A	B	C	D	E	F
1	**Probabilities Calculations**					
2						
3	**Sample Space**		**Event B**			
4			B1	B2	Totals	
5	**Event A**	A1	1	1	2	
6		A2	1	1	2	
7		Totals	2	2	4	
8						
9	**Simple Probabilities**					
10	P(A1)	0.50				
11	P(A2)	0.50				
12	P(B1)	0.50				
13	P(B2)	0.50				
14						
15	**Joint Probabilities**					
16	P(A1 and B1)	0.25				
17	P(A1 and B2)	0.25				
18	P(A2 and B1)	0.25				
19	P(A2 and B2)	0.25				
20						
21	**Addition Rule**					
22	P(A1 or B1)	0.75				
23	P(A1 or B2)	0.75				
24	P(A2 or B1)	0.75				
25	P(A2 or B2)	0.75				
26						

PHStat2 Note:
Enter the labels and values for both events in cells A5, C3, and the cell ranges B5:B6, C4:C5, and C5:D6.

Enter replacement labels for A1, A2, B1, and B2, in cell ranges B5:B6 and C4:D4, if desired.

(Before continuing, press the Delete key to delete this note.)

52

The worksheet that appears is actually a template that the user can manipulate in order to derive several types of probability scenarios.

Step 5: Replace **Event A** and **Event B** in the **Sample Space** table with the event names. For this example that will be **Smoker** and **Developing Cancer**.

Step 6: Next, replace the outcomes that are listed in the **Sample Space** table as A1, A2, B1, and B2 with their titles. For this example that will be respectively **Smoker, Nonsmoker, Cancer, Non Cancer**. The events for the simple and joint probabilities as well as those associated with the addition rule will be modified automatically.

Step 7: Change the numbers within the template to reflect the probabilities associated with the example in the text. The simple and joint probabilities as well as those associated with the addition rule will be modified automatically.

	A	B	C	D	E	F	G
1	Probabilities Calculations						
2							
3	Sample Space			Developing Cancer			
4			Cancer	Non Cancer	Totals		
5	Smoker	Smoker	0.05	0.2	0.25		
6		Nonsmoker	0.03	0.72	0.75		
7		Totals	0.08	0.92	1		
8							
9	Simple Probabilities						
10	P(Smoker)	0.25					
11	P(Nonsmoker)	0.75					
12	P(Cancer)	0.08					
13	P(Non Cancer)	0.92					
14							
15	Joint Probabilities						
16	P(Smoker and Cancer)	0.05					
17	P(Smoker and Non Cancer)	0.20					
18	P(Nonsmoker and Cancer)	0.03					
19	P(Nonsmoker and Non Cance	0.72					
20							
21	Addition Rule						
22	P(Smoker or Cancer)	0.28					
23	P(Smoker or Non Cancer)	0.97					
24	P(Nonsmoker or Cancer)	0.80					
25	P(Nonsmoker or Non Cancer)	0.95					
26							

PHStat2 Note:
Enter the labels and values for both events in cells A5, C3, and the cell ranges B5:B6, C4:C5, and C5:D6.

Enter replacement labels for A1, A2, B1, and B2, in cell ranges B5:B6 and C4:D4, if desired.

(Before continuing, press the Delete key to delete this note.)

Solutions:

Now the revised template can be used to answer the questions from Example 3.15 (page 139) in the text. Before the solutions are presented, remember the symbolism used in the text and how that symbolism is correlated with the template. Remember that "and" probabilities correspond to multiplication rules whereas "or" probabilities correspond to addition rules:

$P(A \cap B) = P(A \text{ and } B)$

$P(A \cup B) = P(A \text{ or } B)$

$P(B|A) = P(A \cap B)/P(A) = P(A \text{ and } B)/P(A)$

Solution to part a: Find P(A), the probability that an adult male smokes. From the template, P(Smoker) is 0.25.

Solution to part b: Find P(B|A), the probability that an adult male smoker develops cancer. To calculate this probability, remember that the $P(B|A) = P(A \cap B)/P(A)$. So, P(Smoker and Cancer)/ P(Smoker) = $0.05/.25 = 0.20$.

Solution to part c: Find $P(B^c|A)$, the probability that an adult male smoker does not develop cancer. To calculate this probability, remember that $P(B^c|A) = P(A \cap B^c)/P(A)$. So, P(Smoker and No Cancer)/P(Smoker) = $0.20/0.25 = 0.80$.

Solution to part d: Find $P(B|A^c)$, the probability that an adult male non-smoker develops cancer. To calculate this probability, remember that $P(B|A^c) = P(A^c \cap B)/P(A^c)$. So, P(Nonsmoker and Cancer)/P(Nonsmoker) = $0.03/0.75 = 0.04$.

Solution to part e: Find $P(B^c|A^c)$, the probability that an adult male nonsmoker does not develop cancer. To calculate this probability, remember that $P(B^c|A^c) = P(A^c \cap B^c)/P(A^c)$. So, P(Nonsmoker and Non Cancer)/P(Nonsmoker) = $0.72/0.75 = 0.96$.

3.3   Random Sampling

In a random sample, each individual, object, or measurement has an equal probability of being selected from a population. Although a random number table does appear in an appendix of the text, PHStat provides the user with two alternative methods of selecting samples randomly from a known population. The first method assumes the user assigns the labels 1 through N to the N units of the population.

The following example from the text will be used to illustrate the use of random sampling. Please refer to page 157 of the text, Exercise 3.99.

Suppose you wish to sample n = 3 elements from a total of N = 600 elements. Use the computer to generate a random sample of 3 from the population of 600 elements.

Steps for choosing a random sample of units labeled 1 through N:

Step 1: Open a clean workbook.

Step 2: Click on the **Add-Ins** tab.

Step 3: Click on **PHStat**.

Step 4: Drop down to **Sampling** and click on **Random Sample Generation....**

Step 5: A box for the **Random Sample Generation** options will appear.

- In the box next to **Sample Size**, enter the size of the desired sample which in this case is 3.
- **Generate a list of random numbers** is the default setting and that is what we want for this scenario.
- In the box next to **Population Size**, enter the total population size, which in this case is 600.
- There is a blank box to enter a title. For this example enter "Ex3.99".

- Click the **OK** button and the random sample will appear in a new sheet labeled **RandomNumbers**. This example shows one possible sample of 3 elements.

	A	B
1	Ex3.99	
2	102	
3	329	
4	189	
5		

Sometimes the user already has a list of items in a spreadsheet. It is possible to select a sample from a set of data that already are contained within the Excel spreadsheet.

Please refer page 42 of the text, Table 2.4 which contains raw data for student loan default rates. Use this data to randomly select 5 states.

	A	B	C
1	AL	12	
2	AK	19.7	
3	AZ	12.1	
4	AR	12.9	
5	CA	11.4	
6	CO	9.5	
7	CT	8.8	
8	DE	10.9	
9	DC	14.7	
10	FL	11.8	
11	GA	14.8	
12	HI	12.8	
13	ID	7.1	
14	IL	9.3	
15	IN	6.7	
16	IA	6.2	
17	KS	5.7	
18	KY	10.3	
19	LA	13.5	
20	ME	9.7	
21	MD	16.6	
22	MA	8.3	
23	MI	11.4	
24	MN	6.6	
25	MS	15.6	

Steps for choosing a random sample of items from a list:

Step 1: Open a clean workbook and enter the data.

Step 2: Click on the **Add-Ins** tab.

Step 3: Click on **PHStat**.

Step 4: Drop down to **Sampling** and click on **Random Sample Generation....**

Step 5: A box for the **Random Sample Generation** options will appear.

- In the box next to **Sample Size** enter 5.
- Click on the circle in front of **Select values from range**.
- Click on the box next to **Values cell range** and either manually enter the range for the data or click and drag in order to enter it. For this example it is Sheet1!$A$1:$A$51.
- For this example, you would need to click on the box in front of **First cell contains label** in order to unselect this option since there are not labels in the first row.
- There is a blank box to enter a title. For this example enter "States".
- Click the **OK** button and the random sample will appear in a new sheet labeled **RandomNumbers**.

	A	B
1	States	
2	AR	
3	RI	
4	WI	
5	MT	
6	NM	

Notice that if the user were to repeatedly sample the same size of n items from a population of size N, the output will differ from trial to trial. The output presented in this example corresponds to the five states within the population that have been selected randomly.

3.4   Sample Exercises

The following exercises were selected from the textbook to test the reader's understanding of PHStat and its usage within the chapter. Try to replicate the output in PHStat for the following problems.

Problem 3.162, page 174 (Study of aggressiveness and birth order)

Psychologists tend to believe that there is a relationship between aggressiveness and order of birth. To test this belief, a psychologist chose 500 elementary school students at random and administered each a test designed to measure the student's aggressiveness. Each student was classified according to one of four categories. The percentage of students falling in the categories is shown here.

a. If one student is chosen at random from the 500, what is the probability that the student is firstborn?
b. What is the probability that the student is aggressive?
c. What is the probability that the student is aggressive, given the student was firstborn?

	Firstborn	Not Firstborn
Aggressive	15%	15%
Not Aggressive	25%	45%

Solution to Problem 3.162

Probabilities Calculations

Sample Space

		Event B		
		Firstborn	not Firstborn	Totals
Event A	Aggressive	0.15	0.15	0.3
	Not Aggress	0.25	0.45	0.7
	Totals	0.4	0.6	1

Simple Probabilities

P(Aggressive)	0.30
P(Not Aggressive)	0.70
P(Firstborn)	0.40
P(not Firstborn)	0.60

PHStat User Note:
Enter the values for both events in the cell range C5:D6.

Enter replacement labels for A1, A2, B1, and B2, in cell ranges B5:B6 and C4:D4, if desired.

Select this note and then select Edit | Cut to delete this note from the worksheet.

Joint Probabilities

P(Aggressive and Firstborn)	0.15
P(Aggressive and not Firstborn)	0.15
P(Not Aggressive and Firstborn)	0.25
P(Not Aggressive and not Firstborn)	0.45

Addition Rule

P(Aggressive or Firstborn)	0.55
P(Aggressive or not Firstborn)	0.75
P(Not Aggressive or Firstborn)	0.85
P(Not Aggressive or not Firstborn)	0.85

Problem 3.104, page 157 (Auditing an accounting system)

In auditing a firm's financial statements, an auditor is required to assess the operational effectiveness of the accounting system. In performing the assessment, the auditor frequently relies on a random sample of actual transactions (Stickney and Weil, *Financial Accounting: An Introduction to Concepts, Methods, and Uses*, 1994). A particular firm has 5,382 customer accounts that are number 0001 to 5382. Draw a random sample of 10 accounts and explain in detail the procedure you used.

One possible solution to Problem 3.104

Randomly Selected Values		
576		
679		
2212		
5202		
4055		
5193		
4303		
1740		
1935		
1986		

4.1   Introduction

In chapter 4 of the textbook, the authors present the two major types of random variables: discrete and continuous variables. By definition, discrete variables are countable and can assume only certain distinct values whereas continuous variables can assume any value within a specified range.   When analyzing discrete variables, specific discrete probability distributions (such as the binomial, Poisson, and hypergeometric distributions) may be utilized in order to calculate the probabilities associated with a specific variable.   Notice that the appendices in the textbook present cumulative probability tables for only specified discrete values for the binomial and Poisson distributions.

Excel 2007 and PHStat can be used to find cumulative and exact probabilities for both the binomial and the Poisson distribution. The major difference between PHStat and Excel 2007 is that PHStat provides an easy-to-use method for interpreting "language" asymmetries most often encountered by students first learning about probabilities. PHStat allows the user to interpret "at least", "at most", "equal to", "more than" and "less than" questions for both the binomial and Poisson distributions by generating all possible cumulative and exact probabilities in a single table contained within the generated output.   The hypergeometric distribution will be demonstrated in Excel only.

The following examples will be used to illustrate techniques in PHStat and Excel.
Example 4.9, page 198 (Heart Association)
Example 4.10, page 200 (Heart Association)
Example 4.11, page 202 (Heart Association)
Example 4.13, page 211 (Whale Sightings)
Example 4.14, page 217 (New Teaching Assistants)
Sample Exercise #1: Problem 4.68, page 209 (Chickens with fecal contamination)
Sample Exercise #2: Problem 4.83, page 215 (Eye fixation problem)
Sample Exercise #3: Problem 4.101, page 219 (On-site disposal of waste)

4.2   Calculating Binomial Probabilities Using PHStat.

Please refer to Example 4.9, page 198, Example 4.10, page 200, and Example 4.11, page 202 (Heart Association) of the text.

The Heart Association claims that only 10% of U.S. adults over 30 can pass the President's Physical Fitness Commission's minimum requirements. Suppose four adults are randomly selected, and each is given the fitness test.
   a.   Find the probability that none of the four adults pass the test.
   b.   Find the probability that three of the four adults pass the test.
   c.   Determine the probability distribution.
   d.   Calculate the mean and standard deviation of the number of the four adults who pass the test.

Steps for calculating binomial probabilities:

Step 1: Open a clean workbook.

Step 2: Click on the **Add-Ins** tab.

Step 3: Click on **PHStat**.

Step 4: Drop down to **Probability & Prob. Distributions** and click on **Binomial....**

Step 5: A box for the **Binomial Probability Distribution** options will appear.

- In the box next to **Sample Size**, enter the number of trials. In this case it is 4.
- In the box next to **Probability of Success**, enter the probability of success on a single trial in decimal form. In this case it is 0.10.
- In the boxes next to **Outcomes From**, enter 0 in the first box and the value of n, for this example 4, into the second box. This way the total probability distribution is provided.

- There is a blank box to enter a title. For this example enter "Heart Association".
- Click on the box in front of **Cumulative Probabilities** so that a check appears; this will provide the cumulative probabilities.
- Click the **OK** button and the results will appear in a new sheet labeled **Binomial**.

	A	B	C	D	E	F	G	
1	Heart Association							
2								
3	Data							
4	Sample size	4						
5	Probability of success	0.1						
6								
7	Statistics							
8	Mean	0.4						
9	Variance	0.36						
10	Standard deviation	0.6						
11								
12	Binomial Probabilities Table							
13			X	P(X)	P(<=X)	P(<X)	P(>X)	P(>=X)
14			0	0.6561	0.6561	0	0.3439	1
15			1	0.2916	0.9477	0.6561	0.0523	0.3439
16			2	0.0486	0.9963	0.9477	0.0037	0.0523
17			3	0.0036	0.9999	0.9963	1E-04	0.0037
18			4	0.0001	1	0.9999	0	1E-04
19								

After reviewing the output, the following answers are derived:
  a. The probability that none of the four adults pass the test is $P(0) = 0.6561$.
  b. The probability that three of the four adults pass the test is $P(3) = 0.0036$.
  c. The column for $P(X)$ corresponds to the probability distribution table on page 201 of the text.
  d. The mean number of individuals that passed the test is 0.4 and the standard deviation is 0.6.

4.3   Calculating Binomial Probabilities using Excel

It is also possible to calculate binomial probabilities without using the PHStat option in Excel. In order to do so, please follow the steps provided.

Steps for calculating binomial probabilities:

Step 1: Open a clean workbook.

Step 2: Click on the **Formulas** tab.

Step 3: Click on **More Functions**.

Step 4: Click on **Statistical** and drop down to **BINOMDIST** and click on it.

Step 5: A box for **Function Arguments** options will appear.

There are four boxes in which to enter data:
- **Number_s** is the number of successes in n trials.
- **Trials** is n, the number of independent trials.
- **Probability_s** is the probability of success for each trial. Enter this in decimal form.
- **Cumulative** is either true or false. Enter **TRUE** if the probability between 0 and **Number_s**, $P(0 \leq X \leq Number\_s)$, is desired. Enter **FALSE** if only the probability of **Number_s**, $P(X = Number\_s)$, is desired.

After all four values are entered, the answer will appear within the **Function Arguments** box. Click the **OK** button and Excel will also enter the answer into the active cell of the spreadsheet.

Now refer to Example 4.12, page 204, of the text to illustrate determining binomial probabilities.

Suppose a poll of voters is taken in a large city. The purpose is to determine X, the number who favor a certain candidate for mayor. Suppose that 60% of all the city's voters favor the candidate.
   a. Find the probability that $X \leq 10$.
   b. Find the probability that $X > 12$.
   c. Find the probability that $X = 11$.

Solutions:

For each of parts (a) through (c) complete **Step 1** through **Step 4** from above and complete **Step 5** as follows:

Solution to part a:

Step 5: A box for **Function Arguments** options will appear. There are four boxes in which to enter data:
- **Number_s** is 10.
- **Trials** is 20.
- **Probability_s** is 0.60.
- **Cumulative** is **TRUE**.

The answer is $P(X \leq 10) = 0.244662797$.

Solution to part b:

Step 5: A box for **Function Arguments** options will appear. There are four boxes in which to enter data:
- **Number_s** is 12.
- **Trials** is 20.
- **Probability_s** is 0.60.
- **Cumulative** is **TRUE**.

The answer given by Excel is 0.58410062, which is $P(0 \leq X \leq 12)$. Since $P(X > 12) = 1 - P(0 \leq X \leq 12)$ the desired answer is $1 - 0.584107062 = 0.415892938$.

Solution to part c:

Step 5: A box for **Function Arguments** options will appear. There are four boxes in which to enter data:

- **Number_s** is 11.
- **Trials** is 20.
- **Probability_s** is 0.60.
- **Cumulative** is **FALSE**.

The answer is P(X = 11) = 0.159738478.

4.4   Calculating Poisson Probabilities using PHStat

According to the authors, Poisson distributions are best used to describe the number of "successes" or events that will occur within a specified time period. Page 210 of the text provides examples of random variables for which the Poisson random variable distribution provides a good model. The following example will help illustrate how to calculate Poisson probabilities in PHStat and Excel.

Please refer to Example 4.13, page 211 (Whale Sightings) of the textbook.

Ecologists often use the number of reported sightings of a rare species of animal to estimate the remaining population size. For example, suppose the number, X, of reported sightings per week of blue whales is recorded. Assume that X has (approximately) a Poisson probability distribution. Furthermore, assume that the average number of weekly sightings is 2.6.

    a.   Find the probability that fewer than two sightings are made during a given week.
    b.   Find the probability that more than five sightings are made during a given week.
    c.   Find the probability that exactly five sightings are made during a given week.

Steps for calculating Poisson probabilities:

Step 1: Open a clean workbook.

Step 2: Click on the **Add-Ins** tab.

Step 3: Click on **PHStat**.

Step 4: Drop down to **Probability & Prob. Distributions** and click on **Poisson....**

Step 5: A box for the **Poisson Probability Distribution** options will appear.

- Next to the box for **Average/Expected No. of Successes** enter the given mean of the distribution. For this example it is 2.6.
- Next to the box for **Title** enter "Whale Sightings".
- Click on the box in front of **Cumulative Probabilities** so that the cumulative probabilities are provided.
- Click the **OK** button and the results will appear in a new sheet labeled **Poisson**.

	A	B	C	D	E	F	G	H	I
1	Whale Sightings								
2									
3			Data						
4	Average/Expected number of successes:				2.6				
5									
6	Poisson Probabilities Table								
7		X	P(X)	P(<=X)	P(<X)	P(>X)	P(>=X)		
8		0	0.074274	0.074274	0.000000	0.925726	1.000000		
9		1	0.193111	0.267385	0.074274	0.732615	0.925726		
10		2	0.251045	0.518430	0.267385	0.481570	0.732615		
11		3	0.217572	0.736002	0.518430	0.263998	0.481570		
12		4	0.141422	0.877423	0.736002	0.122577	0.263998		
13		5	0.073539	0.950963	0.877423	0.049037	0.122577		
14		6	0.031867	0.982830	0.950963	0.017170	0.049037		
15		7	0.011836	0.994666	0.982830	0.005334	0.017170		
16		8	0.003847	0.998513	0.994666	0.001487	0.005334		
17		9	0.001111	0.999624	0.998513	0.000376	0.001487		
18		10	0.000289	0.999913	0.999624	0.000087	0.000376		
19		11	0.000068	0.999982	0.999913	0.000018	0.000087		
20		12	0.000015	0.999996	0.999982	0.000004	0.000018		
21		13	0.000003	0.999999	0.999996	0.000001	0.000004		
22		14	0.000001	1.000000	0.999999	0.000000	0.000001		
23		15	0.000000	1.000000	1.000000	0.000000	0.000000		
24		16	0.000000	1.000000	1.000000	0.000000	0.000000		
25		17	0.000000	1.000000	1.000000	0.000000	0.000000		
26		18	0.000000	1.000000	1.000000	0.000000	0.000000		
27		19	0.000000	1.000000	1.000000	0.000000	0.000000		
28		20	0.000000	1.000000	1.000000	0.000000	0.000000		
29									

PHStat generates a Poisson distribution similar to the one displayed in the appendix at the back of the textbook. Keep in mind that PHStat will only provide outcomes for random variable values between 0 and 20, so it is better suited for some problems than others.

Notice that the labels for Poisson probabilities P (X = number of successes), P (number of successes $\leq$ X), P (number of successes < X), P (number of successes > X), and P (X $\geq$ number of successes) are all listed in the sixth row within the Excel worksheet. The number of successes is listed in the column under X. Please refer to the appropriate column and row when answering the preceding questions.

The solutions to the questions are:
    a. Find the probability that fewer than two sightings are made during a given week:
        $P(X < 2) = P(X \leq 1) = 0.267385$
    b. Find the probability that more than five sightings are made during a given week:
        $P(X > 5) = 0.049037$
    c. Find the probability that exactly five sightings are made during a given week:
        $P(X = 5) = 0.073539$

4.5   Calculating Poisson Probabilities using Excel

Using the same example, recalculate the Poisson probabilities using Excel 2007.

Steps for calculating Poisson probabilities:

Step 1: Open a clean workbook.

Step 2: Click on the **Formulas** tab.

Step 3: Click on **More Functions**.

Step 4: Click on **Statistical** and drop down to **POISSON** and click on it.

Step 5: A box for **Function Arguments** options will appear.

There are three boxes in which to enter data:
- **X** is the number of successes.
- **Mean** is the given mean of the distribution.
- **Cumulative** is either true or false. Enter **TRUE** if the probability between 0 and **X**, $P(0 \leq X \leq X)$, is desired. Enter **FALSE** if only the probability of **X**, $P(X = X)$, is desired.

After all three values are entered, the answer will appear within the **Function Arguments** box. Click the **OK** button and Excel will also enter the answer into the active cell of the spreadsheet.

Please refer again to Example 4.13, page 211 (Whale Sightings).

Solutions:

For each of parts (a) through (c) complete **Step 1** through **Step 4** from above and complete **Step 5** as follows:

Solution to part a:

Step 5: A box for **Function Arguments** options will appear. There are three boxes in which to enter data:
- **X** is 1.
- **Mean** is 2.6.
- **Cumulative** is TRUE.

Note that X = 1 is entered here since P(X < 2) = P(X ≤ 1) for the discrete Poisson random variable. So the answer is 0.267384882.

B	C	D	E	F	G	H	I	J

Function Arguments   ? ✕

POISSON

X	1	🔲	= 1
Mean	2.6	🔲	= 2.6
Cumulative	TRUE	🔲	= TRUE

= 0.267384882

Returns the Poisson distribution.

Cumulative   is a logical value: for the cumulative Poisson probability, use TRUE; for the Poisson probability mass function, use FALSE.

Formula result = 0.267384882

Help on this function   OK   Cancel

Solution to part b:

Step 5: A box for **Function Arguments** options will appear. There are three boxes in which to enter data:

- **X** is 5.
- **Mean** is 2.6.
- **Cumulative** is TRUE.

The answer given by Excel is 0.950962848, which is $P(X \leq 5)$. Since $P(X > 5) = 1 - P(X \leq 5)$ the desired answer is $1 - 0.950962848 = 0.049037152$.

Function Arguments   ? ✕

POISSON

X	5	🔲	= 5
Mean	2.6	🔲	= 2.6
Cumulative	TRUE	🔲	= TRUE

= 0.950962848

Returns the Poisson distribution.

X   is the number of events.

Formula result = 0.950962848

Help on this function   OK   Cancel

Solution to part c:

Step 5: A box for **Function Arguments** options will appear. There are three boxes in which to enter data:

- **X** is 5.
- **Mean** is 2.6.
- **Cumulative** is FALSE.

The answer is $P(X = 5) = 0.073539359$.

4.6   Calculating Hypergeometric Distributions using Excel

Please refer to Example 4.14, page 217 (New Teaching Assistants) of the text.

Suppose a professor randomly selects three new teaching assistants from a total of 10 applicants, six males and four female students. Let x be the number of females who are hired. Find the probability that no females are hired.

Steps for calculating hypergeometric probabilities:

Step 1: Open a clean workbook.

Step 2: Click on the **Formulas** tab.

Step 3: Click on **More Functions**.

Step 4: Click on **Statistical** and drop down to **HYPGEOMDIST** and click on it.

Step 5: A box for **Function Arguments** options will appear.

There are four boxes in which to enter data:

- **Sample_s** is the number of successes in the sample. For this example enter 0 since the problem asks for the probability of 0 successes.
- **Number_sample** is the size of the sample. In this example enter 3 since there will be 3 applicants chosen.
- **Population_s** is the number of successes in the population. In this example enter 4 since there are 4 female applicants to choose from.
- **Number_pop** is the population size. In this example enter 10 since there are 10 total applicants to choose from.

After all four values are entered, the answer will appear within the **Function Arguments** box. Click the **OK** button and Excel will also enter the answer into the active cell of the spreadsheet.

The answer is given as 0.16666667. This is the equivalent of 1/6, which is shown in the text.

4.7 Sample Exercises

The following exercises were selected from the textbook to test the reader's understanding of PHStat and its usage within the chapter. Try to replicate the output in PHStat/Excel for the following problems.

Problem 4.68, page 209 (Chickens with fecal contamination)

The United States Department of Agricultural (USDA) reports that, under its standard inspection system, one in every 100 slaughtered chickens pass inspection with fecal contamination (*Tampa Tribune*, Mar. 31, 2000). Find the probability that, in a random sample of five slaughtered chickens, at least one passes inspection with fecal contamination.

Solution to Problem 4.68, page 209 (Chickens with fecal contamination)

Binomial Probabilities						
Sample size	5					
Probability of success	0.01					
Mean	0.05					
Variance	0.0495					
Standard deviation	0.222486					
Binomial Probabilities Table						
X	P(X)	P(<=X)	P(<X)	P(>X)	P(>=X)	
0	0.95099	0.95099	0	0.04901	1	
1	0.04803	0.99902	0.95099	0.00098	0.04901	
2	0.00097	0.99999	0.99902	9.85E-06	0.00098	
3	9.8E-06	1	0.99999	4.96E-08	9.85E-06	
4	4.95E-08	1	1	1E-10	4.96E-08	
5	1E-10	1	1	0	1E-10	

Problem 4.83, page 215 (Eye fixation problem)

Cognitive scientists at the University of Massachusetts designed an experiment to measure x, the number of times a reader's eye fixated on a single word before moving past that word (*Memory and Cognition*, Sept. 1997). For this experiment, X was found to have a mean of 1. Suppose one of the readers in the experiment is randomly selected and assume that X has a Poisson distribution.

    a.   Find $P(x = 0)$.
    b.   Find $P(x > 1)$.
    c.   Find $P(x \le 2)$.

Solution to Problem 4.83, page 215 (Eye fixation problem), part a.

Function Arguments dialog box (POISSON):
- X = 0 → = 0
- Mean = 1 → = 1
- Cumulative = FALSE → = FALSE
- = 0.367879441

Returns the Poisson distribution.

Cumulative is a logical value: for the cumulative Poisson probability, use TRUE; for the Poisson probability mass function, use FALSE.

Formula result = 0.367879441

Help on this function   [OK]   [Cancel]

Solution to Problem 4.83, part b

$P(x > 1) = 1 - 0.735759 = 0.264241$

	B	C	D	E	F	G	H	I	J

Function Arguments ? ×

POISSON

X	1	= 1
Mean	1	= 1
Cumulative	TRUE	= TRUE

= 0.735758882

Returns the Poisson distribution.

Cumulative is a logical value: for the cumulative Poisson probability, use TRUE; for the Poisson probability mass function, use FALSE.

Formula result = 0.735758882

Help on this function   OK   Cancel

Solution to Problem 4.83, part c.

	B	C	D	E	F	G	H	I	J

Function Arguments ? ×

POISSON

X	2	= 2
Mean	1	= 1
Cumulative	TRUE	= TRUE

= 0.919698603

Returns the Poisson distribution.

Cumulative is a logical value: for the cumulative Poisson probability, use TRUE; for the Poisson probability mass function, use FALSE.

Formula result = 0.919698603

Help on this function   OK   Cancel

Problem 4.101, page 219 (On-site disposal of waste)

The Resource Conservation and Recovery Act mandates the tracking and disposal of hazardous waste produced at U.S. facilities. *Professional Geographer* (Feb. 2000) reported the hazardous waste generation and disposal characteristics of 209 facilities. Only eight of these facilities treated hazardous waste on-site. Find the probability that 4 of the 10 selected facilities treat hazardous waste.

Solution to Problem 4.101, page 219 (On-site disposal of waste)

	B	C	D	E	F	G	H	I	J

Function Arguments [?] [X]

HYPGEOMDIST

Sample_s   `4`   = 4

Number_sample   `10`   = 10

Population_s   `8`   = 8

Number_pop   `209`   = 209

= 0.000168846

Returns the hypergeometric distribution.

        Number_pop is the population size.

Formula result = 0.000168846

Help on this function   [OK]   [Cancel]

5.1   Introduction

In Chapter 4 of the text, discrete random variables were introduced and the distributions necessary to compute probabilities for these variables (binomial and Poisson distributions) were discussed in detail. In Chapter 5 the continuous random variable will be discussed. The most commonly used continuous random variables are the uniform, normal, and exponential distributions. The conditions under which these may be utilized are described in the text. Remember that a continuous variable is defined as one that can assume an infinite number of values contained in one or more intervals and a continuous distribution is graphically depicted as a smooth curve in which the area under the curve is equivalent to probabilities in the distribution.

Although the text does present statistical tables for the normal and exponential distribution that may be utilized to solve for probabilities, PHStat offers an alternative method for calculating those probabilities.

The following examples will be used to illustrate techniques in PHStat and Excel.

> Example 5.6, page 238 (Cell Phone Application)
> Example 5.7, page 240 (Normal Distribution – Automobile Manufacturer)
> Example 5.10, page 244 (College Entrance Exam)
> Example 5.11, page 250 (EPA Gas Mileage Ratings for 100 Cars)
> Example 5.13, page 264 (Emergency Arrivals at Hospitals)
> Sample Exercise #1: Problem 5.39, page 247 (Transmission delays)
> Sample Exercise #2: Problem 5.62, page 256 (Dart throwing errors)

5.2   Calculating Normal Probabilities using PHStat

The normal probability distribution is the most important probability distribution in statistics. This distribution is used throughout the remainder of the text, so it is a good idea to have a solid understanding of these concepts.

Steps for calculating Normal probabilities:

Step 1: Open a clean workbook.

Step 2: Click on the **Add-Ins** tab.

Step 3: Click on **PHStat**.

Step 4: Drop down to **Probability & Prob. Distributions** and click on **Normal....**

Step 5: A box for the **Normal Probability Distribution** options will appear.
- In the box next to **Mean** enter the mean, μ, of the distribution.
- In the box next to **Standard Deviation** enter the standard deviation, σ, of the distribution.
- There are four choices under **Input Options:**
  - The first choice corresponds to $P(X < specified\ value)$. This is the default choice.
  - The second choice corresponds to $P(X > specified\ value)$.
  - The third choice corresponds to $P(first\ specified\ value < X < second\ specified\ value)$.
  - The fourth choice corresponds to the text examples of "using the normal table in reverse". Enter the percent of area under the normal curve to the left (the percentile) of the point of interest.
- There is a blank box to enter a title. Use this when calculating several probabilities in order to keep the output organized.
- Click the **OK** button and the results will appear in a new sheet labeled **Normal**.

The following examples from the textbook will illustrate the use of the normal distribution.

Please refer to Example 5.7, page 240 (Normal Distribution – Automobile Manufacturer) from the textbook.

Suppose an automobile manufacturer introduces a new model that has an advertised mean in-city mileage of 27 miles per gallon. Although such advertisements seldom report any measure of variability, suppose you write the manufacturer for the details of the tests and you find that the standard deviation is 3 miles per gallon. This information leads you to formulate a probability model for the random variable, X, the in-city mileage for this car model. You believe that the probability distribution of X can be approximated by a normal distribution with a mean of 27 and a standard deviation of 3. If you were to buy this model of automobile, what is the probability that you would purchase one that averages less than 20 miles per gallon for in-city driving? In other words, find P(X < 20).

Solution:

Complete **Step 1** through **Step 4**.

Step 5: A box for the **Normal Probability Distribution** options will appear.

- In the box next to **Mean** enter 27.
- In the box next to **Standard Deviation** enter 3.
- There are four choices under **Input Options**; the first one is the appropriate choice for this example since the problem requests the probability of "less than" 20 miles per gallon. Since it is the default, it will already have a check in front of it. Enter 20 in the box next to it.
- Enter the title as "Text Exercise 5.7".
- Click the **OK** button and the results will appear in a new sheet labeled **Normal**. The answer is given as 0.0098153.

	A	B	C
1	Text Example 5.7		
2			
3	Common Data		
4	Mean	27	
5	Standard Deviation	3	
6			
7	Probability for X <=		
8	X Value	20	
9	Z Value	-2.333333	
10	P(X<=20)	0.0098153	
11			
12			

Please refer to Example 5.6, page 238 (Cell Phone Application) from the textbook.

Assume that the length of time, X, between charges of a cellular phone is normally distributed with a mean of 10 hours and a standard deviation of 1.5 hours. Find the probability that the cell phone will last between 8 and 12 hours between charges.

Solution:

Complete **Step 1** through **Step 4**.

Step 5: A box for the **Normal Probability Distribution** options will appear.

- In the box next to **Mean** enter 10.
- In the box next to **Standard Deviation** enter 1.5.
- There are four choices under **Input Options**; the third one is the appropriate choice for this example since the problem requests the probability of "between" 8 and 12 hours. Click on the box in front of this option so that it has a check by it. Enter 8 in the first box next to it and 12 in the second. Since the first option is the default, you will click on the box in front of it in order to unselect this option.
- Enter the title as "Text Exercise 5.6".
- Click the **OK** button and the results will appear in a new sheet labeled **Normal**. The answer is given as 0.8176.

▲	A	B	C	D	E
1	Text Example 5.6				
2					
3	Common Data				
4	Mean	10			
5	Standard Deviation	1.5			
6				Probability for a Range	
7				From X Value	8
8				To X Value	12
9				Z Value for 8	-1.333333
10				Z Value for 12	1.333333
11				P(X<=8)	0.0912
12				P(X<=12)	0.9088
13				P(8<=X<=12)	0.8176
14					

Please refer to Example 5.10, page 244, (College Entrance Exam) from the textbook.

Suppose the scores, *X*, on a college entrance examination are normally distributed with a mean of 550 and a standard deviation of 100. A certain prestigious university will consider for admission only those applicants whose scores exceed the 90th percentile of the distribution. Find the minimum score an applicant must achieve in order to receive consideration for admission to the university.

Solution:

Complete **Step 1** through **Step 4**.

Step 5: A box for the **Normal Probability Distribution** options will appear.

- In the box next to **Mean** enter 550.
- In the box next to **Standard Deviation** enter 100.
- There are four choices under **Input Options**; the fourth one is the appropriate choice for this example since the problem requests a random variable value and not a probability. Click on the

box in front of this option so that it has a check by it. Enter 90, the given percentile, in box next to it. Since the first option is the default, click on the box in front of it in order to unselect this option.

- Enter the title as "Text Exercise 5.10".
- Click the **OK** button and the results will appear in a new sheet labeled **Normal**. The answer is given as 678.1552. So the minimum score an applicant must achieve in order to receive consideration for admission into a prestigious university is 679.

	A	B	C	D	E
1	Text Example 5.10				
2					
3	Common Data				
4	Mean	550			
5	Standard Deviation	100			
6				Find X and Z Given Cum. Pctage.	
7				Cumulative Percentage	90.00%
8				Z Value	1.281552
9				X Value	678.1552
10					

5.3 Calculating Normal Probabilities using Excel

Again please refer to Example 5.7, page 240 (Normal Distribution – Automobile Manufacturer) from the textbook.

Steps for calculating Normal probabilities:

Step 1: Open a clean workbook.

Step 2: Click on the **Formulas** tab.

Step 3: Click on **More Functions**.

Step 4: Click on **Statistical** and drop down to **NORMDIST** and click on it.

Step 5: A box for **Function Arguments** options will appear.

There are four boxes in which to enter data:

- **X** is value of interest. For this example this is 20.
- **Mean** is the given mean of the distribution. For this example this is 27.
- **Standard_dev** is the standard deviation of the distribution. For this example this is 3.
- **Cumulative** is either true or false. Enter **TRUE** if the cumulative probability, *P(X < specified value of interest)*, is to be used. Enter **FALSE** if only the probability mass function is desired. For exercises such as those in Chapter 5 of the text, only cumulative probabilities would be useful, so enter TRUE.
- After all four values are entered, the answer will appear within the **Function Arguments** box. Click the **OK** button and Excel will also enter the answer into the active cell of the spreadsheet.

The answer is P(X < 20) = 0.009815329.

Note that in this example the cumulative probability P(X < 20) was desired, so no additional calculations were necessary. If instead an exercise requested P(X > some value of interest), then the user would have to subtract the Excel output from 1 to arrive at the correct answer. Similarly, if the exercise requested P(first value of interest < X < second value interest), then the user would actually have to compute P(X < second value of interest) and P(X< first value of interest) and then subtract one from the other. Excel also has an **NORMINV** function that will solve problems such as the text Exercise 5.10, but when computing normal probabilities, using the PHStat function is easier to use.

5.4   Assessing the Normality of a Data Set using PHStat

In order to apply the normal distribution to calculate probabilities of continuous random variables, it is necessary to determine if the data are approximately normal. One may never assume normality. PHStat contains a method for determining whether the data, once plotted, is normal or approximately normal. Assessing normality should always be done prior to utilizing the normal distribution to solve for probabilities.

Please refer to Example 5.11, page 250 (EPA Gas Mileage Ratings for 100 Cars) from the text. The example contains data on EPA gas mileage ratings for 100 cars. Using this data set, a normal probability plot will be constructed. Are the data normal or approximately normal?

Steps for assessing normality:

Step 1: Open a new workbook and place the cursor in the upper left-hand corner of the spreadsheet.

Step 2: Input the data manually into the first column. Include a label in the first row.

	A	B
1	MPG	
2	36.3	
3	41	
4	36.9	
5	37.1	
6	44.9	
7	36.8	
8	30	
9	37.2	

Step 3: Click on the **Add-Ins** tab.

Step 4: Click on **PHStat**.

Step 5: Drop down to **Probability & Prob. Distributions** and click on **Normal Probability Plot....**.

Step 6: A box for the **Normal Probability Plot** options will appear.

- Click on the entry box by **Variable Cell Range**. Enter the rows and column(s) where the data is located. This may be done by typing the range in manually or by clicking and dragging the appropriate range of data cells. In this example the data is located in Sheet1!$A$1:$A$101.
- Within the options box, the default setting is **First cell contains label** (one would click on this box to unselect this option if there were no label for the data set).
- There is a blank box to enter a title. This is recommended, particularly when working with several different data sets.

- Click the **OK** button and the results will appear in a new sheet labeled **NormalPlot**.

The following figure presents the normality plot for the EPA mileage ratings on 100 cars. Notice that PHStat arranged the data from smallest to largest and then plotted each data point against its expected z score.

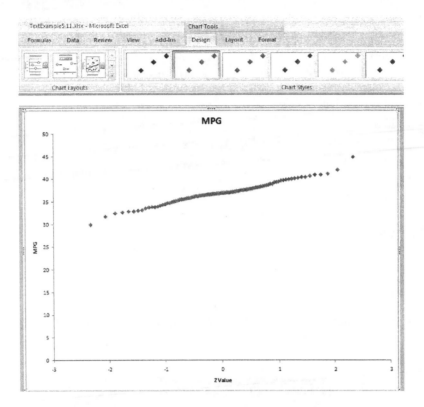

Refer to pages 250 to 252 of the text for an explanation on how to interpret the plot.

5.5   Calculating Exponential Probabilities using PHStat

Often referred to as the "waiting time distribution", the exponential probability distribution is best used to describe the amount of time or distance between occurrences of random events.

Please refer to Example 5.13, page 264, (Emergency Arrivals at Hospitals) from the text.

Suppose the length of time (in hours) between emergency arrivals at a certain hospital is modeled as an exponential distribution with $\theta = 2$. What is the probability that more than 5 hours pass without an emergency room arrival?

Steps for calculating exponential probabilities:

Step 1: Open a clean workbook.

Step 2: Click on the **Add-Ins** tab.

Step 3: Click on **PHStat**.

Step 4: Drop down to **Probability & Prob. Distributions** and click on **Exponential…**.

Step 5: A box for the **Exponential Probability Distribution** options will appear.

- In the box next to **Mean per unit (Lambda)** enter the inverse of the mean ($1/\theta$ by the text's notation) of the distribution.
- In the box next to **X Value** enter the value of interest.
- There is a blank box to enter a title. Use this when calculating several probabilities in order to keep the output organized.
- Click the **OK** button and the results will appear in a new sheet labeled **Exponential**. The answer given is the cumulative probability from 0 to the value of interest.

◢	A	B
1	Text Example 5.13	
2		
3	Data	
4	Mean	0.5
5	X Value	5
6		
7	Results	
8	P(<=X)	0.9179
9		

PHStat calculated the probability for P $(X \leq 5)$ = 0.9179. In order to find the probability of P $(X > 5)$; subtract this value from one. Thus, P $(X > 5)$ = 1 - P $(X \leq 5)$ = 1-0.9179 = 0.0821.

5.6   Sample Exercises

The following exercises were selected from the textbook to test the reader's understanding of PHStat and its usage within the chapter. Try to replicate the output in PHStat for the following problems.

Problem 5.39, page 247 (Transmission delays in wireless technology)

Resource Reservation Protocol (RSVP) was originally designed to establish signaling links for stationary networks. In *Mobile Networks and Applications* (Dec. 2003), RSVP was applied to mobile wireless technology (e.g., a PC notebook with wireless LAN card for Internet access). A simulation study revealed that the transmission delay (measured in milliseconds) of an RSVP linked wireless device has an approximate normal distribution with mean μ = 48.5 milliseconds and σ = 8.5 milliseconds.
a. What is the probability that the transmission delay is less than 57 milliseconds?
b. What is the probability that the transmission delay is between 40 and 60 milliseconds?

Solution to Problem 5.39 part a:

◢	A	B	C
1	Exercise 5.39a		
2			
3	Common Data		
4	Mean	48.5	
5	Standard Deviation	8.5	
6			
7	Probability for X <=		
8	X Value	57	
9	Z Value	1	
10	P(X<=57)	0.8413447	
11			

Solution to Problem 5.39 part b:

	A	B	C	D	E
1	Exercise 5.39b				
2					
3	Common Data				
4	Mean	48.5			
5	Standard Deviation	8.5			
6				Probability for a Range	
7				From X Value	40
8				To X Value	60
9				Z Value for 40	-1
10				Z Value for 60	1.352941
11				P(X<=40)	0.1587
12				P(X<=60)	0.9120
13				P(40<=X<=60)	0.7533
14					

Problem 5.62, page 256 (Dart throwing errors)

How accurate are you at the game of darts? Researchers at Iowa State University attempted to develop a probability model for dart throws (*Chance*, Summer 1997). For each of 590 throws made at certain targets on a dartboard, the distance from the dart to the target point was measured (to the nearest millimeter). The error distribution for the dart throws is described by the frequency table shown in the text. Construct a normal probability plot for the data and use it to assess whether the error distribution is approximately normal.

Solution to Problem 5.62

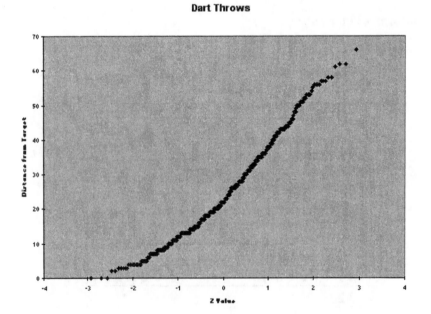

Dart Throws

90

6.1   Introduction

Chapter 6 in the textbook begins to introduce the basic properties of a sampling distribution.   The sampling distribution is used to estimate a population parameter such as the population mean (μ) or the population proportion (p). In this section, PHStat will be used to derive various probabilities using the sampling distribution of the sample mean.   In the previous chapter, the Normal Probability Distribution was discussed and probabilities were calculated using specific examples from the textbook. Be sure to reread that section before continuing with the following examples.

The following examples will be used to illustrate techniques in PHStat and Excel.

> Example 6.8, page 292, (Automobile Batteries)
> Sample Exercise #1: Problem 6.36, page 296 (Research on eating disorders)

6.2   Calculating Probabilities Involving the Sampling Distribution of $\bar{x}$ using PHStat

Recall from the text that the sampling distribution of the sample mean is a probability distribution of possible sample means of a given sample size. The most important tool from thus chapter is the Central Limit Theorem (see page 290 of the text). According to this theorem, if samples of size n are sufficiently large, then the sampling distribution of the sample mean is approximately a normal distribution. Although the approximation improves as the sample size increases, this theorem allows the researcher to make statements about the population using a sample, without any information about the shape of the original distribution from which the sample is taken.

Using the Central Limit Theorem, probabilities for sample distributions may be estimated using the **Normal Probability Distribution** function within PHStat. The steps for calculating these probabilities are basically the same as those for calculating normal probabilities from Chapter 5, but with one very important difference: the user must first determine the parameters for the sampling distribution.

Steps for calculating probabilities for problems that make use of the Central Limit Theorem:

Step 1: Open a clean workbook.

Step 2: Click on the **Add-Ins** tab.

Step 3: Click on **PHStat**.

Step 4: Drop down to **Probability & Prob. Distributions** and click on **Normal....**

Step 5: A box for the **Normal Probability Distribution** options will appear.
- In the box next to **Mean** enter the mean of the distribution. For these types of problems that will be μ since the mean of the sampling distribution will be the same as the mean of the population.
- In the box next to **Standard Deviation** enter the standard deviation of the distribution. For these types of problems that will be $\sigma/\sqrt{n}$ since the standard deviation of the sampling distribution will be the standard deviation of the population divided by the square root of n.

- There are four choices under **Input Options:**
  - The first choice corresponds to *P(X < specified value)*. This is the default choice.
  - The second choice corresponds to *P(X > specified value)*.
  - The third choice corresponds to *P(first specified value < X < second specified value)*.
  - The fourth choice corresponds to the text examples of "using the normal table in reverse". Enter the percent of area under the normal curve to the left (the percentile) of the point of interest.
- There is a blank box to enter a title. Use this when calculating several probabilities in order to keep the output organized.
- Click the **OK** button and the results will appear in a new sheet labeled **Normal**.

The following examples from the textbook will illustrate the use of the normal distribution.

Please refer to Example 6.8, page 292, (Automobile Batteries) from the textbook.

A manufacturer of automobile batteries claims that the distribution of the lengths of life of its best battery has a mean of 54 months and a standard deviation of 6 months. Suppose a consumer group decides to check the claim by purchasing a sample of 50 of these batteries and subjecting them to tests that determine battery life. Assuming that the manufacturer's claim is true, what is the probability the consumer group's sample has a mean life of 52 or fewer months?

Solution:

Complete **Step 1** through **Step 4**.

Step 5: A box for the **Normal Probability Distribution** options will appear.

- In the box next to **Mean** enter 54.
- In the box next to **Standard Deviation** enter 0.848528 since $6/\sqrt{50}$ is approximately 0.848528.
- There are four choices under **Input Options**; the first one is the appropriate choice for this example since the problem requests the probability of 52 "or fewer". Since it is the default, it will already have a check in front of it. Enter 52 in the box next to it.
- Enter the title as "Text Exercise 6.8".

- Click the **OK** button and the results will appear in a new sheet labeled **Normal**. The answer is given as 0.0092111. Thus, the probability the consumer group will observe a sample mean of 52 or less is only 0.0092111 if the manufacturer's claims is true.

	A	B
	Book1	
1	Text Exercise 6.8	
2		
3	**Common Data**	
4	Mean	54
5	Standard Deviation	0.848528
6		
7	**Probability for X <=**	
8	X Value	52
9	Z Value	-2.357023
10	P(X<=52)	0.0092111
11		
12		
13		

Note that as in Chapter 5, the **NORMDIST** function in Excel can be used to solve the problems in Chapter 6, but using PHStat is easier.

6.3   Sample Exercises

The following exercises were selected from the textbook to test the reader's understanding of PHStat and its usage within the chapter. Try to replicate the output in PHStat for the following problems.

Problem 6.36, page 296 (Research on eating disorders)

Refer to *The American Statistician* (May, 2001) study of female students who suffer from bulimia. Recall that each student completed a questionnaire from which a "fear of negative evaluation" (FNE) score was produced. (The higher the score, the greater the fear of negative evaluation.) Suppose the FNE scores of bulimic students have a distribution with mean of 18 and a standard deviation of 5. Now consider a random sample of 45 female students with bulimia.
   a. What is the probability that the sample mean FNE score is greater than 17.5?
   b. What is the probability that the sample mean FNE score is between 18 and 18.5?
   c. What is the probability that the sample mean FNE score is less than 18.5?

Solution to Problem 6.36

Solution for Exercise 6.36 part a	
Mean	18
Standard Deviation	0.7454
Probability for X>	17.5
Z Value	-0.670780789
P(X>17.5)	0.748819977

Solution for Exercise 6.36 part b		
Mean	18	
Standard Deviation	0.7454	
Probability for range	18	<= X <= 18.5
Z Value for 18	0	
Z Value for 18.5	0.670780789	
P(X<=18)	0.5	
P(X<=18.5)	0.748819977	
P(18<=X<=18.5)	0.248819977	

Solution for Exercise 6.36 part c	
Mean	18
Standard Deviation	0.7454
Probability for X<=	18.5
Z Value	0.670780789
P(X<=18.5)	0.748819977

7.1   Introduction

In Chapter 7 of the textbook, population means and proportions are estimated using statistics generated from a single sample. Instead of simply stating the point estimates for a population parameter, the user may also quantify the "confidence" in the estimate. According to the Central Limit Theorem, the sampling distribution of the sample mean is normally distributed (for large sample sizes). Based on this assumption, an interval can be created to determine how "close" the sample statistic is to the population parameter. In this chapter, examples will be used to demonstrate

1. Estimating population means
2. Estimating population proportions
3. Calculating the sample size necessary to estimate population means and proportions

PHStat can be used to calculate confidence intervals for population means, μ, and population proportions, p. When calculating intervals estimating population means, remember to determine whether:

1. The population standard deviation, σ, is known
2. The population standard deviation, σ, is unknown

When the population standard deviation is unknown, the confidence interval for the population mean can be estimated using statistics generated from the sample data.

The following examples will be used to demonstrate techniques in PHStat and Excel.

Exercise 7.13, page 314 (Latex Allergy)
Example 7.2, page 320 (Printhead Failure)
Example 7.4, page 328 (BEBR Survey).
Example 7.6, page 335 (NFL Footballs)
Example 7.7, page 337 (Telephone Manufacturer – Customer Complaints)
Sample Exercise #1: Problem 7.43, page 325 (Study on waking sleepers early)
Sample Exercise #2: Problem 7.100, page 344 (Role importance for the elderly)
Sample Exercise #3: Problem 7.77, page 339 (Bacteria in bottled water)

7.2   Estimation of the Population Mean: Sigma Known (or large sample size)

When calculating the confidence interval for a population mean, one must determine whether the Normal distribution or the t-distribution is appropriate. PHStat gives the user two options when computing a confidence interval for a mean: **sigma known** (uses the Normal distribution) or **sigma unknown** (uses the t-distribution). Typically the user knows s, but does not know σ, however, if n is large enough one may assume that the sampling distribution is Normal and use the z-score in the formula for the confidence interval (see the note at the bottom of page 309 of the text).

The following example from the text will be used to illustrate how to calculate a confidence interval for a mean. Please refer to page 314 of the text, Exercise 7.13.

Health-care workers who use latex gloves with glove powder on a daily basis are particularly susceptible to developing a latex allergy. Symptoms of a latex allergy include conjunctivitis, hand eczema, nasal congestion, a skin rash, and shortness of breath. Each in a sample of 46 hospital employees who were diagnosed with latex allergy based on a skin-prick test reported on their exposure to latex gloves. (*Current Allergy & Clinical Immunology*, March 2004.) Summary statistics for the number of latex gloves used per week are $\bar{x} = 19.3$ and $s = 11.9$.

Steps for calculating a confidence interval for a population mean, sigma known:

Step 1: Open a clean workbook.

Step 2: Click on the **Add-Ins** tab.

Step 3: Click on **PHStat**.

Step 4: Drop down to **Confidence Intervals** and click on **Estimate for the Mean, sigma known....**

Step 5: A box for the **Estimate for the Mean, sigma known** options will appear.

- In the box next to **Population Standard Deviation** enter the standard deviation, σ, of the population. If n is large, then the sample standard deviation may be used if σ is not given. For this problem use 11.9, the sample standard deviation.
- In the box next to **Confidence Level** the default is 95%. Other values of the confidence level may be entered if needed. For this example it is 95%.
- There are two choices under **Input Options:**
    - The first choice is **Sample Statistics Known**. This is the default choice. Since the summary statistics are given in this problem, enter the **Sample Size** as 46 and the **Sample Mean** as 19.3.
    - The second choice is **Sample Statistics Unknown**. This would be used if the problem provided raw data as opposed to summary statistics. The example presented in the next section demonstrates use of this option.
- There is a blank box to enter a title. Use this when calculating several confidence intervals in order to keep the output organized.
- Click the **OK** button and the results will appear in a new sheet labeled **Confidence**.

	A	B
1	Exercise 7.13	
2		
3	Data	
4	Population Standard Deviation	11.9
5	Sample Mean	19.3
6	Sample Size	46
7	Confidence Level	95%
8		
9	Intermediate Calculations	
10	Standard Error of the Mean	1.754559278
11	Z Value	-1.95996398
12	Interval Half Width	3.438872994
13		
14	Confidence Interval	
15	Interval Lower Limit	15.86112701
16	Interval Upper Limit	22.73887299
17		

The 95% confidence interval for the mean number of latex gloves used per week by health-care workers is (15.86112701, 22.73887299)

7.3   Estimation of the Population Mean: Sigma Unknown

Although certain conditions must still be met (refer to Section 7.3 of the text), the confidence interval for a mean may still be constructed if σ is unknown and n is small.

The following example from the text will be used to illustrate how to calculate a confidence interval for a mean if σ is unknown. Please refer to page 320 of the text, Example 7.2 (Printhead Failure).

Some quality control experiments require *destructive sampling* (i.e. the test to determine whether the item is defective destroys the item) in order to measure a particular characteristic of the product. The cost of destructive sampling often dictates small samples.   Suppose a manufacturer of printers for personal computers wishes to estimate the mean number of characters printed before the printhead fails.   The printer manufacturer tests n = 15 printheads and records the number of characters printed until failure for each. These 15 measurements (in millions of characters) are listed in Table 7.5. Form a 99% confidence interval for the mean number of characters printed before the printhead fails.

Steps for calculating a confidence interval for a population mean, sigma unknown:

Step 1: Open a clean workbook and enter the raw data into column A.

	A
1	1.13
2	1.55
3	1.43
4	0.92
5	1.25
6	1.36
7	1.32
8	0.85
9	1.07
10	1.48
11	1.2
12	1.33
13	1.18
14	1.22
15	1.29
16	
17	

Step 2: Click on the **Add-Ins** tab.

Step 3: Click on **PHStat**.

Step 4: Drop down to **Confidence Intervals** and click on **Estimate for the Mean, sigma unknown**....

Step 5: A box for the **Estimate for the Mean, sigma unknown** options will appear.

- In the box next to **Confidence Level** the default is 95%. Other values of the confidence level may be entered if needed. For this example it is 99%.
- There are two choices under **Input Options:**
    - The first choice is **Sample Statistics Known**. This is the default choice.
    - The second choice is **Sample Statistics Unknown**. This option will be used for this example since the raw data is given instead of the summary statistics. Click on the circle in front of this option. Click on the entry box by **Sample Statistics Unknown**. Enter the rows and column(s) where the data are located. This may be done by typing the range in manually or by clicking and dragging the appropriate range of data cells. In this example the data are located in Sheet1!$A$1:$A$15. Notice that the box next to **First cell contains label** will be checked. Click on this box to unselect this option.
- There is a blank box to enter a title. Use this when calculating several confidence intervals in order to keep the output organized.
- Click the **OK** button and the results will appear in a new sheet labeled **Confidence**.

	A	B
1	Text Example 7.2	
2		
3	Data	
4	Sample Standard Deviation	0.19316413
5	Sample Mean	1.238666667
6	Sample Size	15
7	Confidence Level	99%
8		
9	Intermediate Calculations	
10	Standard Error of the Mean	0.049874764
11	Degrees of Freedom	14
12	t Value	2.976842734
13	Interval Half Width	0.148469328
14		
15	Confidence Interval	
16	Interval Lower Limit	1.09
17	Interval Upper Limit	1.39
18		

The confidence interval for the mean number of characters is (1.09, 1.39).

7.4   Estimation of the Population Proportion

When certain conditions are met (see page 327 of the text), one may calculate the confidence interval for a population proportion.

The following example from the text will be used to illustrate how to calculate a confidence interval for a proportion. Please refer to page 328 of the text, Example 7.4 (BEBR Survey).

Many polling agencies conduct surveys to determine the current consumer sentiment concerning the state of the economy. For example, the Bureau of Economic and Business Research (BEBR) at the University of Florida conduct s quarterly surveys to gauge consumer sentiment in the Sunshine State. Suppose that BEBR randomly samples 484 consumers and finds that 257 are optimistic about the state of the economy. Use a 90% confidence interval to estimate the proportion of all consumers in Florida who are optimistic about the state of the economy.

Steps for calculating a confidence interval for a population proportion:

Step 1: Open a clean workbook.

Step 2: Click on the **Add-Ins** tab.

Step 3: Click on **PHStat**.

Step 4: Drop down to **Confidence Intervals** and click on **Estimate for the Proportion....**

Step 5: A box for the **Estimate for the Proportion** options will appear.

- In the box next to **Sample Size** enter n. For this problem enter 484, the number of randomly chosen consumers.
- In the box next to **Number of Successes** enter x. For this problem enter 257, the number of optimistic consumers.
- In the box next to **Confidence Level** the default is 95%. Other values of the confidence level may be entered if needed. For this example it is 90%.

- There is a blank box to enter a title. Use this when calculating several confidence intervals in order to keep the output organized.
- Click the **OK** button and the results will appear in a new sheet labeled **Confidence**.

	A	B
1	**Text Example 7.4**	
2		
3	**Data**	
4	**Sample Size**	484
5	**Number of Successes**	257
6	**Confidence Level**	90%
7		
8	Intermediate Calculations	
9	Sample Proportion	0.530991736
10	Z Value	-1.64485363
11	Standard Error of the Proportion	0.022683572
12	Interval Half Width	0.037311156
13		
14	**Confidence Interval**	
15	**Interval Lower Limit**	0.49368058
16	**Interval Upper Limit**	0.568302892
17		

The 90% confidence interval for the proportion of Florida consumers who are optimistic about the economy is (0.49368058, 0.568302892).

Note that if it is necessary to use Wilson's adjustment for estimating p (see page 329 of the text), the PHStat steps outlined above may still be used. Two substitutions would have to be made in **Step 5**: enter $n + 4$ instead of n and $x + 2$ instead of x.

7.5   Determining the Sample Size Necessary for Estimating a Mean

In Section 7.5 of the textbook, the authors discuss the importance of determining the sample size necessary for making an inference about a population parameter if a specific level of confidence is required.

The following example from the text will be used to illustrate how to determine the sample size necessary to estimate the population mean to within a certain specification. Please refer to page 335 of the text, Example 7.6 (NFL Footballs).

Suppose the manufacturer of official NFL footballs uses a machine to inflate the new balls to a pressure of 13.5 pounds. When the machine is properly calibrated, the mean inflation pressure is 13.5 pounds but uncontrollable factors cause the pressures of individual footballs to vary randomly from about 13.3 to 13.7 pounds.   For quality control purposes, the manufacturer wishes to estimate the mean inflation pressure to within 0.025 pound of its true value with a 99% confidence interval. What sample size should be specified for this experiment?

Steps for determining the sample size for a population mean:

Step 1: Open a clean workbook.

Step 2: Click on the **Add-Ins** tab.

Step 3: Click on **PHStat**.

Step 4: Drop down to **Sample Size** and click on **Determination for the Mean…**.

Step 5: A box for the **Determination for the Mean** options will appear.

- In the box next to **Population Standard Deviation** enter the standard deviation, σ, of the population. Most of the time this value will be unknown (see page 335 of the text). In this problem it is estimated by calculating the range and dividing that by 4, therefore enter 0.1.
- In the box next to **Sampling Error** enter the SE given in the problem. In this problem it is 0.025.
- In the box next to **Confidence Level** the default is 95%. Other values of the confidence level may be entered if needed. For this example it is 99%.
- There is a blank box to enter a title. Use this when calculating several sample sizes in order to keep the output organized.
- Click the **OK** button and the results will appear in a new sheet labeled **Sample Size**.

	A	B
1	Text Example 7.6	
2		
3	Data	
4	Population Standard Deviation	0.1
5	Sampling Error	0.025
6	Confidence Level	99%
7		
8	Intemediate Calculations	
9	Z Value	-2.5758293
10	Calculated Sample Size	106.1583456
11		
12	Result	
13	Sample Size Needed	107
14		

Therefore, the sample size needed to estimate the mean inflation pressure to within 0.025 pounds of its true value with 99% confidence is 107 footballs. Note that "rounding up" as the text does in this example is not necessary; PHStat has already calculated the necessary sample size.

7.6 Determining the Sample Size Necessary for Estimating a Proportion

The following example from the text will be used to illustrate how to determine the sample size necessary to estimate the population proportion to within a certain specification. Please refer to page 337 of the text, Example 7.7 (Telephone Manufacturer – Customer Complaints).

Suppose a large telephone manufacturer that entered the post-regulation market quickly has an initial problem with excessive customer complaints and consequent returns of the phones for repair or replacement. The manufacturer wants to estimate the magnitude of the problem in order to design a quality control program. How many telephones should be sampled and checked in order to estimate the fraction defective, p, to be within 0.01 with 90% confidence?

Steps for determining the sample size for a population proportion:

Step 1: Open a clean workbook.

Step 2: Click on the **Add-Ins** tab.

Step 3: Click on **PHStat**.

Step 4: Drop down to **Sample Size** and click on **Determination for the Proportion....**

Step 5: A box for the **Determination for the Proportion** options will appear.

- In the box next to **Estimate of True Proportion** enter an estimate for p. Although not stated in the original problem, in the solution portion in the text it is stated that a reasonably conservative estimate would be 10% defective. Therefore enter 0.10.
- In the box next to **Sampling Error** enter the SE given in the problem. In this problem it is 0.01.
- In the box next to **Confidence Level** the default is 95%. Other values of the confidence level may be entered if needed. For this example it is 90%.
- There is a blank box to enter a title. Use this when calculating several sample sizes in order to keep the output organized.
- Click the **OK** button and the results will appear in a new sheet labeled **Sample Size**.

	A	B
1	Text Example 7.7	
2		
3	Data	
4	Estimate of True Proportion	0.1
5	Sampling Error	0.01
6	Confidence Level	90%
7		
8	Intermediate Calculations	
9	Z Value	-1.64485363
10	Calculated Sample Size	2434.989109
11		
12	Result	
13	Sample Size Needed	2435
14		
15		

PHStat calculates the necessary sample size to be 2,435. Notice, however, that the book solution is 2,436. This slight difference is attributed to the fact that PHStat calculates a more accurate z-value (1.64485363) than the textbook table ($z = 1.645$). Therefore, 2,435 telephones should be sampled and checked in order to estimate the proportion of defectives to be within 0.01 with 90% confidence.

7.7 Sample Exercises

The following exercises were selected from the textbook to test the reader's understanding of PHStat and its usage within the chapter. Try to replicate the output in PHStat for the following problems.

Problem 7.43, page 325 (Study on waking sleepers early)

Scientists have discovered increased levels of the hormone adrenocorticotropin in people just before they awake from sleeping (*Nature*, Jan. 7, 1999). In the study, 15 subjects were monitored during their sleep after being told they would be woken at a particular time. One hour prior to the designated wake-up time, the adrenocorticotropin level (pg/ml) was measured in each, with a mean of 37.3 and a standard deviation of 13.9. Estimate the true mean adrenocorticotropin level of sleepers one-hour prior to waking using a 95% confidence level.

Solution for Problem 7.43

Solution for Exercise 7.43	
Sample Standard Deviation	13.9
Sample Mean	37.3
Sample Size	15
Confidence Level	95%
Standard Error of the Mean	3.588964567
Degrees of Freedom	14
t Value	2.144788596
Interval Half Width	7.697570274
Interval Lower Limit	29.60
Interval Upper Limit	45.00

Problem 7.100, page 325 (Role importance for the elderly)

Refer to the *Psychology and Aging* (December 2000) study of the roles that elderly people feel are the most important to them, presented in Exercise 2.10 (p. 35). Recall that in a national sample of 1,102 adults 65 years or older, 424 identified "spouse" as their most salient role. Use a 95% confidence interval to estimate the true percentage of adults 65 years or older who feel that being a spouse is their most salient role. Give a practical interpretation of this interval.

	A	B
1	Text Exercise 7.100	
2		
3	Data	
4	Sample Size	1102
5	Number of Successes	424
6	Confidence Level	95%
7		
8	Intermediate Calculations	
9	Sample Proportion	0.384754991
10	Z Value	-1.95996398
11	Standard Error of the Proportion	0.014656335
12	Interval Half Width	0.028725889
13		
14	Confidence Interval	
15	Interval Lower Limit	0.356029102
16	Interval Upper Limit	0.41348088
17		

Problem 7.77, page 339 (Bacteria in bottled water)

Is the bottled water you drink safe? According to an article in *U.S. News & World Report* (April 12, 1999), the Natural Resources Defense Council warns that the bottled water you are drinking may contain more bacteria and other potentially carcinogenic chemicals than allowed by state and federal regulators. Of the more than 1,000 bottles studied, nearly one-third exceeded government levels. Suppose that the

Natural Resources Defense Council wants an updated estimate of the population proportion of bottled water that violates at least one government standard. Determine the sample size (number of bottles) needed to estimate this proportion to be within ±0.01 with 99% confidence.

Solution for Problem 7.77

Solution for Exercise 7.77	
Estimate of True Proportion	**0.3333**
Sampling Error	**0.01**
Confidence Level	**99%**
Z Value	-2.575834515
Calculated Sample Size	14743.53704
Sample Size Needed	**14744**

8.1   Introduction

In Chapter 8 of the textbook, the authors define hypothesis testing as the utilization of sample information in order to test what the true value of the population mean, μ, population proportion, p, or population variance, σ^2, may be. From a single sample, a researcher may infer characteristics about the population using a specific level of error or level of significance. Remember that the procedures for conducting hypothesis tests on large or small samples are basically the same; however, the probability distributions necessary for determining the critical values that separate the rejection or not rejection regions are different.

PHStat can be used to conduct hypothesis tests for population means, μ, population proportion, p, and population variance, σ^2. When calculating intervals estimating population means, remember to determine whether:
1.   The population standard deviation, σ, is known
2.   The population standard deviation, σ, is unknown

When the population standard deviation is unknown, the test statistic can be estimated using statistics generated from the sample data.

In order to estimate a population proportion, p, enter both the number of successes and the sample size into the PHStat menu. At this time, PHStat does not provide an option for entering a specific data set to use to estimate proportions.

The following examples will be used to illustrate techniques in PHStat and Excel.

Exercise 8.27, page 364 (Heart rate during laughter)
Example 8.5, page 374 (Air pollution standards)
Example 8.7 and 8.8, pages 381&382 (Alkaline batteries and defective shipments)
Exercise 8.113, page 397 (Latex allergy in health-care workers)
Sample Exercise #1: Problem 8.35, page 365 (Social interaction of mental patients)
Sample Exercise #2: Problem 8.91, page 386 (Testing the placebo effect)

8.2   Tests of Hypothesis of a Population Mean: Sigma Known (or large sample size)

When conducting a hypothesis test of the population mean, one must determine whether it is appropriate to use the Normal distribution or the t-distribution. PHStat gives the user two options when conducting a one sample hypothesis test of the mean: **sigma known** (uses the Normal distribution) or **sigma unknown** (uses the t-distribution). Typically the user knows s, but does not know σ. However, if n is large enough one may assume that the sampling distribution is Normal (see the note at the bottom of page 362 of the text). In the majority of cases the population standard deviation, σ, is not known and it will be necessary to substitute the sample standard deviation, s, as its estimate.

The following example from the text will be used to illustrate how to conduct a one sample hypothesis test of the mean. Please refer to page 364 of the text, Exercise 8.27.

Laughter is often called "the best medicine" since studies have shown that laughter can reduce muscle tension and increase oxygenation of the blood. In the *International Journal of Obesity* (Jan. 2007), researchers at Vanderbilt University investigated the physiological changes that accompany laughter. Ninety subjects (18 – 34 years old) watched film clips designed to evoke laughter. During the laughing period, the researchers measured the heart rate (beats per minute) of each subject with the following summary results: $\bar{x} = 73.5$, $s = 6$. It is well known that the mean resting heart rate is 71 beats per minute. At $\alpha = 0.05$, is there sufficient evidence to indicate the true mean heart rate during laughter exceeds 71 beats per minute?

Steps for a conducting a one sample hypothesis test of the population mean, sigma known:

Step 1: Open a clean workbook.

Step 2: Click on the **Add-Ins** tab.

Step 3: Click on **PHStat**.

Step 4: Drop down to **One-Sample Tests** and click on **Z Test for the Mean, sigma known....**

Step 5: A box for the **Z Test for the Mean, sigma known** options will appear.

- In the box next to **Null Hypothesis** enter the hypothesized mean. For this problem that is 71.
- In the box next to **Level of Significance** enter the value of α. For this example it is 0.05.
- In the box next to **Population Standard Deviation** enter the standard deviation, σ, of the population. If n is large, then the sample standard deviation may be used if σ is not given. For this problem enter 6, the sample standard deviation.
- There are two choices under **Sample Statistics Options:**
  - The first choice is **Sample Statistics Known**. This is the default choice. Since the summary statistics are given in this problem, enter the **Sample Size** as 90 and the **Sample Mean** as 73.5.
  - The second choice is **Sample Statistics Unknown**. This would be used if the problem provided raw data as opposed to summary statistics. The example presented in the next section demonstrates use of this option.
- There are three choices under **Test Options** (refer to page 359 of the text for a description):
  - The first choice is **Two-Tail Test**. This is the default choice.
  - The second choice is **Upper-Tail Test**. Click on the circle in front of this option since this problem attempts to determine whether the true mean *exceeds* 71.
  - The third choice is **Lower-Tail Test**.
- There is a blank box to enter a title. Use this when calculating several confidence intervals in order to keep the output organized.
- Click the **OK** button and the results will appear in a new sheet labeled **Hypothesis**.

	A	B	C
1	Text Exercise 8.27		
2			
3	Data		
4	Null Hypothesis μ=	71	
5	Level of Significance	0.05	
6	Population Standard Deviation	6	
7	Sample Size	90	
8	Sample Mean	73.5	
9			
10	Intermediate Calculations		
11	Standard Error of the Mean	0.632455532	
12	Z Test Statistic	3.952847075	
13			
14	Upper-Tail Test		
15	Upper Critical Value	1.644853627	
16	p-Value	3.86134E-05	
17	Reject the null hypothesis		
18			

The test statistic is z = 3.952847075 and it exceeds the critical value of 1.644853627.   Since the test statistic lies within the critical region, the conclusion is to reject the null hypothesis.   Therefore, it is determined that the true mean heart rate during laughter does exceed 71 beats per minute. Notice that the PHStat output also provides the p-value. In this problem it is calculated to be 0.0000386134. Since this is less than the level of significance used in this case (α = 0.05), the same conclusion (of course!) is reached: reject the null hypothesis.

8.3   Tests of Hypothesis of a Population Mean: Sigma Unknown

Although certain conditions must still be met (refer to Section 8.4 of the text), hypothesis testing for a mean may still be conducted if σ is unknown and n is small.

The following example from the text will be used to illustrate how to conduct a one sample hypothesis test of the mean. Please refer to page 374 of the text, Example 8.5.

A major car manufacturer wants to test a new engine to determine whether it meets new air-pollution standards. The mean emission μ of all engines of this type must be less than 20 parts per million of carbon. Ten engines are manufactured for testing purposes, and the emission level of each is determined. The data (in parts per million) are listed in Table 8.5. Do the data supply sufficient evidence to allow the manufacturer to conclude that this type of engine meets the pollution standard?   Assume that the manufacturer is willing to risk a Type I error with probability α = 0.01.

Steps for a conducting a one sample hypothesis test of the population mean, sigma unknown:

Step 1: Open a clean workbook.

Step 2: Click on the **Add-Ins** tab.

Step 3: Click on **PHStat**.

Step 4: Drop down to **One-Sample Tests** and click on **t Test for the Mean, sigma unknown....**

Step 5: A box for the **t Test for the Mean, sigma unknown** options will appear.

- In the box next to **Null Hypothesis** enter the hypothesized mean. For this problem that is 20.
- In the box next to **Level of Significance** enter the value of α. For this example it is 0.01.
- There are two choices under **Sample Statistics Options:**
  - The first choice is **Sample Statistics Known**. This is the default choice.
  - The second choice is **Sample Statistics Unknown**. This option will be used for this example since the raw data is given instead of the summary statistics. Click on the circle in front of this option. Click on the entry box by **Sample Statistics Unknown**. Enter the rows and column(s) where the data are located. This may be done by typing the range in manually or by clicking and dragging the appropriate range of data cells. In this example the data are located in Sheet1!$A$1:$A$10. Notice that the box next to **First cell contains label** will be checked. Click on this box to unselect this option.
- There are three choices under **Test Options** (refer to pages 373 and 359 of the text for a description):
  - The first choice is **Two-Tail Test**. This is the default choice.
  - The second choice is **Upper-Tail Test.**
  - The third choice is **Lower-Tail Test.** Click on the circle in front of this option since this problem attempts to determine whether this type of engine meets the pollution standard (in other words, mean is *less than* 20).
- There is a blank box to enter a title. Use this when calculating several confidence intervals in order to keep the output organized.
- Click the **OK** button and the results will appear in a new sheet labeled **Hypothesis**.

	A	B	C	D	E
1	Text Example 8.5				
2					
3	Data				
4	Null Hypothesis μ=	20			
5	Level of Significance	0.01			
6	Sample Size	10			
7	Sample Mean	17.57			
8	Sample Standard Deviation	2.952230795			
9					
10	Intermediate Calculations				
11	Standard Error of the Mean	0.933577349			
12	Degrees of Freedom	9			
13	*t* Test Statistic	-2.602890915			
14					
15	Lower-Tail Test			Calculations Area	
16	Lower Critical Value	-2.821437921		For one-tailed tests:	
17	*p*-Value	0.014301146		TDIST value	0.014301
18	Do not reject the null hypothesis			1-TDIST value	0.985699
19					

The test statistic is t = -2.602890915 and it is not more extreme than the critical value of -2.821437921. Since the test statistic does not lie within the critical region, the conclusion is to fail to reject the null hypothesis. Therefore, there is not enough evidence to conclude that this type of engine meets the pollution standard. Notice that the PHStat output also provides the p-value. In this problem it is calculated to be 0.014301146. Since this is greater than the level of significance used in this case (α = 0.01), the same conclusion (of course!) is reached: fail to reject the null hypothesis.

8.4   Tests of Hypothesis of a Population Proportion

If certain conditions are met (refer to page 381 of the text), then a hypothesis test for the population proportion may be performed.

The following example from the text will be used to illustrate how to conduct a one sample hypothesis test of the proportion. Please refer to pages 381 and 382 of the text, Examples 8.7 and 8.8.

The reputations (and hence sales) of many businesses can be severely damaged by shipments of manufactured items that contain a large percentage of defectives. For example, a manufacturer of alkaline batteries may want to be reasonably certain that fewer than 5% of its batteries are defective. Suppose 300 batteries are randomly selected from a very large shipment; each is tested and 10 defective batteries are found. Does this provide sufficient evidence for the manufacturer to conclude that the fraction defective in the entire shipment is less than 0.05? Use α = 0.01.

Steps for a conducting a one sample hypothesis test of the population proportion:

Step 1: Open a clean workbook.

Step 2: Click on the **Add-Ins** tab.

Step 3: Click on **PHStat**.

Step 4: Drop down to **One-Sample Tests** and click on **Z Test for the Proportion....**

Step 5: A box for the **Z Test for the Proportion** options will appear.

- In the box next to **Null Hypothesis** enter the hypothesized proportion. For this problem that is 0.05.
- In the box next to **Level of Significance** enter the value of α. For this example it is 0.01.
- In the box next to **Number of Successes** enter x. For this problem enter 10, the number of defective batteries.
- In the box next to **Sample Size** enter n. For this problem enter 300, the number of randomly chosen batteries.
- There are three choices under **Test Options** (refer to page 381 of the text for a description):
    - The first choice is **Two-Tail Test**. This is the default choice.
    - The second choice is **Upper-Tail Test.**
    - The third choice is **Lower-Tail Test.** Click on the circle in front of this option since this problem attempts to determine whether the proportion of defective batteries is *less than* 5%.
- There is a blank box to enter a title. Use this when calculating several confidence intervals in order to keep the output organized.
- Click the **OK** button and the results will appear in a new sheet labeled **Hypothesis**.

	A	B	C
1	Text Example 8.7 & 8.8		
2			
3	Data		
4	Null Hypothesis π =	0.05	
5	Level of Significance	0.01	
6	Number of Successes	10	
7	Sample Size	300	
8			
9	Intermediate Calculations		
10	Sample Proportion	0.033333333	
11	Standard Error	0.012583057	
12	Z Test Statistic	-1.324532357	
13			
14	Lower-Tail Test		
15	Lower Critical Value	-2.326347874	
16	p-Value	0.092663152	
17	Do not reject the null hypothesis		
18			

The test statistic is z = -1.324532357 and it is not more extreme than the critical value of -2.326347874. Since the test statistic does not lie within the critical region, the conclusion is to fail to reject the null hypothesis. Even though the sample proportion is 3%, there is not enough evidence to conclude that the population proportion of defective batteries is less than 5%. Notice that the PHStat output also provides the p-value. In this problem it is calculated to be 0.092663152. Since this is greater than the level of significance used in this case (α = 0.01), the same conclusion (of course!) is reached: fail to reject the null hypothesis.

8.5   Tests of Hypothesis of a Population Variance

The following example from the text will be used to illustrate how to conduct a one sample hypothesis test of the variance. Please refer to page 397 of the text, Exercise 8.113.

Refer to the *Current Allergy & Clinical Immunology* (March 2004) study of n = 46 hospital employees who were diagnosed with latex allergy from the exposure to the powder on latex gloves, presented in Exercise 8.26 (p. 364). Recall that the number of latex gloves used per week by the sampled workers is summarized as follows: $\bar{x} = 19.3$ and $s = 11.9$. Let σ^2 represent the variance in the number of latex gloves used per week by all hospital employees. Consider testing $H_0: \sigma^2 = 100$ against $H_a: \sigma^2 \neq 100$. (Note that in part **a.** of this problem a significance level of $\alpha = 0.01$ is given.)

Steps for a conducting a one sample hypothesis test of the population variance:

Step 1: Open a clean workbook.

Step 2: Click on the **Add-Ins** tab.

Step 3: Click on **PHStat**.

Step 4: Drop down to **One-Sample Tests** and click on **Chi-Square Test for the Variance....**

Step 5: A box for the **Chi-Square Test for the Variance** options will appear.

- In the box next to **Null Hypothesis** enter the hypothesized variance. For this problem that is 100.
- In the box next to **Level of Significance** enter the value of α. For this example it is 0.01.
- In the box next to **Sample Size** enter n. For this problem enter 46, the number of hospital workers with latex allergies.
- In the box next to **Sample Standard Deviation** enter the standard deviation, s, of the sample. For this problem enter 11.9.
- There are three choices under **Test Options** (refer to page 396 of the text for a description):
  - The first choice is **Two-Tail Test**. This is the default choice. This is the appropriate choice for this example since the alternative hypothesis contains ≠.
  - The second choice is **Upper-Tail Test.**
  - The third choice is **Lower-Tail Test.**
- There is a blank box to enter a title. Use this when calculating several confidence intervals in order to keep the output organized.
- Click the **OK** button and the results will appear in a new sheet labeled **Hypothesis**.

	A	B	C
1	Text Exercise 8.113		
2			
3	Data		
4	Null Hypothesis σ^2=	100	
5	Level of Significance	0.01	
6	Sample Size	46	
7	Sample Standard Deviation	11.9	
8			
9	Intermediate Calculations		
10	Degrees of Freedom	45	
11	Half Area	0.005	
12	Chi-Square Statistic	63.7245	
13			
14	Two-Tail Test		
15	Lower Critical Value	24.31101434	
16	Upper Critical Value	73.1660608	
17	p-Value	0.034382801	
18	Do not reject the null hypothesis		
19			

The test statistic is $\chi^2 = 63.7245$ and it is not more extreme than the critical value of 73.1660608. Since the test statistic does not lie within the critical region, the conclusion is to fail to reject the null hypothesis. One cannot dispute that the population variance is 100. Notice that the PHStat output also provides the p-value. In this problem it is calculated to be 0.034382801. Since this is greater than the level of significance used in this case ($\alpha = 0.01$), the same conclusion (of course!) is reached: fail to reject the null hypothesis.

8.6 Sample Exercises

The following exercises were selected from the textbook to test the reader's understanding of PHStat and its usage within the chapter. Try to replicate the output in PHStat for the following problems.

Problem 8.35, page 365 (Social interaction of mental patients)

The *Community Mental Health Journal* (Aug. 2000) presented the results of a survey of over 6,000 clients of the Department of Mental Health and Addiction Services (DMHAS) in Connecticut. One of the many variables measured for each mental health patient was frequency of social interaction (on a 5-point scale, where 1 = very infrequently, 3 = occasionally, and 5 = very frequently). The 6,681 clients who were evaluated had a mean social interaction score of 2.95 with a standard deviation of 1.10. Conduct a hypothesis test ($\alpha = 0.01$) to determine if the true mean social interaction score of all Connecticut mental health patients differs from 3. (Note: Although it is numerically possible to conduct a hypothesis test of the mean, given the nature of the data, a test of proportion may be more appropriate: i.e. test the claim that the proportion of patients that gave the response "occasionally" is greater than 50%)

Solution for Problem 8.35:

Solution for Exercise 8.35		
Null Hypothesis $\mu=$		3
Level of Significance		0.01
Population Standard Deviation		1.1
Sample Size		6
Sample Mean		2.95
Standard Error of the Mean		0.44907312
Z Test Statistic		-0.111340443
Two-Tailed Test		
Lower Critical Value		-2.575834515
Upper Critical Value		2.575834515
p-Value		0.911346284
Do not reject the null hypothesis		

Problem 8.91, page 386 (Testing the placebo effect)

The *placebo effect* describes the phenomenon of improvement in the condition of a patient taking a placebo - a pill that looks and tastes real but contains no medically active chemicals. Physicians at a clinic in La Jolla, California, gave what they thought were drugs to 7,000 asthma, ulcer, and herpes patients. Although the doctors later learned that the drugs were really placebos, 70% of the patients reported an improved condition (*Forbes*, May 22, 1995). Use this information (at $\alpha = 0.05$) to test the placebo effect at the clinic. Assume that if the placebo is ineffective, the probability of a patient's condition improving is 0.5.

Solution for Problem 8.91

Solution for Exercise 8.91		
Null Hypothesis $p=$		0.5
Level of Significance		0.05
Number of Successes		4900
Sample Size		7000
Sample Proportion		0.7
Standard Error		0.005976143
Z Test Statistic		33.46640106
Upper-Tail Test		
Upper Critical Value		1.644853
p-Value		0
Reject the null hypothesis		

9.1   Introduction

In Chapters 7 and 8 of the text, inferences about the population from a single sample were conducted by developing confidence intervals and conducting hypothesis tests. Using these techniques, inferences were made about the population mean, μ, the population parameter, p, and the population variance, σ^2. In this chapter inferences will be made about populations using multiple samples. Differences in population parameters will be the points of interest.

PHStat and Excel will perform the following.
1. Tests of hypothesis for comparing population means (independent and dependent samples)
2. Tests of hypothesis for comparing population proportions.
3. Tests of hypothesis for comparing population variances.

Before conducting inferences on two samples, PHStat requires that descriptive statistics for the sample data already be provided. The required information may be input directly into the main menus necessary for conducting analysis.

The following examples will be used to demonstrate techniques in PHStat and Excel.

Examples 9.2 and 9.3, page 414 (Low-fat versus Regular Diet)
Exercise 9.22, page 428 (Pig castration study)
Exercise 9.48, page 442 (Homophone confusion)
Examples 9.6 and 9.7, pages 446&447, (Antismoking Campaigns)
Example 9.10 and 9.11, pages 458-460 (Metabolic Rates of White Mice)
Sample Exercise #1: Problem 9.67, page 452 ("Tip of the Tongue" study)
Sample Exercise #2: Problem 9.101, page 465 (Gender differences in math test scores)

9.2   Tests for Differences in Two Means: Independent Sampling (n_1, n_2 large)

The following example from the text will be used to illustrate how to conduct a two sample hypothesis test for comparing means. Please refer to page 414 of the text, Examples 9.2 and 9.3.

A dietitian has developed a diet that is low in fats, carbohydrates, and cholesterol. Although the diet was initially intended to be used by people with heart disease, the dietitian wishes to examine the effect this diet has on the weights of obese people. Two random samples of 100 obese people each are selected and one group of 100 is placed on the low-fat diet. The other 100 are placed on a diet that contains approximately the same quantity of food but is not as low in fats, carbohydrates, and cholesterol. For each person, the amount of weight lost (or gained) in a 3-week period is recorded. Using a 0.05 level of significance, is there a difference in the mean weight loss between these two groups?

For the steps below, summary statistics will be required. The text does provide them in this example, but recall that for any raw data Excel can calculate summary statistics (see Chapter 2).

	Low-Fat Diet	Regular Diet
Sample Size	100	100
Sample Mean	9.31	7.40
Sample Standard Deviation	4.67	4.04

In the majority of cases the population standard deviation, σ, is not known and it will be necessary to substitute the sample standard deviation, s, as its estimate. Refer to page 413 of the text for when this is appropriate.

Steps for a conducting a two sample hypothesis test comparing population means:

Step 1: Open a clean workbook.

Step 2: Click on the **Add-Ins** tab.

Step 3: Click on **PHStat**.

Step 4: Drop down to **Two-Sample Tests** and click on **Z Test for the Differences in Two Means....**

Step 5: A box for the **Z Test for the Differences in Two Means** options will appear.

- In the box next to **Hypothesized Difference** enter D_0. For this problem that is 0.
- In the box next to **Level of Significance** enter the value of α. For this example it is 0.05.
- There are three pieces of information needed for the **Population 1 Sample:**
  - In the box next to **Sample Size** enter n_1. For this example that is 100, the number of people on the low-fat diet.
  - In the box next to **Sample Mean** enter $\bar{x}_1$. For this example that is 9.31, the average weight loss on the low-fat diet.
  - In the box next to **Population Std. Deviation** enter σ_1 if available. For this example enter 4.67, the sample standard deviation for the low-fat diet, as an estimate for the population standard deviation.
- There are three pieces of information needed for the **Population 2 Sample:**
  - In the box next to **Sample Size** enter n_2. For this example that is 100, the number of people on the regular diet.
  - In the box next to **Sample Mean** enter $\bar{x}_2$. For this example that is 7.40, the average weight loss on the regular diet.
  - In the box next to **Population Std. Deviation** enter σ_2 if available. For this example enter 4.04, the sample standard deviation for the regular diet, as an estimate for the population standard deviation.
- There are three choices under **Test Options** (refer to page 413 of the text for a description):
  - The first choice is **Two-Tail Test**. This is the default choice. This is the appropriate choice for this example since the alternative hypothesis contains $\neq$.
  - The second choice is **Upper-Tail Test.**
  - The third choice is **Lower-Tail Test.**

125

- There is a blank box to enter a title. Use this when calculating several confidence intervals in order to keep the output organized.
- Click the **OK** button and the results will appear in a new sheet labeled **Hypothesis**.

	A	B	C
1	Text Example 9.2 & 9.3		
2			
3	Data		
4	Hypothesized Difference	0	
5	Level of Significance	0.05	
6	Population 1 Sample		
7	Sample Size	100	
8	Sample Mean	9.31	
9	Population Standard Deviation	4.67	
10	Population 2 Sample		
11	Sample Size	100	
12	Sample Mean	7.4	
13	Population Standard Deviation	4.04	
14			
15	Intermediate Calculations		
16	Difference in Sample Means	1.91	
17	Standard Error of the Difference in Means	0.617499	
18	Z-Test Statistic	3.093122	
19			
20	Two-Tail Test		
21	Lower Critical Value	-1.95996	
22	Upper Critical Value	1.959964	
23	p-Value	0.001981	
24	Reject the null hypothesis		
25			

The test statistic is $z = 3.093122$ and it exceeds the critical value of 1.959964. Since the test statistic lies within the critical region, the conclusion is to reject the null hypothesis. Therefore, it is determined that there is a difference in the mean weight loss for the two diets. Notice that the PHStat output also provides the p-value. In this problem it is calculated to be 0.001981. Since this is less than the level of significance used in this case ($\alpha = 0.05$), the same conclusion (of course!) is reached: reject the null hypothesis.

9.3 Tests for Differences in Two Means: Independent Sampling (n_1, n_2 small)

Although certain conditions must still be met (see Section 9.2 of the text), hypothesis testing for comparing two means may still be conducted if σ_1 and σ_2 are unknown and n_1 and n_2 are small.

The following example from the text will be used to illustrate how to conduct a two sample hypothesis test for comparing means. Please refer to page 428 of the text, Exercise 9.22.

Two methods of castrating male piglets were investigated in *Applied Animal Behaviour Science* (Nov. 1, 2000). Method 1 involved an incision in the spermatic cords, while Method 2 involved pulling and severing the cords. Forty-nine male piglets were randomly allocated to one of the two methods. During castration, the researchers measured the number of high-frequency vocal responses (squeals) per second

126

over a 5-second period. The data are summarized in the accompanying table. Conduct a test of hypothesis to determine whether the population mean number of high-frequency vocal responses differs for piglets castrated by the two methods. Use $\alpha = 0.05$.

Steps for a conducting a two sample hypothesis test comparing population means:

Step 1: Open a clean workbook.

Step 2: Click on the **Add-Ins** tab.

Step 3: Click on **PHStat**.

Step 4: Drop down to **Two-Sample Tests** and click on **t Test for the Differences in Two Means....**

Step 5: A box for the **t Test for the Differences in Two Means** options will appear.

- In the box next to **Hypothesized Difference** enter D_0. For this problem that is 0 since the null hypotheses is $H_0: \mu_1 = \mu_2$.
- In the box next to **Level of Significance** enter the value of α. For this example it is 0.05.
- There are three pieces of information needed for the **Population 1 Sample:**
  - In the box next to **Sample Size** enter n_1. For this example that is 24, the number of piglets randomly assigned to Method 1.
  - In the box next to **Sample Mean** enter $\bar{x}_1$. For this example that is 0.74, the average number of squeals per second using Method 1.
  - In the box next to **Sample Standard Deviation** enter s_1. For this example enter 0.09, the sample standard deviation for Method 1.
- There are three pieces of information needed for the **Population 2 Sample:**
  - In the box next to **Sample Size** enter n_2. For this example that is 25, the number of piglets randomly assigned to Method 2.
  - In the box next to **Sample Mean** enter $\bar{x}_2$. For this example that is 0.70, the average number of squeals per second using Method 2.
  - In the box next to **Sample Standard Deviation** enter s_2. For this example enter 0.09, the sample standard deviation for Method 2.
- There are three choices under **Test Options** (refer to page 416 of the text for a description):
  - The first choice is **Two-Tail Test**. This is the default choice. This is the appropriate choice for this example since the problem is to detect whether the two methods differ (in *either* direction).
  - The second choice is **Upper-Tail Test.**

- The third choice is **Lower-Tail Test.**
- There is a blank box to enter a title. Use this when calculating several confidence intervals in order to keep the output organized.
- Click on the box in front of **Confidence Interval Estimate** to obtain the confidence interval for $\mu_1 - \mu_2$. Once this option is chosen, there is a box to enter the **Confidence level.** In this case use the default value of 95% since the hypothesis test uses $\alpha = 0.05$. Refer to page 418 of the text for an explanation on how to interpret the confidence interval for the difference of two means.
- Click the **OK** button and the results will appear in a new sheet labeled **Hypothesis**.

	A	B	C	D	E
1	Text Exercise 9.22				
2	(assumes equal population variances)				
3	Data			Confidence Interval Estimate	
4	Hypothesized Difference	0		for the Difference Between Two Means	
5	Level of Significance	0.05			
6	Population 1 Sample			Data	
7	Sample Size	24		Confidence Level	95%
8	Sample Mean	0.74			
9	Sample Standard Deviation	0.09		Intermediate Calculations	
10	Population 2 Sample			Degrees of Freedom	47
11	Sample Size	25		t Value	2.01174048
12	Sample Mean	0.7		Interval Half Width	0.051741246
13	Sample Standard Deviation	0.09			
14				Confidence Interval	
15	Intermediate Calculations			Interval Lower Limit	-0.011741246
16	Population 1 Sample Degrees of Freedom	23		Interval Upper Limit	0.091741246
17	Population 2 Sample Degrees of Freedom	24			
18	Total Degrees of Freedom	47			
19	Pooled Variance	0.0081			
20	Difference in Sample Means	0.04			
21	t Test Statistic	1.555232			
22					
23	Two-Tail Test				
24	Lower Critical Value	-2.01174			
25	Upper Critical Value	2.01174			
26	p-Value	0.1266			
27	Do not reject the null hypothesis				
28					

The test statistic is $t = 1.555232$ and it is not more extreme than the critical value of 2.01174. Since the test statistic does not lie within the critical region, the conclusion is to fail to reject the null hypothesis. Therefore, there is not enough evidence to conclude that the mean number of high-frequency vocal responses differs between Method 1 and Method 2. Notice that the PHStat output also provides the p-value. In this problem it is calculated to be 0.1266. Since this is greater than the level of significance used in this case ($\alpha = 0.05$), the same conclusion (of course!) is reached: fail to reject the null hypothesis. PHStat also provides the confidence interval. In this example it is believed that the difference in the mean number of squeals for Method 1 and Method 2 is between (-0.011741246, 0.091741246) with 95% confidence.

9.4   Tests for Differences in Two Means: Paired Data

If it is desired to compare two means, but the data for two groups is paired, then the preceding tests are not appropriate.

The following example from the text will be used to illustrate how to conduct a two sample hypothesis test for comparing means for paired data. Please refer to page 442 of the text, Exercise 9.48.

A *homophone* is a word whose pronunciation is the same as that of another word having a different meaning and spelling (e.g., *nun* and *none*, *doe* and *dough*, etc.) *Brain and Language* (Apr. 1995) reported on a study of homophone spelling in patients with Alzheimer's disease. Twenty Alzheimer's patients were asked to spell 24 homophone pairs given in random order. Then the number of homophone confusions (e.g., spelling *doe* given the context, *bake bread dough*) was recorded for each patient. One year later, the same test was given to the same patients. The data for the study are provided in the accompanying table (refer to text). The researchers posed the following question: "Do Alzheimer's patients show a significant increase in the mean homophone confusion errors over time?" Perform an analysis of the data to answer the researchers' question.

Steps for a conducting a two sample hypothesis test comparing population means for paired data:

Step 1: Open a clean workbook and enter the raw data for the first group into column A and the raw data for the second group into column B. It is important that the data be entered such that each row represents a matched pair. Include labels in the first row.

	A	B
1	time1	time2
2	5	5
3	1	3
4	0	0
5	1	1
6	0	1
7	2	1
8	5	6
9	1	2
10	0	9
11	5	8
12	7	10

Step 2: Click on the **Data** tab.

Step 3: Click on **Data Analysis**.

130

Step 4: A box for **Data Analysis** options will appear. Click on **t-test: Paired Two Sample for Means** and then click the **OK** button.

Step 5: A box for the **t-test: Paired Two Sample for Means** options will appear.

- Click on the entry box by **Variable 1 Range**. Enter the rows and column(s) where the data for the first group are located. This may be done by typing the range in manually or by clicking and dragging the appropriate range of data cells. In this example the data are located in Sheet1!$A$1:$A$21.
- Click on the entry box by **Variable 2 Range**. Enter the rows and column(s) where the data for the second group are located. This may be done by typing the range in manually or by clicking and dragging the appropriate range of data cells. In this example the data are located in Sheet1!$B$1:$B$21.

- In the box next to **Hypothesized Mean Difference** enter D_0. In this case that is 0. Note that the alternative hypothesis could be written as H_a: $\mu_d < 0$.
- Since there are labels in the first row, click on the box in front of **Labels** so that there is a check mark beside it.
- In the box next to **Alpha** enter α. No level of significance was specified in this problem, so use the default value of 0.05.
- Click the **OK** button and the test results will appear in a new sheet. The width of columns A, B, and C may need to be adjusted in order to read all of the output. The test statistic, p-value, and critical value(s) are included in the output.

	A	B	C	D
A1		f_x	t-Test: Paired Two Sample for Means	

TextExercise9.48.xlsx

	A	B	C	D
1	t-Test: Paired Two Sample for Means			
2				
3		time1	time2	
4	Mean	4.15	5.8	
5	Variance	12.23947368	17.74736842	
6	Observations	20	20	
7	Pearson Correlation	0.669932241		
8	Hypothesized Mean Difference	0		
9	df	19		
10	t Stat	-2.306004365		
11	P(T<=t) one-tail	0.016274706		
12	t Critical one-tail	1.729132792		
13	P(T<=t) two-tail	0.032549412		
14	t Critical two-tail	2.09302405		
15				

The test statistic is t = -2.036004365 and it is more extreme than the critical value of -1.729132792. Notice that interpreting the Excel output for hypothesis testing requires more of the user than the PHStat output. The user must realize that this is a lower-tailed test (H_a: $\mu_1 < \mu_2$ since the researchers wish to determine if the mean homophone confusion *increases* over time), so the critical value is negative. Since the test statistic lies within the critical region, the conclusion is to reject the null hypothesis. Therefore, it is determined that there is a difference in the mean homophone confusion over time. In this problem the p-value is calculated to be 0.016274706. Since this is less than the level of significance used in this case ($\alpha = 0.05$), the same conclusion (of course!) is reached: reject the null hypothesis.

9.5 Tests for Differences in Two Proportions

In this section, PHStat will be used to compare two population proportions using two random and independent samples. Remember that the two samples represent binomial experiments.

The following example from the text will be used to illustrate how to conduct a two sample hypothesis test for comparing proportions. Please refer to pages 446 and 447 of the text, Examples 9.6 and 9.7.

In the past decade, intensive antismoking campaigns have been sponsored by both federal and private agencies. Suppose the American Cancer Society randomly sampled 1,500 adults in 1997 and then sampled 1,750 adults in 2007 to determine whether there was evidence that the percentage of smokers had decreased. The results of the two sample surveys are shown in the table below, where x_1 and x_2 represent the numbers of smokers in the 1997 and 2007 samples, respectively. Do these data indicate that the fraction of smokers decreased over this ten-year period? Use $\alpha = 0.05$.

Year	1997	2007
Sample Size	1,500	1,750
x_i	555	578
Sample Proportion	555/1,500	578/1,750

Steps for a conducting a two sample hypothesis test comparing population proportions:

Step 1: Open a clean workbook.

Step 2: Click on the **Add-Ins** tab.

Step 3: Click on **PHStat**.

Step 4: Drop down to **Two-Sample Tests** and click on **Z Test for the Differences in Two Proportions…**.

Step 5: A box for the **Z Test for the Differences in Two Proportions** options will appear.

- In the box next to **Hypothesized Difference** enter D_0. For this problem that is 0.
- In the box next to **Level of Significance** enter the value of α. For this example it is 0.05.
- There are two pieces of information needed for the **Population 1 Sample:**
  - In the box next to **Number of Successes** enter x_1. For this problem enter 555, the number of smokers in the 1997 sample.
  - In the box next to **Sample Size** enter n_1. For this example that is 1500, the number of randomly sampled adults in 1997.
- There are two pieces of information needed for the **Population 2 Sample:**
  - In the box next to **Number of Successes** enter x_2. For this problem enter 578, the number of smokers in the 2007 sample.
  - In the box next to **Sample Size** enter n_2. For this example that is 1750, the number of randomly sampled adults in 2007.
- There are three choices under **Test Options** (refer to page 445 of the text for a description):
  - The first choice is **Two-Tail Test**. This is the default choice.
  - The second choice is **Upper-Tail Test.** Click on this option for this example since it is desired to determine whether the proportion *decreased* from 1997 to 2007.
  - The third choice is **Lower-Tail Test.**
- There is a blank box to enter a title. Use this when calculating several confidence intervals in order to keep the output organized.
- Click on the box in front of **Confidence Interval Estimate** to obtain the confidence interval for $p_1 - p_2$. Once this option is chosen, there is a box to enter the **Confidence level.** In this case use the default value of 95% since the hypothesis test uses $\alpha = 0.05$. Refer to page 418 of the text for an explanation on how to interpret the confidence interval for the difference of two means.
- Click the **OK** button and the results will appear in a new sheet labeled **Hypothesis**.

134

	A	B	C	D	E	F
4	Hypothesized Difference	0		of the Difference Between Two Proportions		
5	Level of Significance	0.05				
6	Group 1			Data		
7	Number of Successes	555		Confidence Level	95%	
8	Sample Size	1500				
9	Group 2			Intermediate Calculations		
10	Number of Successes	578		Z Value	-1.959963985	
11	Sample Size	1750		Std. Error of the Diff. between two Proportions	0.01678685	
12				Interval Half Width	0.032901621	
13	Intermediate Calculations					
14	Group 1 Proportion	0.37		Confidence Interval		
15	Group 2 Proportion	0.330285714		Interval Lower Limit	0.006812665	
16	Difference in Two Proportions	0.039714286		Interval Upper Limit	0.072615906	
17	Average Proportion	0.348615385				
18	Z Test Statistic	2.368523545				
19						
20	Upper-Tail Test					
21	Upper Critical Value	1.644853627				
22	p-Value	0.008929622				
23	Reject the null hypothesis					
24						

The test statistic is z = 2.368523545 and it exceeds the critical value of 1.644853627. Since the test statistic lies within the critical region, the conclusion is to reject the null hypothesis. Therefore, it is determined that the proportion of smokers did decrease from 1997 to 2007. Notice that the PHStat output also provides the p-value. In this problem it is calculated to be 0.008929622. Since this is less than the level of significance used in this case ($\alpha = 0.05$), the same conclusion (of course!) is reached: reject the null hypothesis.

Although it was not requested in these problems, PHStat can calculate the confidence interval for $p_1 - p_2$. In this example the confidence interval is (0.006812665, 0.072615906). Refer to page 449 of the text for an example of how to interpret this output.

9.6   Tests for Differences in Two Variances

In this section, PHStat will be used to compare population variances using two random and independent samples. Remember that in order to conduct this test, the two sampled populations must be normally distributed and the samples must be randomly and independently selected from their respective populations.

The following example from the text will be used to illustrate how to conduct a two sample hypothesis test for comparing variances. Please refer to page 458-460 of the text, Examples 9.10 and 9.11.

An experimenter wants to compare the metabolic rates of white mice subjected to different drugs. The weights of the mice may affect their metabolic rates, and thus the experimenter wishes to obtain mice that are relatively homogeneous with respect to weight. Five hundred mice will be needed to complete the study. Currently, 13 mice from supplier 1 and another 18 mice from supplier 2 are available for comparison. The experimenter weighs these mice and obtains the raw data in the text. For the steps below, summary statistics will be required. The text does provide them in this example, but recall that for any raw data Excel can calculate summary statistics (see Chapter 2).

	Supplier 1	Supplier 2
Sample Size	13	18
Sample Standard Deviation	0.2021	0.0982

Do these data provide sufficient evidence to indicate a difference in the variability of weights of mice obtained from the two suppliers? (Use $\alpha = 0.10$)

Steps for a conducting a two sample hypothesis test comparing population variances:

Step 1: Open a clean workbook.

Step 2: Click on the **Add-Ins** tab.

Step 3: Click on **PHStat**.

Step 4: Drop down to **Two-Sample Tests** and click on **F Test for the Differences in Two Variances…**.

Step 5: A box for the **F Test for the Differences in Two Variances** options will appear.

- In the box next to **Level of Significance** enter the value of α. For this example it is 0.10.
- There are two pieces of information needed for the **Population 1 Sample:**
  - In the box next to **Sample Standard Deviation** enter s_1. For this example enter 0.2021, the sample standard deviation for Supplier 1.
  - In the box next to **Sample Size** enter n_1. For this example that is 13, the number of rats from Supplier 1.
- There are two pieces of information needed for the **Population 2 Sample:**
  - In the box next to **Sample Standard Deviation** enter s_2. For this example enter 0.0982, the sample standard deviation for Supplier 2.
  - In the box next to **Sample Size** enter n_2. For this example that is 18, the number of rats from Supplier 2.
- There are three choices under **Test Options** (refer to page 461 of the text for a description):
  - The first choice is **Two-Tail Test**. This is the default choice. Use this option for this example since it is desired to determine whether there is a difference (in either direction) in variability between the two groups.
  - The second choice is **Upper-Tail Test.**
  - The third choice is **Lower-Tail Test.**
- There is a blank box to enter a title. Use this when calculating several confidence intervals in order to keep the output organized.
- Click the **OK** button and the results will appear in a new sheet labeled **Hypothesis**.

	A	B	C	D	E	F
1	Text Examples 9.10 and 9.11					
2						
3	Data					
4	Level of Significance	0.1				
5	Population 1 Sample					
6	Sample Size	13				
7	Sample Standard Deviation	0.2021				
8	Population 2 Sample					
9	Sample Size	18				
10	Sample Standard Deviation	0.0982				
11						
12	Intermediate Calculations					
13	F Test Statistic	4.235548				
14	Population 1 Sample Degrees of Freedom	12				
15	Population 2 Sample Degrees of Freedom	17				
16				Calculations Area		
17	Two-Tail Test			FDIST value	0.003568	
18	Lower Critical Value	0.387171		1-FDIST value	0.996432	
19	Upper Critical Value	2.380654				
20	p-Value	0.007136				
21	Reject the null hypothesis					
22						

The test statistic is $F = 4.235548$ and it exceeds the critical value of 2.380654. Since the test statistic lies within the critical region, the conclusion is to reject the null hypothesis. Therefore, it is determined that the population variances of weights of mice obtained from the two suppliers do differ. Notice that the PHStat output also provides the p-value. In this problem it is calculated to be 0.007136. Since this is less than the level of significance used in this case ($\alpha = 0.10$), the same conclusion (of course!) is reached: reject the null hypothesis.

9.7 Sample Exercises

The following exercises were selected from the textbook to test the reader's understanding of PHStat and its usage within the chapter. Try to replicate the output in PHStat for the following problems.

Problem 9.67, "Tip of the tongue" Study (page 452)

Trying to think of a word you know, but can't instantly retrieve is called the "tip of the tongue" phenomenon. *Psychology and Aging* (Sep. 2001) published a study of this phenomenon in senior citizens. The researchers compared 40 people between 60 and 72 years of age with 40 between 73 and 83 years of age. When primed with initial syllable of a missing word (e.g., seeing the word "include" to help recall the word "incisor"), the younger seniors had a higher recall rate. Suppose 31 of the 40 seniors in the younger group could recall the word when primed with the initial syllable, while only 22 of the 40 seniors could recall the word. The recall rates of the two groups were compared. Using the 0.05 level of significance, does one group of elderly have a significantly higher recall rate than the other? (Is there a difference in the recall rate between groups?)

Solution for Problem 9.67

Solution to Exercise 9.67	
Hypothesized Difference	0
Level of Significance	0.05
Group 1	
Number of Successes	31
Sample Size	40
Group 2	
Number of Successes	22
Sample Size	40
Group 1 Proportion	0.775
Group 2 Proportion	0.55
Difference in Two Proportions	0.225
Average Proportion	0.6625
Z Test Statistic	2.127980706
Two-Tailed Test	
Lower Critical Value	-1.959961082
Upper Critical Value	1.959961082
p-Value	0.033338564
Reject the null hypothesis	

Problem 9.101, Gender differences in math test scores (page 465)

Refer to the *American Educational Journal* (Fall, 1998) study to compare the mathematics achievement test scores of male and female students. The researchers hypothesized that the distribution of test scores for males is more variable than the corresponding distribution for females. Use the summary information reproduced in the table to test this claim at $\alpha = 0.01$.

	Males	Females
Sample Size	1,764	1,739
Mean	48.9	48.4
Standard Deviation	12.96	11.85

Solution for Problem 9.101

Solution to Exercise 9.101			
Level of Significance	**0.01**		
Population 1 Sample			
Sample Size	**1764**		
Sample Standard Deviation	**12.96**		
Population 2 Sample			
Sample Size	**1739**		
Sample Standard Deviation	**11.85**		
F-Test Statistic	1.196116007		
Population 1 Sample Degrees of Freedom	1763		
Population 2 Sample Degrees of Freedom	1738		
		Calculations Area	
Upper-Tail Test		FDIST value	9.1202E-05
Upper Critical Value	**1.117699711**	1-FDIST value	0.999908798
p-Value	**9.1202E-05**		
Reject the null hypothesis			

140

10.1 Introduction

In Chapter 10 of the textbook, the authors discuss how to design and analyze experiments in order to compare more than two population means. The goal of ANOVA (analysis of variance) is to identify if there is a difference in the mean response of the various treatment groups. Although it is not available in PHStat/Excel, the text presents several methods for multiple comparisons of means. These tests are beneficial when it has been determined by ANOVA that at least one of the treatment means differs from the others.

Although analysis of variance software does not currently exist in PHStat, Excel will be used in order to solve problems within this chapter. In particular, Excel may be used in the analysis of completely randomized and factorial designs. Notice that when regression analysis is discussed in subsequent chapters, basic ANOVA calculations will be performed within the regression analysis.

Recall from Section 10.2 of the textbook that a completely randomized design is one for which independent random samples of experimental units are selected for each treatment. In ANOVA, the null hypothesis tests whether the treatment means are all equal. The alternative hypothesis is that at least one of the treatment means differs from the others.

The following examples will be used to illustrate techniques in PHStat and Excel.

Example 10.4, page 493 (USGA and golf balls)
Example 10.10, page 531 (USGA and golf balls)
Sample Exercise #1: Problem 10.32, Hair Color and Pain, page 5xx
Sample Exercise #2: Problem 10.93, page 543 (Impact of Vitamin B Supplement)

10.2 Completely Randomized Design

The completely randomized design is a design in which treatments are randomly assigned to the experimental units. Please refer to page 492 for the conditions for which using a one-way ANOVA is appropriate.

The following example from the text will be used to illustrate how to conduct a one-way ANOVA test. Please refer to page 493 of the text, Example 10.4.

Suppose the USGA wants to compare the mean distances associated with four different brands of golf balls when struck with a driver. A completely randomized design is employed, with Iron Byron, the USGA's robotic golfer, using a driver to hit a random sample of 10 balls of each brand in a random sequence. The distance is recorded for each hit, and the results are shown in the table below, organized by brand. Set up a test to compare the mean distances for the four brands. Use $\alpha = 0.10$

	Brand A	Brand B	Brand C	Brand D
1	251.2	263.2	269.7	251.6
2	245.1	262.9	263.2	248.6
3	248.0	265.0	277.5	249.4
4	251.1	254.5	267.4	242.0
5	260.5	264.3	270.5	246.5
6	250.0	257.0	265.5	251.3
7	253.9	262.8	270.7	261.8
8	244.6	264.4	272.9	249.0
9	254.6	260.6	275.6	247.1
10	248.8	255.9	266.5	245.9

Steps for a conducting a One-way ANOVA:

Step 1: Open a clean workbook and enter the raw data for each treatment into a separate column. Include labels in the first row.

	A	B	C	D	E
1	Brand A	Brand B	Brand C	Brand D	
2	251.2	263.2	269.7	251.6	
3	245.1	262.9	263.2	248.6	
4	248	265	277.5	249.4	
5	251.1	254.5	267.4	242	
6	260.5	264.3	270.5	246.5	
7	250	257	265.5	251.3	
8	253.9	262.8	270.7	261.8	
9	244.6	264.4	272.9	249	
10	254.6	260.6	275.6	247.1	
11	248.8	255.9	266.5	245.9	
12					
13					

Step 2: Click on the **Data** tab.

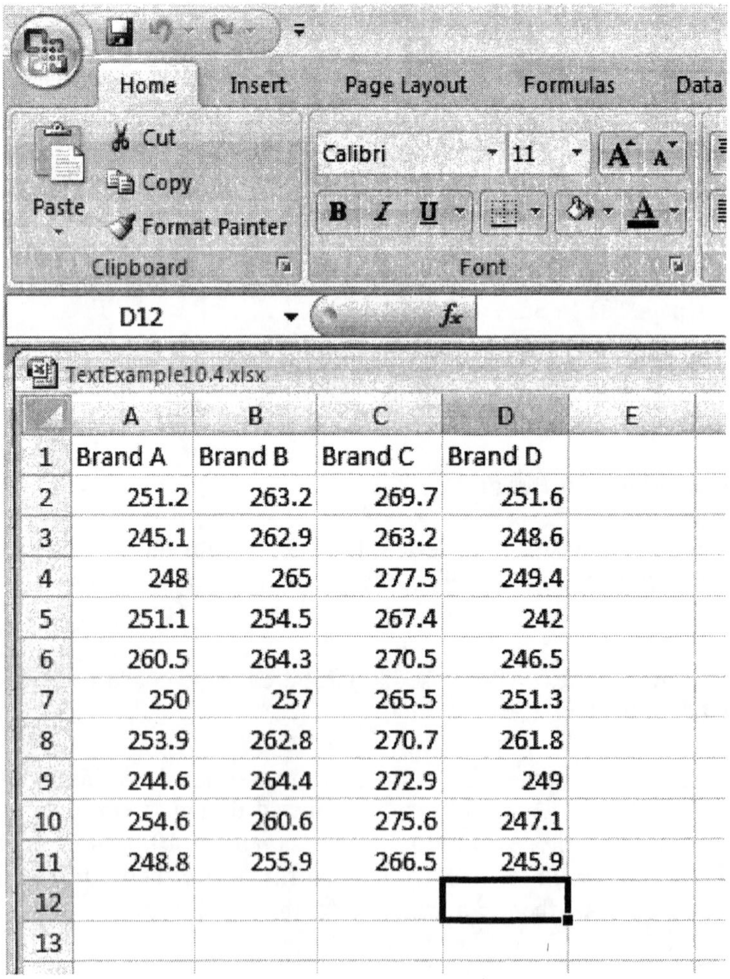

Step 3: Click on **Data Analysis**.

Step 4: A box for **Data Analysis** options will appear. Click on **ANOVA: Single Factor** and then click the **OK** button.

Step 5: A box for the **ANOVA: Single Factor** options will appear.

- Click on the entry box by **Input Range**. Enter the rows and columns where the data are located. This may be done by typing the range in manually or by clicking and dragging the appropriate range of data cells. In this example the data are located in $A$1:$D$11.

- Next to **Grouped By** there are two options: **Columns** and **Rows.** In this example, use the default choice of **Columns** since in Step 1 the data were entered by treatments into columns.
- Since there are labels in the first row, click on the box in front of **Labels in First Row** so that there is a check mark beside it.
- In the box next to **Alpha** enter α. The default value is 0.05, but in this problem it is specified that $\alpha = 0.10$.
- Click the **OK** button and the test results will appear in a new sheet. The width of column A may need to be adjusted in order to read all of the output. The test statistic, p-value, and critical value are included in the output.

TextExample10.4.xlsx

	A	B	C	D	E	F	G	H
1	Anova: Single Factor							
2								
3	SUMMARY							
4	*Groups*	*Count*	*Sum*	*Average*	*Variance*			
5	Brand A	10	2507.8	250.78	22.42178			
6	Brand B	10	2610.6	261.06	14.94711			
7	Brand C	10	2699.5	269.95	20.25833			
8	Brand D	10	2493.2	249.32	27.07289			
9								
10								
11	ANOVA							
12	*Source of Variation*	*SS*	*df*	*MS*	*F*	*P-value*	*F crit*	
13	Between Groups	2794.389	3	931.4629	43.98875	3.97E-12	2.242605	
14	Within Groups	762.301	36	21.17503				
15								
16	Total	3556.69	39					
17								

The test statistic is $F = 43.98875$ and it is more extreme than the critical value of 2.242605. Since the test statistic lies within the critical region, the conclusion is to reject the null hypothesis. Therefore, it is determined that the mean distances of the golf balls differ for at least two of the brands. In this problem the p-value is calculated to be 3.97×10^{-12}. Since this is less than the level of significance used in this case ($\alpha = 0.10$), the same conclusion (of course!) is reached: reject the null hypothesis.

10.3 The Factorial Design

In the factorial design, an additional factor is added to the simple one-way design of experiment.

The following example from the text will be used to illustrate how to conduct a two-factor ANOVA test. Please refer to page 531 of the text, Example 10.10.

Suppose the USGA tests four different brands (A, B, C, D) of golf balls and two different clubs (driver, five-iron) in a completely randomized design. Each of the eight Brand-Club combinations (treatments) is randomly and independently assigned to four experimental units, each experimental unit consisting of a specific position in the sequence of hits by Iron Byron. The distance response is recorded for each of the 32 hits, and the results are shown in the table below.

    a. Use Excel to partition the Total Sum of Squares into the components necessary to analyze this 4x2 factorial experiment.

    b. Follow the steps for analyzing a two-factor factorial experiment and interpret the results of your analysis. Use $\alpha = 0.10$ for the tests conducted.

		Brand			
		A	B	C	D
	Driver	226.4	238.3	240.5	219.8
	Driver	232.6	231.7	246.9	228.7
	Driver	234.0	227.7	240.3	232.9
Club	Driver	220.7	237.2	244.7	237.6
	Five-Iron	163.8	184.4	179.0	157.8
	Five-Iron	179.4	180.6	168.0	161.8
	Five-Iron	168.6	179.5	165.2	162.1
	Five-Iron	173.4	186.2	156.5	160.3

Steps for a conducting a Two-way ANOVA:

Step 1: Open a clean workbook and enter the raw data for each of the different levels of the first factor into separate columns. Each column will then be comprised of groupings of the second factor. Include labels in the first row and first column.

	A	B	C	D	E	F
1		Brand A	Brand B	Brand C	Brand D	
2	DRIVER	226.4	238.3	240.5	219.8	
3	DRIVER	232.6	231.7	246.9	228.7	
4	DRIVER	234	227.7	240.3	232.9	
5	DRIVER	220.7	237.2	244.7	237.6	
6	5IRON	163.8	184.4	179	157.8	
7	5IRON	179.4	180.6	168	161.8	
8	5IRON	168.6	179.5	165.2	162.1	
9	5IRON	173.4	186.2	156.5	160.3	

Step 2: Click on the **Data** tab.

Step 3: Click on **Data Analysis**.

Step 4: A box for **Data Analysis** options will appear. Click on **ANOVA: Two-Factor With Replication** and then click the **OK** button.

Step 5: A box for the **ANOVA: Two-Factor With Replication** options will appear.

- Click on the entry box by **Input Range**. Enter the rows and columns where the data are located. This may be done by typing the range in manually or by clicking and dragging the appropriate range of data cells. In this example the data are located in $A$1:$E$9.
- Click on the entry box by **Rows per sample**. Enter the number of observations for each treatment. For this example enter 4.
- In the box next to **Alpha** enter α. The default value is 0.05, but in this problem it is specified to use $\alpha = 0.10$.
- Click the **OK** button and the test results will appear in a new sheet. The width of columns A and B may need to be adjusted in order to read all of the output. The test statistics, p-values, and critical values are included in the output.

The ANOVA output generated for factorial design is separated into two sections. First, a statistical summary is provided for the various treatments. For each of the eight Brand-Club treatments, Excel provides the average distance and variance of each drive. Please refer to the textbook to interpret this data.

	A	B	C	D	E	F	G
		Brand A	Brand B	Brand C	Brand D	Total	
1	Anova: Two-Factor With Replication						
2							
3	SUMMARY	Brand A	Brand B	Brand C	Brand D	Total	
4	DRIVER						
5	Count	4	4	4	4	16	
6	Sum	913.7	934.9	972.4	919	3740	
7	Average	228.425	233.725	243.1	229.75	233.75	
8	Variance	37.42916667	24.46917	10.53333	57.21667	61.07067	
9							
10	SIRON						
11	Count	4	4	4	4	16	
12	Sum	685.2	730.7	668.7	642	2726.6	
13	Average	171.3	182.675	167.175	160.5	170.4125	
14	Variance	44.52	9.929167	86.1225	3.86	98.19183	
15							
16	Total						
17	Count	8	8	8	8		
18	Sum	1598.9	1665.6	1641.1	1561		
19	Average	199.8625	208.2	205.1375	195.125		
20	Variance	967.4826786	759.3429	1688.454	1396.336		
21							

The second section contains the analysis of variance output necessary to answer the question. The first test to conduct is:

H_0: The eight treatment means are equal.
H_a: At least two of the means differ.

The test statistic necessary for this test is $F = \frac{MST}{MSE}$, however, neither the F test statistic nor the MST is given in the Excel output (Note that MSE is MS(Within) in the Excel output). It is possible to obtain SST by adding SS(Sample), SS(Columns), and SS(Interaction) and since MST $= \frac{SST}{df = ab-1}$, the F test statistic may be obtained. Therefore, in this example SST $= 32093.11125 + 800.73625 + 765.96125 = 33659.8087$, and $ab - 1 = 4 \times 2 - 1 = 7$, so MST $= \frac{33659.8087}{7} = 4808.544$. Therefore, the necessary $F = \frac{4808.544}{34.26} = 140.35$. Since Excel does not perform this test explicitly, this F test statistic will have to be compared to the critical value of F with 7 and 24 degrees of freedom. The result is to reject the null hypothesis and conclude that at least two of the Brand-Club combinations differ in mean distance.

21						
22						
23	ANOVA					
24	*Source of Variation*	*SS*	*df*	*MS*	*F*	*P-value*	*F crit*
25	Sample	32093.11125	1	32093.11	936.7516	9.63E-21	2.927117
26	Columns	800.73625	3	266.9121	7.790779	0.00084	2.32739
27	Interaction	765.96125	3	255.3204	7.452435	0.001079	2.32739
28	Within	822.24	24	34.26			
29							
30	Total	34482.04875	31				
31							
32							

Since there is a difference in the mean distance in at least two of the brand/club treatments, it makes sense to test for interaction between brand and club. Excel does explicitly perform this test. Notice that the F test statistic for interaction is 7.452435 and this is more extreme than the critical value of 2.32739. Since the test statistic lies within the critical region, the conclusion is to reject the null hypothesis. Therefore, it is determined that brand and club interact to affect mean response. In this problem the p-value is calculated to be 0.001079. Since this is less than the level of significance used in this case ($\alpha = 0.10$), the same conclusion (of course!) is reached: reject the null hypothesis.

10.4   Sample Exercises

The following exercises were selected from the textbook to test the reader's understanding of Excel and its usage within the chapter. Try to replicate the output for the following problems.

Problem 10.111, Hull failures of oil tankers (page 550)

Refer to the *Marine Technology* (Jan. 1995) study of major ocean oil spills by tanker vessels, presented in Exercise 2.186 (p. 106). The spillage amounts (thousands of metric tons) and cause of accident for 48 tankers are saved in the **OILSPILL** file. (Note: Delete the two tankers with oil spills of unknown cause.) Conduct an analysis of variance (at $\alpha=0.01$) to compare the mean spillage amounts for the for types of accident: (1) collision, (2) grounding, (3) fire/explosion, and (4) hull failure. Interpret your results.

Output Table for Problem 10.111

	A	B	C	D	E	F	G
1	Anova: Single Factor						
2							
3	SUMMARY						
4	Groups	Count	Sum	Average	Variance		
5	collision	10	766	76.6	4950.933		
6	fire	12	851	70.91667	3681.538		
7	grounding	14	669	47.78571	810.3352		
8	hullfail	12	653	54.41667	3179.538		
9							
10							
11	ANOVA						
12	Source of Variation	SS	df	MS	F	P-value	F crit
13	Between Groups	6575.889	3	2191.963	0.738687	0.534652	4.260643
14	Within Groups	130564.6	44	2967.377			
15							
16	Total	137140.5	47				
17							

Problem 10.93, Impact of Vitamin B Supplement (page 543)

In the *Journal of Nutrition* (July 1995), University of Georgia researchers examined the impact of a vitamin-B supplement (nicotinamide) on the kidney. The experimental subjects were 28 Zucker rats-a species that tends to develop kidney problems. Half of the rats were classified as obese and half as lean. Within each group, half were randomly assigned to receive a vitamin-B-supplemented diet and half were not. Thus, a 2x2 factorial experiment was conducted with seven rats assigned to each of the four combinations of size (lean or obese) and diet (supplemental or not). One of the response variables measured was weight (in grams) of the kidney at the end of a 20-week feeding period. The data (simulated from the summary information provided in the journal article) are shown in the table. Conduct an analysis of variance on the data. Summarize the results in an ANOVA table. Conduct the appropriate ANOVA F-tests at $\alpha = 01$.

		Diet	
		Regular	Vitamin B
	Lean	1.62	1.51
	Lean	1.80	1.65
	Lean	1.71	1.45
	Lean	1.81	1.44
	Lean	1.47	1.63
	Lean	1.37	1.35
	Lean	1.71	1.66
Size	Obese	2.35	2.93
	Obese	2.97	2.72
	Obese	2.54	2.99
	Obese	2.93	2.19
	Obese	2.84	2.63
	Obese	2.05	2.61
	Obese	2.82	2.64

Solution for Problem 10.93

	A	B	C	D	E	F	G
1	Anova: Two-Factor With Replication						
2							
3	SUMMARY	regular	vitaminB	Total			
4	*lean*						
5	Count	7	7	14			
6	Sum	11.49	10.69	22.18			
7	Average	1.641429	1.527143	1.584286			
8	Variance	0.027748	0.014757	0.023134			
9							
10	*obese*						
11	Count	7	7	14			
12	Sum	18.5	18.71	37.21			
13	Average	2.642857	2.672857	2.657857			
14	Variance	0.117924	0.068157	0.086126			
15							
16	*Total*						
17	Count	14	14				
18	Sum	29.99	29.4				
19	Average	2.142143	2.1				
20	Variance	0.337234	0.391677				
21							
22							
23	ANOVA						
24	*Source of Variation*	*SS*	*df*	*MS*	*F*	*P-value*	*F crit*
25	Sample	8.067889	1	8.067889	141.1792	1.533E-11	7.822871
26	Columns	0.012432	1	0.012432	0.217549	0.6451199	7.822871

H ◀ ▶ H Sheet4 / Sheet1 / **Sheet6** / Sheet2 / Sheet3 / ⏳

	A	B	C	D	E	F	G	H
8	Variance	0.027748	0.014757	0.023134				
9								
10	obese							
11	Count	7	7	14				
12	Sum	18.5	18.71	37.21				
13	Average	2.642857	2.672857	2.657857				
14	Variance	0.117924	0.068157	0.086126				
15								
16	Total							
17	Count	14	14					
18	Sum	29.99	29.4					
19	Average	2.142143	2.1					
20	Variance	0.337234	0.391677					
21								
22								
23	ANOVA							
24	Source of Variation	SS	df	MS	F	P-value	F crit	
25	Sample	8.067889	1	8.067889	141.1792	1.533E-11	7.822871	
26	Columns	0.012432	1	0.012432	0.217549	0.6451199	7.822871	
27	Interaction	0.036432	1	0.036432	0.637523	0.4324398	7.822871	
28	Within	1.371514	24	0.057146				
29								
30	Total	9.488268	27					
31								
32								
33								

Sheet4 Sheet1 **Sheet6** Sheet2 Sheet3

11.1   Introduction

In Chapters 11-13, McClave and Sincich begin to introduce students to the basics of forecasting, starting with the simple linear regression model presented in Chapter 11.   Students are given methods to investigate relationships between variables. In the simplest case, two variables are studied in which the independent variable, x, is used to predict the dependent variable, y.

PHStat can be used to calculate coefficients of correlation, coefficients of determination, and the output necessary to analyze the usefulness of the regression and the regression coefficients.

The following examples will be used to illustrate techniques in PHStat and Excel.

    Section 11.7, Table 11.7 data, page 609 (Fire Insurance Company)
    Sample Exercise #1: Problem 11.129, page 623 (Study of Fertility Rates)
    Sample Exercise #2: Problem 11.19, page 574 (College Protests of Labor)

11.2   The Scatterplot

When attempting to determine whether there is a linear relationship between two variables, the first step is to plot the data. A visual inspection can help the user determine whether to proceed with the simple linear regression analysis.

Within Chapter 2 of this manual, the steps for constructing a scattergram are outlined; the scatterplot of Chapter 11 is the same as the scattergram of Chapter 2. Refer to those steps to construct a scatterplot for the following example.

The following example from the text will be used to illustrate how to analyze bivariate data using simple linear regression. Please refer to page 610 of the text, Table 11.7.

Suppose a fire insurance company wants to relate the amount of fire damage in major residential fires to the distance between the burning house and the nearest fire station. The study is to be conducted in a large suburb of a major city; a sample of 15 recent fires in this suburb is selected. The amount of damage, y, and the distance between the fire and the nearest fire station, x, are recorded for each fire. Generate regression statistics.

Distance from Fire Station x (miles)	Fire Damage y (thousands of dollars)
3.4	26.2
1.8	17.8
4.6	31.3
2.3	23.1
3.1	27.5
5.5	36.0
0.7	14.1
3.0	22.3
2.6	19.6
4.3	31.3
2.1	24.0
1.1	17.3
6.1	43.2
4.8	36.4
3.8	26.1

The scatterplot for this data is:

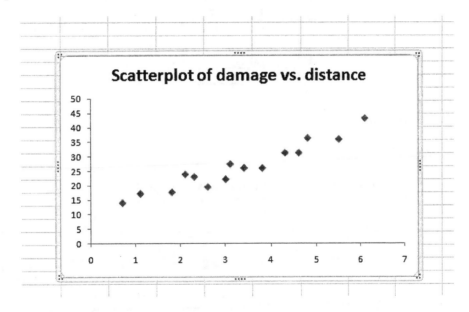

By visual inspection, it looks as if there is a strong positive linear correlation between distance from fire station and fire damage in thousands of dollars.

11.3   Simple Linear Regression: A Complete Model

In this section, a complete analysis of regression will be provided. The goal is first to analyze all elements necessary to determine if a straight-line regression is an appropriate model for the data. Once determining that the model is appropriate, the second goal is to interpret the coefficient of correlation, coefficient of determination, and regression coefficients.

Again refer to page 610 of the text, Table 11.7. This data will be used to demonstrate a complete analysis.

Steps for conducting simple linear regression:

Step 1: Open a clean workbook and enter the data for the x variable into column A and the data for the y variable into column B. It is important that the data be entered such that each row represents one observation. Include labels in the first row.

	A	B	C
1	distance	damage	
2	3.4	26.2	
3	1.8	17.8	
4	4.6	31.3	
5	2.3	23.1	
6	3.1	27.5	
7	5.5	36	
8	0.7	14.1	
9	3	22.3	
10	2.6	19.6	
11	4.3	31.3	
12	2.1	24	
13	1.1	17.3	
14	6.1	43.2	
15	4.8	36.4	
16	3.8	26.1	
17			
18			

Step 2: Click on the **Add-Ins** tab.

Step 3: Click on **PHStat**.

Step 4: Drop down to **Regression** and click on **Simple Linear Regression…**.

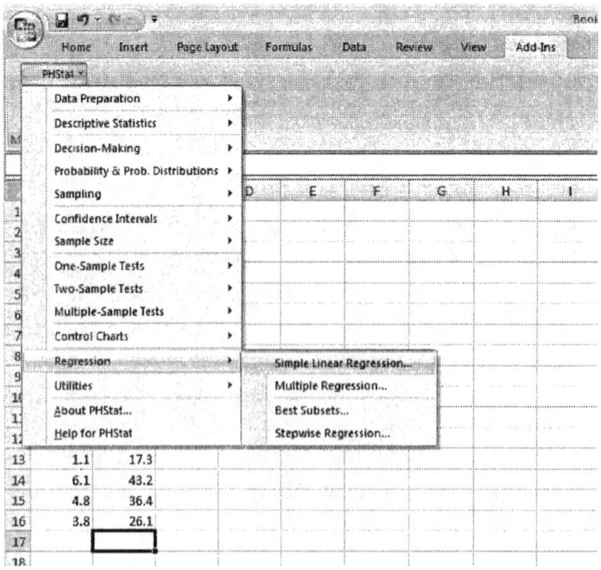

Step 5: A box for the **Simple Linear Regression** options will appear.

- Click on the entry box by **Y Variable Cell Range**. Enter the rows and column where the data for the y variable are located. This may be done by typing the range in manually or by clicking and dragging the appropriate range of data cells. In this example the data are located in Sheet1!$B$1:$B$16.
- Click on the entry box by **X Variable Cell Range**. Enter the rows and column where the data for the x variable are located. This may be done by typing the range in manually or by clicking and dragging the appropriate range of data cells. In this example the data are located in Sheet1!$A$1:$A$16.
- The box in front of **First cells in both ranges contain labels** should have a check in it since this is how the data were entered.
- The default value for the **Confidence level for regression coefficients** is 95%. For this example keep 95%, but it could be changed for other problems.
- The boxes in front of **Regression Statistics Table, ANOVA and Coefficients Table,** and **Residual Plot** should have a check in them since this output will be needed.
- There is a blank box to enter a title. Use this when computing several analyses in order to keep the output organized.
- Click the box in front of **Confidence and Prediction Interval for X =** so that it has a check in it. This option computes the predicted value of y, given a particular x value. Click on the entry box to the right of **Confidence and Prediction Interval for X =** and enter 3.5. On page 612 of the

textbook, the predicted value of y is calculated for x = 3.5. There is also a box to enter the **Confidence level for interval estimates**. Since the text uses 95% in this example, enter 95 into this box.

- Click the **OK** button and the regression results will appear in a new sheet labeled **SLR** and the prediction results will appear in a new sheet labeled **Estimate**.

Use the PHStat output to follow the steps on pages 609 – 612 in the text:

	A	B	C	D	E	F	G
1	Text Table 11.7						
2							
3	Regression Statistics						
4	Multiple R	0.960977715					
5	R Square	0.923478169					
6	Adjusted R Square	0.917591874					
7	Standard Error	2.316346184					
8	Observations	15					
9							
10	ANOVA						
11		df	SS	MS	F	Significance F	
12	Regression	1	841.766358	841.766358	156.8861596	1.2478E-08	
13	Residual	13	69.75097535	5.365459643			
14	Total	14	911.5173333				
15							
16		Coefficients	Standard Error	t Stat	P-value	Lower 95%	Upper 95%
17	Intercept	10.27792855	1.420277811	7.236562082	6.58556E-06	7.20960489	13.34625221
18	distance	4.919330727	0.392747749	12.52542054	1.2478E-08	4.070850801	5.767810653
19							
20							
21							
22	RESIDUAL OUTPUT						
23							
24	Observation	Predicted damage	Residuals				
25	1	27.00365302	-0.803653021				
26	2	19.13272386	-1.332723858				

(Sheet tabs: Estimate, SLR, Sheet1, Sheet2, Sheet3)

distance Residual Plot (chart in top right, Residuals vs distance)

Step 1: The first piece of output to consider is the p-value within the ANOVA table in the sheet labeled **SLR**. In this case it is very small, 1.2478×10^{-8} (refer to the value under *Significance F*). This is smaller than any reasonable alpha level, therefore reject the null hypothesis that there is no linear relationship between distance and damage and determine that a straight line model is appropriate in this case.

Step 2: In this case the slope of the line is 4.91933 and the intercept is 10.2779. These values are listed under *Coefficients*. Therefore, the linear equation for this model is y = 4.91933x + 10.2779.

Step 3: The estimate of the standard deviation is 2.316346. This is labeled as Standard Error.

Step 4: The p-value for testing the slope is given in the table under *P-value* and it is 1.2478×10^{-8}. Clearly the slope is significant. The coefficient of determination is also given. In this case it is $R^2 = 0.923478169$, so 92.35% of the changes in the dollar value of fire damage, y, can be explained by the distance from the fire station, x. The coefficient of correlation, r, is labeled multiple R in the output. In this example r = 0.960977715. This value of r supports that there is a strong linear relationship between distance and damage.

Step 5: Refer to the sheet labeled **Estimate**. Recall that the value x = 3.5 was entered into PHStat. The resulting predicted y, labeled as YHat, is 27.49559 and the 95% confidence interval for this estimate is (22.32394, 32.66723).

◢	A	B	
1	Confidence Interval Estimate		
2			
3	Data		
4	X Value	3.5	
5	Confidence Level	95%	
6			
7	Intermediate Calculations		
8	Sample Size	15	
9	Degrees of Freedom	13	
10	t Value	2.160369	
11	XBar, Sample Mean of X	3.28	
12	Sum of Squared Differences from XBar	34.784	
13	Standard Error of the Estimate	2.316346	
14	h Statistic	0.068058	
15	Predicted Y (YHat)	27.49559	
16			
17	For Average Y		
18	Interval Half Width	1.305483	
19	Confidence Interval Lower Limit	26.1901	
20	Confidence Interval Upper Limit	28.80107	
21			
22	For Individual Response Y		
23	Interval Half Width	5.171646	
24	Prediction Interval Lower Limit	22.32394	
25	Prediction Interval Upper Limit	32.66723	
26			

Also notice that there is a residual plot in the output in the sheet labeled **SLR**. Chapter 12 of the text describes how to interpret a residual plot. The important aspect of this is that the residuals do not show any trends.

11.4   Sample Exercises

The following exercises were selected from the textbook to test the reader's understanding of PHStat and its usage within the chapter. Try to replicate the output in PHStat for the following problems.

Problem 11.129, Study of Fertility Rates, page 623.

The fertility rate of a country is defined as the number of children a woman citizen bears, on average, in her lifetime. *Scientific American* (Dec. 1993) reported on the declining fertility rate in developing countries. The researchers found that family planning can have a great effect on fertility rate. The accompanying table gives the fertility rate, y, and contraceptive prevalence, x, (measured as the percentage of married women who use contraception for each of 27 developing countries. According to the researchers, "the data reveal that differences in contraceptive prevalence explain about 90% of the variation in fertility rates." Do you concur?

Country	Contraceptive Prevalence x	Fertility Rate y
Mauritius	76	2.2
Thailand	69	2.3
Colombia	66	2.9
Costa Rica	71	3.5
Sri Lanka	63	2.7
Turkey	62	3.4
Peru	60	3.5
Mexico	55	4.0
Jamaica	55	2.9
Indonesia	50	3.1
Tunisia	51	4.3
El Salvador	48	4.5
Morocco	42	4.0
Zimbabwe	46	5.4
Egypt	40	4.5
Bangladesh	40	5.5
Botswana	35	4.8
Jordan	35	5.5
Kenya	28	6.5
Guatemala	24	5.5
Cameroon	16	5.8
Ghana	14	6.0
Pakistan	13	5.0
Senegal	13	6.5
Sudan	10	4.8
Yemen	9	7.0
Nigeria	7	5.7

Solution:

Solution for Exercise 11.129 of the text	
Regression Statistics	
Multiple R	0.865019842
R Square	0.748259327
Adjusted R Square	0.738189701
Standard Error	0.695710716
Observations	27

Problem 11.21, College protests of labor exploitations, page 574

Refer to the *Journal of World-Systems Research* (Winter 2004) study of student "sit-ins" for a "sweat free campus" at universities in Exercise 2.145 (p.93). Recall that the **SITIN** file contains data on the duration (in days) of each sit-in as well as the number of student arrests. The data for 5 sit-ins where there was at least one arrest are shown below. Let y = number of arrests and x = duration.
   a. Give the equation of a straight-line model relating y to x.
   b. Fit the model to the data for the 5 sit-ins using the methods of least squares. Give the equation of the least squares prediction equation.

Sit-In	University	Duration	Number of Arrests
12	Wisconsin	4	54
14	SUNY Albany	1	11
15	Oregon	3	14
17	Iowa	4	16
18	Kentucky	1	12

Solution:

SUMMARY OUTPUT

Regression Statistics	
Multiple R	0.6009
R Square	0.3611
Adjusted R Square	0.1481
Standard Error	16.9132
Observations	5

ANOVA

	df	SS	MS	F	Significance F
Regression	1	485.0261	485.0261	1.6956	0.2838
Residual	3	858.1739	286.0580		
Total	4	1343.2000			

	Coefficients	Standard Error	t Stat	P-value	Lower 95%	Upper 95%
Intercept	2.52	16.35	0.15	0.89	-49.52	54.56
Duration	7.26	5.58	1.30	0.28	-10.48	25.01

12.1 Introduction

In Chapter 11, a straight-line regression model was developed in order to define the relationship between a dependent variable, y, and a single independent variable, x. In Chapter 12, the basic regression model is extended in order to analyze the relationship between the dependent variable, y, and multiple independent variables. PHStat and Excel will be used to illustrate how to build models using quantitative independent variables and to test the utility of the model.

The following examples will be used to illustrate techniques in PHStat and Excel.

Example 12.1, page 631 (Antique Grandfather Clocks)
Example 12.2, page 635 (Antique Grandfather Clocks)
Example 12.3, page 636 (Antique Grandfather Clocks)
Example 12.4, page 641 (Antique Grandfather Clocks)
Example 12.5, page 649 (Antique Grandfather Clocks)
Example 12.6, page 655 (Antique Grandfather Clocks)
Example 12.7, , page 661 (Energy Conservation)
Example 12.13, page 699 (Executive Salary)
Sample Exercise #1: Problem 12.20, page 645 (Study of contaminated fish)
Sample Exercise #2: Problem 12.33, page 652 (Study of contaminated fish)
Sample Exercise #3: Problem 12.156, page 733 (Carp Experiment)

12.2   Multiple Regression Models

Please refer to the process in the textbook to review how to analyze a multiple regression model. In this section, a first-order model will be developed for independent variables that are quantitative in nature.

The following example from the text will be used to illustrate how to analyze multivariate data using multiple regression. Several concepts will be presented using the same example. Please refer to page 631 of the text, Example 12.1; page 635, Example 12.2; page 636, Example 12.3; and page 641, Example 12.4.

A collector of antique grandfather clocks knows that the price received for the clocks increases linearly with the age of the clock. Moreover, the collector hypothesizes that the auction price of the clocks will increase linearly as the number of bidders increases. Thus, the following first-order model is hypothesized:

$$y = \beta_0 + \beta_1 x_1 + \beta_2 x_2 + \varepsilon$$

where
y = Auction price
x_1 = Age of clock (years)
x_2 = Number of bidders

A sample of 32 auction prices of grandfather clocks, and the number of bidders, is given in Table 12.1.

Steps for conducting multiple regression:

Step 1: Open a clean workbook and enter the data for the y variable into column C and the data for the x variables into columns A and B. For this example enter the x_1 values into column A and the x_2 values into column B. It is important that the data be entered such that each row represents one observation. Include labels in the first row.

Step 2: Click on the **Add-Ins** tab.

Step 3: Click on **PHStat**.

Step 4: Drop down to **Regression** and click on **Multiple Regression….**

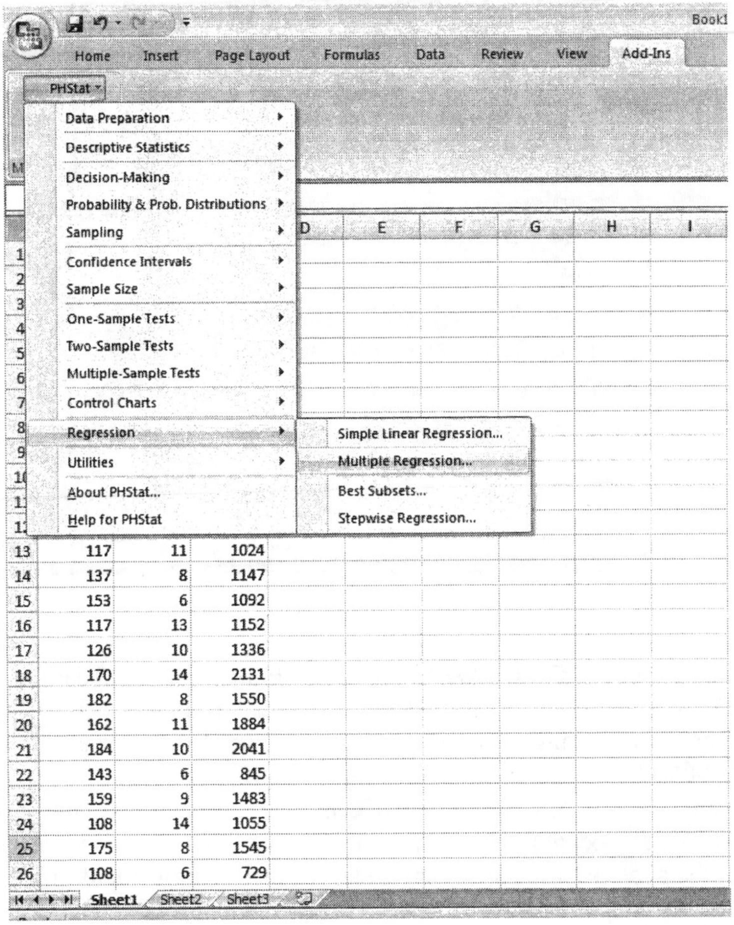

Step 5: A box for the **Multiple Regression** options will appear.

- Click on the entry box by **Y Variable Cell Range**. Enter the rows and column where the data for the y variable are located. This may be done by typing the range in manually or by clicking and dragging the appropriate range of data cells. In this example the data are located in Sheet1!$C$1:$C$33.
- Click on the entry box by **X Variables Cell Range**. Enter the rows and column where the data for the x variables are located. This may be done by typing the range in manually or by clicking and dragging the appropriate range of data cells. In this example the data are located in Sheet1!$A$1:$B$33.
- The box in front of **First cells in both ranges contain labels** should have a check in it since this is how the data were entered.
- The default value for the **Confidence level for regression coefficients** is 95%. For this example enter 90 since the 90% confidence interval is used Example 12.3.
- The boxes in front of **Regression Statistics Table** and **ANOVA and Coefficients Table** should have a check in them since this output will be needed.
- There is a blank box to enter a title. Use this when computing several analyses in order to keep the output organized.
- Click the **OK** button and the regression results will appear in a new sheet labeled **MR**

	A	B	C	D	E	F	G	H	I	J
1	Text Example 12.1									
2										
3	Regression Statistics									
4	Multiple R	0.94463957								
5	R Square	0.892343916								
6	Adjusted R Square	0.884919359								
7	Standard Error	133.4846678								
8	Observations	32								
9										
10	ANOVA									
11		df	SS	MS	F	Significance F				
12	Regression	2	4283062.96	2141531.48	120.1881617	9.21636E-15				
13	Residual	29	516726.5399	17818.15655						
14	Total	31	4799789.5							
15										
16		Coefficients	Standard Error	t Stat	P-value	Lower 95%	Upper 95%	Lower 90.0%	Upper 90.0%	
17	Intercept	-1338.95134	173.8094707	-7.703558013	1.70581E-08	-1694.431617	-983.4710644	-1634.2757	-1043.62698	
18	x1	12.7405741	0.904740307	14.08202331	1.69276E-14	10.89017243	14.59097576	11.2033054	14.2778428	
19	x2	85.95298437	8.728523289	9.847368395	9.34495E-11	68.10115008	103.8048187	71.1221148	100.783854	
20										
21										

Example 12.1 in the text presents scatterplots for y and x_1 and for y and x_2. Although not presented in this section of the manual, the steps for plotting scatterplots (scattergrams) may be found in Chapter 2.

In order to conduct the overall F-test (as in Example 12.4), notice that in the output the F test statistic is 120.1881617 and the p-value (found under the label *significance F* in the output) is 9.21636×10^{-15}. Therefore it is believed that at least one of the model coefficients is nonzero and the model appears to be useful in predicting auction prices.

Notice that the adjusted R^2 is 0.884919359, which is located under the heading *Regression Statistics*. Therefore, the model has explained about 88.49% of the total sample variation in auction prices after adjusting for sample size and the number of independent variables in the model.

Since the hypothesized model is determined to be useful, it makes sense to determine the estimates of the coefficients (as in Example 12.1). These estimates may be found in the PHStat output under *Coefficients*. In this case $\hat{\beta}_0 = -1338.95134$, $\hat{\beta}_1 = 12.7405741$, and $\hat{\beta}_2 = 85.95298437$. Therefore the least squares prediction equation is $\hat{y} = -1338.95 + 12.74x_1 + 85.95x_2$. Refer to Example 12.2 for the interpretation of these estimates.

Example 12.1 also shows how to determine the SSE and the estimate of σ. In the PHStat output the SSE is located under the heading of *SS* for Residual, so it is 516726.5399 and estimate of σ^2 may be found under the heading of *MS* for Residual, so it is 17818.15655. Take the square root to get the estimate of σ = 133.5

In order to test the hypothesis that the mean auction price of a clock increases as the number of bidders increases when age is held constant (as in Example 12.3), notice that the t test statistic for β_2 may be found in the PHStat output under the heading t *Stat* for x_2. In this case the test statistic is 9.847368395. This is more extreme than the critical value of 1.699 presented in the text. Therefore reject the null hypothesis and conclude that the mean auction price of a clock increases with the number of bidders

The 90% confidence interval for β_1 and its interpretation may be found in Example 12.4. The upper and lower limits of the interval may be found in the PHStat output under the heading *Lower 90%* and *Upper*

90% for x_1. The clock collector is 90% confident that β_1 falls between 11.2033054 and 14.2778428. Notice that the 95% interval is always generated; in order to generate another interval, specify the percentage interval in Data Options. The PHStat output also displays the confidence interval limits for the other coefficients.

Notice that when entering the options to produce the output for the multiple regression output, the options for **Durbin-Watson statistics**, **Coefficients of Partial Determination**, **VIF**, and **Confidence & Prediction Interval Estimates** were not chosen. There are many useful tools for model building and analysis, but many of them are beyond the scope of this book. Chapter 12 of the text is an introduction to the basics of model building; this area of analysis is very complex.

12.3   Using the Model for Estimation and Prediction

In this section, PHStat and Excel will be used to estimate the mean value of the dependent variable once particular values of the independent variables are substituted into the prediction equation. Please refer to page 649 of the text, Example 12.5.

Refer to Examples 12.1 – 12.4 and the first order model $E(y) = \beta_0 + \beta_1 x_1 + \beta_2 x_2$, where y = auction price of a grandfather clock, x_1 = age of the clock, and x_2 = number of bidders.
   a. Estimate the average auction price for all 150-year-old clocks sold at an auction with 10 bidders. Use a 95% confidence interval. Interpret the result.
   b. Predict the auction price for a single 150-year-old clock sold at an auction with 10 bidders. Use a 95% prediction interval. Interpret the result.

Steps for estimating E(y) and predicting y in multiple regression:

Step 1: Open a clean workbook and enter the data for the y variable into column C and the data for the x variables into columns A and B. For this example enter the x_1 values into column A and the x_2 values into column B. It is important that the data be entered such that each row represents one observation. Include labels in the first row.

Step 2: Click on the **Add-Ins** tab.

Step 3: Click on **PHStat**.

Step 4: Drop down to **Regression** and click on **Multiple Regression…**.

Step 5: A box for the **Multiple Regression** options will appear.

- Click on the entry box by **Y Variable Cell Range**. Enter the rows and column where the data for the y variable are located. This may be done by typing the range in manually or by clicking and dragging the appropriate range of data cells. In this example the data are located in Sheet1!$C$1:$C$33.
- Click on the entry box by **X Variables Cell Range**. Enter the rows and column where the data for the x variables are located. This may be done by typing the range in manually or by clicking and dragging the appropriate range of data cells. In this example the data are located in Sheet1!$A$1:$B$33.
- The box in front of **First cells in both ranges contain labels** should have a check in it since this is how the data were entered.
- The default value for the **Confidence level for regression coefficients** is 95%.
- The boxes in front of **Regression Statistics Table** and **ANOVA and Coefficients Table** should have a check in them. The output will not be explicitly used for this example, but PHStat uses some of the calculations in these to generate the desired output.
- There is a blank box to enter a title. Use this when computing several analyses in order to keep the output organized.

- Click on the box in front of **Confidence Interval Estimate & Prediction Interval** so that there is a check in it. In the box next to **Confidence level for interval estimates** enter 95 since 95% is used in Example 12.5.
- Click the **OK** button and the regression results will appear in a new sheet labeled **MR**. Another new sheet labeled **Intervals** will also appear.

Step 6: The new sheet, **Intervals**, contains a PHStat Users Note; press the Delete key on the computer keyboard to eliminate this message. Notice that values on this sheet are already calculated assuming that the x values of interest are 0. Click on the cell to the right of **x1 given value** and replace the 0 with 150, the x_1 of interest in this example. Click on the cell to the right of **x2 given value** and replace the 0 with 10, the x_2 of interest in this example.

	A	B	C	D	E	F	G
1	Confidence Interval Estimate and Prediction Interval						
2							
3	Data						
4	Confidence Level	95%					
5		1					
6	x1 given value	0					
7	x2 given value	0					
8							
9	X'X	32					
10		4638					
11		305					
12							
13	Inverse of X'X	1.695447					
14		-0.00773	4.59E-05	0.000112			
15		-0.05705	0.000112	0.004276			
16							
17	X'G times Inverse of X'X	1.695447	-0.00773	-0.05705			
18							
19	[X'G times Inverse of X'X] times XG	1.695447					
20	t Statistic	2.04523					
21	Predicted Y (YHat)	-1338.95					
22							
23	For Average Predicted Y (YHat)						
24	Interval Half Width	355.4803					
25	Confidence Interval Lower Limit	-1694.43					
26	Confidence Interval Upper Limit	-983.471					

PHStat2 User Note:
Enter the values for the given X's in the cell range B6:B7. (You can interactively change these values at any time.)

(Before continuing, press the Delete key to delete this note.)

Intervals / MR / Sheet1 / Sheet2 / Sheet3
Ready

The 95% confidence interval for the average auction price for all 150-year-old clocks sold at an auction with 10 bidders can be found under the heading **For Average Predicted Y (YHat)**. The upper and lower limits for the interval are 1381.398 and 1481.931.

	A	B	C	D	E	F
1	Confidence Interval Estimate and Prediction Interval					
2						
3	Data					
4	Confidence Level	95%				
5		1				
6	x1 given value	150				
7	x2 given value	10				
8						
9	X'X	32	4638	305		
10		4638	695486	43594		
11		305	43594	3157		
12						
13	Inverse of X'X	1.695447	-0.00773	-0.05705		
14		-0.00773	4.59E-05	0.000112		
15		-0.05705	0.000112	0.004276		
16						
17	X'G times Inverse of X'X	-0.03463	0.000285	0.002574		
18						
19	[X'G times Inverse of X'X] times XG	0.033901				
20	t Statistic	2.04523				
21	Predicted Y (YHat)	1431.665				
22						
23	For Average Predicted Y (YHat)					
24	Interval Half Width	50.26636				
25	Confidence Interval Lower Limit	1381.398				
26	Confidence Interval Upper Limit	1481.931				

H ◀ ▶ H Intervals MR Sheet1 Sheet2 Sheet3
Ready

The 95% prediction interval for the auction price for a single 150-year-old clock sold at an auction with 10 bidders can be found under the heading **For Individual Response Y**. The upper and lower limits for the interval are 1154.069 and 1709.26.

		B
20	t Statistic	2.04523
21	Predicted Y (YHat)	1431.665
22		
23	For Average Predicted Y (YHat)	
24	Interval Half Width	50.26636
25	Confidence Interval Lower Limit	1381.398
26	Confidence Interval Upper Limit	1481.931
27		
28	For Individual Response Y	
29	Interval Half Width	277.5958
30	Prediction Interval Lower Limit	1154.069
31	Prediction Interval Upper Limit	1709.26

H ◀ ▶ H Intervals MR Sheet1 Sheet2 Sheet3
Ready

The interpretation for these values can be found on page 649 of the text.

12.4   Model Building: Interaction Models

The following example from the text will be used to illustrate how to analyze multivariate data using multiple regression with interaction terms. Please refer to page 655 of the text, Example 12.6. The raw data for this example is taken from Table 12.1 from the text.

Suppose the collector of grandfather clocks, having observed many auctions, believes that the rate of increase of the auction price with age will be driven upward by a large number of bidders. Consequently, the interaction model is proposed:

$$y = \beta_0 + \beta_1 x_1 + \beta_2 x_2 + \beta_3 x_1 x_2 + \varepsilon$$

a.   Use the global F-test at $\alpha = 0.05$ to test the overall utility of the model.
b.   Test the hypothesis that the price-age slope increases as the number of bidders increases.
c.   Estimate the change in auction price of a 150-year-old grandfather clock, y, for each additional bidder.

Steps for conducting multiple regression with interaction terms:

Step 1: Open a clean workbook and enter the data for the y variable into column D and the data for the x variables into columns A and B. For this example enter the x_1 values into column A and the x_2 values into column B. It is important that the data be entered such that each row represents one observation. Include labels in the first row.

Step 2: Highlight the cells C2:C33. Click on the function box and enter = **A2:A33*B2:B33** and then press the CTRL, SHIFT, and ENTER keys on the computer keyboard simultaneously. This will multiply the values in column A with the corresponding values of column B. This new column is needed since the model contains an interaction term ($x_1 x_2$) that is made up of the two given x variables. Enter a label for this column into the first row. See pictures labeled **Steps 2a-2d**.

Step 2c					Step 2d			

Home | Insert | Page Layout | Formulas | Data

Cut
Copy
Paste Format Painter
Clipboard

Calibri 11 A A
B I U

Font

C2 fx {= A2:A33*B2:B33}

Book2

	A	B	C	D	E
1	x1	x2		y	
2	127	13	1651	1235	
3	115	12	1380	1080	
4	127	7	889	845	
5	150	9	1350	1522	

Home | Insert | Page Layout | Formulas

Cut
Copy
Paste Format Painter
Clipboard

Calibri 11
B I U

Font

C2 fx {= A2:A33*B

Book2

	A	B	C	D
1	x1	x2	x1x2	y
2	127	13	1651	1235
3	115	12	1380	1080
4	127	7	889	845
5	150	9	1350	1522
6	156	6	936	1047

Step 3: Click on the **Add-Ins** tab.

Step 4: Click on **PHStat**.

Step 5: Drop down to **Regression** and click on **Multiple Regression....**

Step 6: A box for the **Multiple Regression** options will appear.

Multiple Regression

Data

Y Variable Cell Range: Sheet1!$D$1:$D$33

X Variables Cell Range: Sheet1!$A$1:$C$33

☑ First cells in both ranges contain label

Confidence level for regression coefficients: 95 %

Regression Tool Output Options

☑ Regression Statistics Table

☑ ANOVA and Coefficients Table

☐ Residuals Table

☐ Residual Plots

Output Options

Title: Text Example 12.6

☐ Durbin-Watson Statistic

☐ Coefficients of Partial Determination

☐ Variance Inflationary Factor (VIF)

☐ Confidence Interval Estimate & Prediction Interval

Confidence level for interval estimates: %

Help OK Cancel

- Click on the entry box by **Y Variable Cell Range**. Enter the rows and column where the data for the y variable are located. This may be done by typing the range in manually or by clicking and dragging the appropriate range of data cells. In this example the data are located in Sheet1!$D$1:$D$33.
- Click on the entry box by **X Variables Cell Range**. Enter the rows and column where the data for the x variables are located. This may be done by typing the range in manually or by clicking and dragging the appropriate range of data cells. In this example the data are located in Sheet1!$A$1:$C$33.
- The box in front of **First cells in both ranges contain labels** should have a check in it since this is how the data were entered.
- The default value for the **Confidence level for regression coefficients** is 95%. For this example use 95%.
- The boxes in front of **Regression Statistics Table** and **ANOVA and Coefficients Table** should have a check in them since this output will be needed.
- There is a blank box to enter a title. Use this when computing several analyses in order to keep the output organized.
- Click the **OK** button and the regression results will appear in a new sheet labeled **MR**.

	A	B	C	D	E	F	G	H
1	Text Example 12.6							
2								
3	*Regression Statistics*							
4	Multiple R	0.976668248						
5	R Square	0.953880866						
6	Adjusted R Square	0.948939531						
7	Standard Error	88.91451215						
8	Observations	32						
9								
10	ANOVA							
11		*df*	*SS*	*MS*	*F*	*Significance F*		
12	Regression	3	4578427.367	1526142.456	193.0410958	8.34956E-19		
13	Residual	28	221362.1332	7905.790471				
14	Total	31	4799789.5					
15								
16		*Coefficients*	*Standard Error*	*t Stat*	*P-value*	*Lower 95%*	*Upper 95%*	
17	Intercept	320.4579934	295.141285	1.085778268	0.286836952	-284.1115146	925.0275013	
18	x1	0.878142475	2.03215593	0.43212357	0.668961315	-3.284540189	5.04082514	
19	x2	-93.26482436	29.89161615	-3.120099759	0.004164589	-154.4950236	-32.03462517	
20	x1x2	1.297845824	0.212332598	6.112324878	1.35347E-06	0.862902218	1.732789429	
21								
22								

The *F*-test statistic for the global F test is 193.0410958 and the p-value is 8.34956×10^{-19} (this is found under the label *Significance F*). The p-value is certainly less than 0.05. Therefore the model is a statistically useful predictor of auction price.

The t test statistic for the interaction parameter, β_3, is 6.11232 (this is found under the label *t Stat*) with a p-value of 1.35347×10^{-6}. Since the text chooses a one tail test, the p-value in the PHStat output should be divided by 2. Certainly the appropriate p-value is less than 0.05. Thus, as the number of bidders increases, the rate of change of the mean price of the clocks with age increases.

In order to estimate the change in auction price of a 150-year-old grandfather clock for each additional bidder, the values of $\hat{\beta}_2$ and $\hat{\beta}_3$ are needed. These may be found under the label *Coefficients* in the output. For this example $\hat{\beta}_2 = -93.2648$ and $\hat{\beta}_3 = 1.2978$. Notice that in the model the coefficient for x_2 is β_2 and the coefficient for x_1x_2 is β_3. Therefore the estimated rate of change of y for a unit increase in x_2 is $y = -93.2648 + 1.2978(150) = 101.41$.

12.5   Model Building: Quadratic and Other Higher-Order Models

Models that allow for curvature in relationships are presented in this section. PHStat and Excel may be used to construct higher-order models through the creation of additional columns to represent squared or cubed terms of x, the independent variable.

The following example from the text will be used to illustrate how to analyze multivariate data using multiple regression with higher-order terms. Please refer to page 661 of the text, Example 12.7.

In all-electric homes, the amount of electricity expended is of interest to consumers, builders, and groups involved with energy conservation. Suppose we wish to investigate the monthly electrical usage, y, in all-electric homes and its relationship to the size, x, of the home. Moreover, suppose we think that monthly electrical usage in all-electric homes is related to the size of the home by the quadratic model

$$y = \beta_0 + \beta_1 x + \beta_2 x^2 + \varepsilon$$

To fit the model, the values of y and x are collected for 10 homes during a particular month. The data are shown in Table 12.3. Answer parts a through f in the text by generating output in PHStat.

Steps for conducting multiple regression with higher order terms:

Step 1: Open a clean workbook and enter the data for the y variable into column C and the data for the x variable into column A. It is important that the data be entered such that each row represents one observation. Include labels in the first row.

Step 2: Highlight the cells B2:B11. Click on the function box and enter = **A2:A11^2** and then press the CTRL, SHIFT, and ENTER keys on the computer keyboard simultaneously. This will square the values in column A. This new column is needed since the model contains a quadratic term. Enter a label for this column into the first row. See **Step 2** from the previous section for illustrations of a similar process.

Step 3: Click on the **Add-Ins** tab.

Step 4: Click on **PHStat**.

Step 5: Drop down to **Regression** and click on **Multiple Regression….**

Step 6: A box for the **Multiple Regression** options will appear.

- Click on the entry box by **Y Variable Cell Range**. Enter the rows and column where the data for the y variable are located. This may be done by typing the range in manually or by clicking and dragging the appropriate range of data cells. In this example the data are located in Sheet1!$C$1:$C$11.
- Click on the entry box by **X Variables Cell Range**. Enter the rows and column where the data for the x variables are located. This may be done by typing the range in manually or by clicking and dragging the appropriate range of data cells. In this example the data are located in Sheet1!$A$1:$B$11.
- The box in front of **First cells in both ranges contain labels** should have a check in it since this is how the data were entered.
- The default value for the **Confidence level for regression coefficients** is 95%. Use 99% for consistency in this example since the testing is done using $\alpha = 0.01$.
- The boxes in front of **Regression Statistics Table** and **ANOVA and Coefficients Table** should have a check in them since this output will be needed.
- There is a blank box to enter a title. Use this when computing several analyses in order to keep the output organized.

- Click the **OK** button and the regression results will appear in a new sheet labeled **MR**.

	A	B	C	D	E	F	G	H	I	J
1	Text Example 12.7									
2										
3	*Regression Statistics*									
4	Multiple R	0.990901117								
5	R Square	0.981885024								
6	Adjusted R Square	0.976709317								
7	Standard Error	46.80133336								
8	Observations	10								
9										
10	ANOVA									
11		*df*	*SS*	*MS*	*F*	*Significance F*				
12	Regression	2	831069.5464	415534.7732	189.7103041	8.00078E-07				
13	Residual	7	15332.55363	2190.364804						
14	Total	9	846402.1							
15										
16		*Coefficients*	*Standard Error*	*t Stat*	*P-value*	*Lower 95%*	*Upper 95%*	*Lower 99.0%*	*Upper 99.0%*	
17	Intercept	-1216.143887	242.8063685	-5.008698472	0.001550025	-1790.289714	-641.9980598	-2065.8407	-366.44706	
18	x	2.398930177	0.245835602	9.758269998	2.51335E-05	1.817621351	2.980239003	1.5386326	3.25922776	
19	x^2	-0.00045004	5.90766E-05	-7.617907059	0.000124415	-0.000589734	-0.000310346	-0.0006568	-0.0002433	
20										

Example 12.7 in the text presents a scatterplot for y and x. Although not presented in this section of the manual, the steps for plotting scatterplots (scattergrams) may be found in Chapter 2.

The estimates for β_0, β_1, and β_2 may be found in the PHStat output under *Coefficients*. In this example $\hat{\beta}_0$ = -1216.143887, $\hat{\beta}_1$ = 2.398930177, and $\hat{\beta}_2$ = -0.00045004. Therefore the least squares prediction equation is $\hat{y}$ = -1216.14 + 2.3989x - 0.00045x^2. Refer to page 662 - 664 of the text for the interpretation of these estimates.

In order to conduct the global *F*-test, notice that in the output the *F* test statistic is 189.7103041 and the p-value (found under the label *significance F* in the output) is 8.00078×10^{-7}. Therefore it is believed that the model is a useful predictor of electrical usage.

In order to test the hypothesis that there is a concave-downward curvature in the electrical usage-home size relationship, notice that the *t* test statistic for β_2 may be found in the PHStat output under the heading *t Stat* for x_2. In this case the test statistic is -7.617907059 and the p-value is 0.000124415. Since the test presented in the text is one sided and the PHStat output provides a p-value for a two sided test, this p-value must be divided by 2. The p-value is less than 0.01, therefore reject the null hypothesis and conclude that there is evidence of downward curvature in the population.

12.6   Model Building: Stepwise Regression

Stepwise regression is a tool that may assist the researcher when faced with the issue of which of a large set of independent variables to include in a model.

The following example from the text will be used to illustrate how to analyze multivariate data using stepwise regression. Please refer to page 699 of the text, Example 12.13.

An international management consulting company develops multiple-regression models for executive salaries of its client firms. The consulting company has found that models which use the natural logarithm of salary as the dependent variable have better predictive power than those using salary as the dependent variable. A preliminary step in the construction of these models is the determination of the most important independent variables. For one firm, 10 potential independent variables (7 quantitative and 3 qualitative) were measured in a sample of 100 executives. The data, described in Table 12.7, are saved in the **EXECSAL** file. Since it would be very difficult to construct a complete second-order model with all of the 10 independent variables, use stepwise regression to decide which of the 10 variables should be included in the building of the final model for the logarithm of the executive salaries.

Steps for conducting stepwise regression:

Step 1: Open the **EXECSAL** file that came with the textbook.

Step 2: Click on the **Add-Ins** tab.

Step 3: Click on **PHStat**.

Step 4: Drop down to **Regression** and click on **Stepwise Regression...**

Step 5: A box for the **Stepwise Regression** options will appear.

- Click on the entry box by **Y Variable Cell Range**. Enter the rows and column where the data for the y variable are located. This may be done by typing the range in manually or by clicking and dragging the appropriate range of data cells. In this example the data are located in EXECSAL!$A$1:$A$100.
- Click on the entry box by **X Variables Cell Range**. Enter the rows and column where the data for the x variables are located. This may be done by typing the range in manually or by clicking and dragging the appropriate range of data cells. In this example the data are located in EXECSAL!$B$1:$K$100.
- The box in front of **First cells in both ranges contain labels** will have a check by default. Click on this box to remove the check since this file does not include labels for the data.
- The default value for the **Confidence level for regression coefficients** is 95%.
- There are two options under **Stepwise Criteria.** The default option is **p values**. Keep this option; it is the method the text uses in this section.
- There are three options under **Stepwise Options.** The default is **General Stepwise**. Keep this option; it is the method the text uses in this section. In the boxes to the right of **p value to enter** and **p value to remove** enter 0.15, the value used in the text example.

- There is a blank box to enter a title. Use this when computing several analyses in order to keep the output organized.
- Click the **OK** button and the regression results will appear in a new sheet labeled **MR** and **Stepwise**.

	A	B	C	D	E	F	G	H
1	Stepwise Analysis							
2								
3	*Regression Statistics*							
4	Multiple R	0.960679346						
5	R Square	0.922904807						
6	Adjusted R Square	0.914242425						
7	Standard Error	0.076084043						
8	Observations	100						
9								
10	ANOVA							
11		*df*	*SS*	*MS*	*F*	*Significance F*		
12	Regression	10	6.167466203	0.61674662	106.5416975	4.40784E-45		
13	Residual	89	0.515201564	0.005788782				
14	Total	99	6.682667768					
15								
16		*Coefficients*	*Standard Error*	*t Stat*	*P-value*	*Lower 95%*	*Upper 95%*	
17	Intercept	10.0219208	0.148051595	67.6920828	2.44525E-78	9.727745443	10.31609616	
18	X Variable 1	0.027921042	0.001773294	15.74529988	1.83725E-27	0.024397545	0.031444539	
19	X Variable 2	0.029033159	0.00342561	8.475325879	4.56781E-13	0.022226546	0.035839773	
20	X Variable 3	0.22433563	0.017079257	13.13497591	1.52143E-22	0.190399511	0.258271749	
21	X Variable 4	0.00051397	4.92155E-05	10.44323479	3.90898E-17	0.000416179	0.00061176	
22	X Variable 5	0.002048045	0.000524988	3.901129179	0.000185599	0.001004905	0.003091184	
23	X Variable 6	-0.015377012	0.016856939	-0.912206654	0.364124274	-0.048871389	0.018117366	
24	X Variable 7	-0.000509708	0.001437779	-0.354510688	0.723794806	-0.003366545	0.002347129	
25	X Variable 8	-0.002633204	0.005128263	-0.513468944	0.608895685	-0.012822954	0.007556546	
26	X Variable 9	-0.026560066	0.020368906	-1.303951487	0.195612644	-0.067032648	0.013912516	
27	X Variable 10	-0.000977418	0.00295925	-0.330292551	0.741955159	-0.006857384	0.004902548	
28								
29								
30								
31								

The sheet labeled **Stepwise** displays the steps used to determine which x variables will be considered for inclusion. Note that even though the variables were not labeled in the spreadsheet, PHStat automatically labeled the variables X1 – X10 in the output based on order of appearance from the spreadsheet. In the sheet labeled **MR**, there is a column of *P-values* for the intercept and each one of the x variables. If the p-value is greater than $\alpha = 0.15$, then it is suggested that the variable not be considered for the model. In this case, it would be recommended that the user consider x_1 through x_5 for building a model for predicting executive salaries.

		df	SS	MS	F	Significance F	
1	Stepwise Analysis						
2	Table of Results for General Stepwise						
3							
4	X1 entered.						
5							
6		df	SS	MS	F	Significance F	
7	Regression	1	4.136434068	4.136434068	159.2039798	2.97316E-22	
8	Residual	98	2.5462337	0.025981977			
9	Total	99	6.682667768				
10							
11		Coefficients	Standard Error	t Stat	P-value	Lower 95%	Upper 95%
12	Intercept	11.09088709	0.033055393	335.5242852	8.6301E-152	11.02528974	11.15648444
13	X1	0.027838755	0.002206342	12.61760595	2.97316E-22	0.023460341	0.032217169
14							
15							
16	X3 entered.						
17							
18		df	SS	MS	F	Significance F	
19	Regression	2	5.006704553	2.503352277	144.8869335	7.35798E-30	
20	Residual	97	1.675963214	0.017277971			
21	Total	99	6.682667768				
22							
23		Coefficients	Standard Error	t Stat	P-value	Lower 95%	Upper 95%
24	Intercept	10.96837233	0.03200962	342.658624	2.3284E-151	10.9048421	11.03190256
25	X1	0.027258137	0.001801074	15.13437706	2.7955E-27	0.023683503	0.030832771
26	X3	0.197135209	0.027776879	7.097097206	2.10494E-10	0.142005795	0.252264623
27							
28							
29	X4 entered.						
30							

12.7   Sample Exercises

The following exercises were selected from the textbook to test the reader's understanding of PHStat and its usage within the chapter. Try to replicate the output in PHStat for the following problems.

Problem 12.20, Study of contaminated fish, page 645

Refer to the U.S. Army Corps of Engineers data on fish contaminated from the toxic discharges of a chemical plant located on the banks of the Tennessee River in Alabama, shown in the table in the text. Recall that the engineers measured the length (in centimeters), weight (in grams), and DDT level (in parts per million) for 144 captured fish. In addition, the number of miles upstream from the river was recorded. The data are saved in the **DDT** file. (The first and last five observations are shown in the text table.) Answer parts a. – e. in the text.

Solution to Problem 12.20

	A	B	C	D	E	F	G	H
1	Regression Analysis							
2								
3	Regression Statistics							
4	Multiple R	0.197149936						
5	R Square	0.038868097						
6	Adjusted R Square	0.018272414						
7	Standard Error	97.47563451						
8	Observations	144						
9								
10	ANOVA							
11		df	SS	MS	F	Significance F		
12	Regression	3	53793.58245	17931.19415	1.887196277	0.134522621		
13	Residual	140	1330209.905	9501.499324				
14	Total	143	1384003.488					
15								
16		Coefficients	Standard Error	t Stat	P-value	Lower 95%	Upper 95%	
17	Intercept	-108.0676955	62.70035758	-1.723557882	0.086995307	-232.0296688	15.89427778	
18	mile	0.085086799	0.082212774	1.034958371	0.302472765	-0.07745227	0.247625867	
19	length	3.770915444	1.618864937	2.329357661	0.021270588	0.570332547	6.97149834	
20	weight	-0.049414672	0.029261995	-1.68869798	0.093503459	-0.107267207	0.008437863	
21								
22								

Problem 12.33, Study of contaminated fish, page 652

Refer to Exercise 12.20 (p. 645) and the U.S. Army Corps of Engineers data on contaminated fish. You fit the first-order model

$$E(y) = \beta_0 + \beta_1 x_1 + \beta_2 x_2 + \beta_3 x_3$$

to the data, where y=DDT level (parts per million), x_1=number of miles upstream, x_2=length (centimeters), x_3=weight (in grams). Predict, with 95% confidence, the DDT level of a fish caught 100 miles upstream with a length of 40 centimeters and a weight of 800 grams.

Solution to Problem 12.33

	A	B	C	D	E	F
5		1				
6	mile given value	100				
7	length given value	40				
8	weight given value	800				
9						
10	X'X	144	38674	6165	151159	
11		38674	11891410	1639938	40879759	
12		6165	1639938	270712	6714077	
13		151159	40879759	6714077	1.79E+08	
14						
15	Inverse of X'X	0.413759	-0.00029	-0.00924	6.28E-05	
16		-0.00029	7.11E-07	3.52E-06	-5.2E-08	
17		-0.00924	3.52E-06	0.000276	-3.3E-06	
18		6.28E-05	-5.2E-08	-3.3E-06	9.01E-08	
19						
20	X'G times Inverse of X'X	0.065676	-0.00012	-0.00053	-4.3E-06	
21						
22						
23	[X'G times Inverse of X'X] times XG	0.029211				
24	t Statistic	1.977054				
25	**Predicted Y (YHat)**	**11.74586**				
26						
27	**For Average Predicted Y (YHat)**					
28	Interval Half Width	32.93713				
29	Confidence Interval Lower Limit	-21.1913				
30	Confidence Interval Upper Limit	44.683				
31						
32	**For Individual Response Y**					
33	Interval Half Width	195.509				
34	Prediction Interval Lower Limit	-183.763				
35	Prediction Interval Upper Limit	207.2548				

⑊ ◀ ▶ ▶│ **Intervals** / MR / DDT /

Problem 12.156, Carp experiment, page 733

Fisheries Science (Feb. 1995) reported on a study of the variables that affect endogenous nitrogen excretion (ENE) in carp raised in Japan. Carp were divided into groups of 2 to 15 fish each according to body weight and each group placed in a separate tank. The carp were then fed a protein-free diet three times daily for a period of 20 days. One day after terminating the feeding experiment, the amount of ENE in each tank was measured. The accompanying table gives the mean body weight (in grams) and ENE amount (in milligrams per 100 grams of body weight per day) for each carp group. Fit the quadratic model $E(y) = \beta_0 + \beta_1 x_1 + \beta_2 x^2$ to the data. Use the resulting printout to test $H_0: \beta_2 = 0$ against $H_a: \beta_2 \neq 0$ using $\alpha = 0.10$.

	A	B	C	D
1	Tank	Body Weight, x	x squared	ENE, y
2	1	11.7	136.9	15.3
3	2	25.3	640.1	9.3
4	3	90.2	8136.0	6.5
5	4	213.0	45369.0	6.0
6	5	10.2	104.0	15.7
7	6	17.6	309.8	10.0
8	7	32.6	1062.8	8.6
9	8	81.3	6609.7	6.4
10	9	141.5	20022.3	5.6
11	10	285.7	81624.5	6.0

Solution to Problem 12.156, Carp Experiment, page 733

Regression Analysis

Regression Statistics	
Multiple R	0.858733141
R Square	0.737422607
Adjusted R Square	0.662400495
Standard Error	2.194326595
Observations	10

ANOVA

	df	SS	MS	F	Significance F
Regression	2	94.65851555	47.32925778	9.82940343	0.009276853
Residual	7	33.70548445	4.815069207		
Total	9	128.364			

	Coefficients	Standard Error	t Stat	P-value	Lower 95%	Upper 95%
Intercept	13.71273728	1.306249375	10.49779433	1.55178E-05	10.62395054	16.80152402
Body Weight,x	-0.101839013	0.028810911	-3.534737733	0.009536966	-0.169965942	-0.033712084
x squared	0.000273478	0.000101599	2.691740332	0.031007513	3.32348E-05	0.000513721

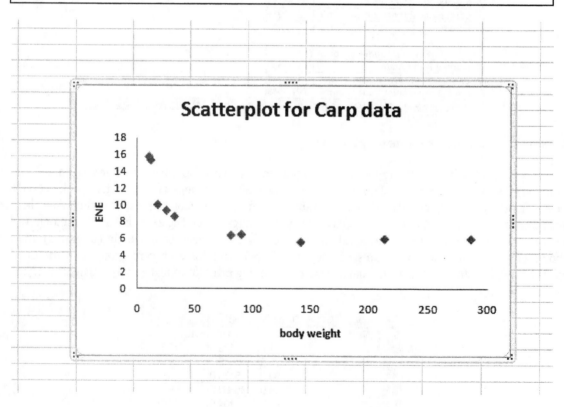

13.1   Introduction

In this chapter, various methods are presented for analyzing categorical data. This chapter considers categorical data that may be classified according to only one factor or two factors. The data are often times presented in contingency tables. The Chi-square test is utilized to analyze data classified according to qualitative variables.

The following examples will be used to illustrate techniques in PHStat and Excel.

> Example 13.2, page 747 (TV Program on marijuana)
> Example 13.3, page 759 (Survey Results regarding Marital Status & Religious Affiliation)
> Sample Exercise #1: Problem 13.13, page 752 (Internet search engines)
> Sample Exercise #2: Problem 13.33, page 766 (Politics and religion)

13.2   Testing Categorical Probabilities: One-Way Table

The following example from the text will be used to illustrate how to conduct a hypothesis test for a one-way table based on a multinomial experiment. Please refer to page 747 of the text, Example 13.2.

Suppose an educational television station has broadcast a series of programs on the physiological and psychological effects of smoking marijuana. Now that the series is finished, the station wants to see whether the citizens within the viewing area have changed their minds about how the possession of marijuana should be considered legally. Before the series was shown, it was determined that 7% of the citizens favored legalization, 18% favored decriminalization, 65% favored the existing law (an offender could be fined or imprisoned), and 10% had no opinion.

A summary of the opinions (after the series was shown) of a random sample of 500 people in the viewing area is given as follows: legalization: 39, decriminalization: 99, existing laws: 336, and no opinion: 26. Test at the $\alpha = 0.01$ level to see whether these data indicate that the distribution of opinions differs significantly from the proportions that existed before the educational series was aired.

Steps for conducting a one-way Chi-square test:

Step 1: Open a clean workbook. Enter the actual counts into column B and enter the corresponding expected cell counts into column C. Notice that for this example labels were placed in the first column and first row. These are not necessary, but help keep track of the data.

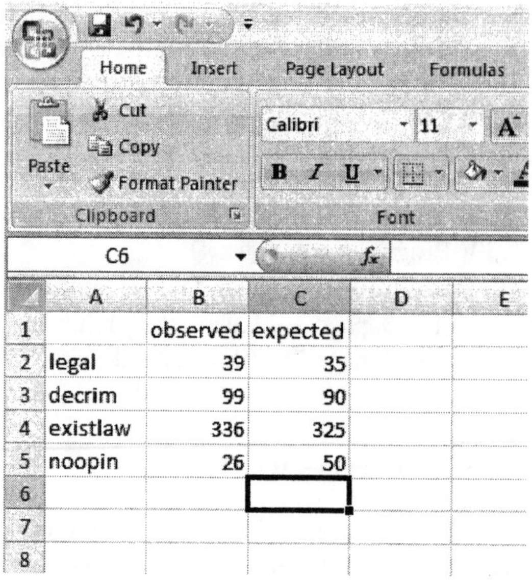

Step 2: Click on the **Formulas** tab.

Step 3: Click on **More Functions**.

Step 4: Click on **Statistical** and drop down to **CHITEST** and click on it.

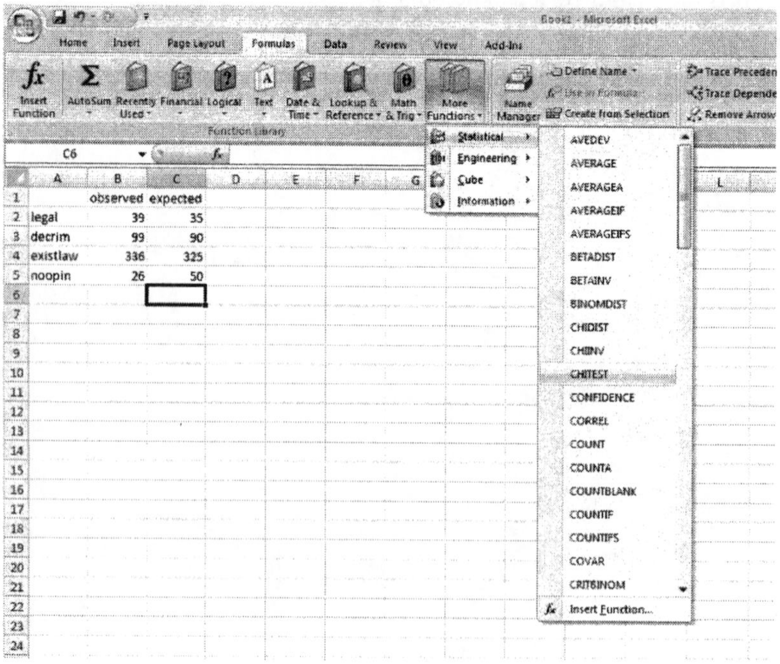

Step 5: A box for **Function Arguments** options will appear. There are two boxes in which to enter data:
- In the box next to **Actual_range** enter the location of the actual data counts. For this example that is B2:B5.
- In the box next to **Expected_range** enter the location of the expected counts. For this example that is C2:C5.

After the two boxes are filled, the answer will appear within the **Function Arguments** box. Click the **OK** button and Excel will also enter the answer into the active cell of the spreadsheet.

The computed output for this problem is 0.004127065; this is the p-value for this hypothesis test. This Excel function does not provide the critical chi-square value based on the sample data, but the p-value is enough information with which to make a conclusion. Since the p-value is smaller than $\alpha = 0.01$, the conclusion is to reject the null hypothesis. Therefore, it is determined that at least one of the proportions does not equal its hypothesized value.

13.3   Testing Categorical Probabilities: Two-way Table

The following example from the text will be used to illustrate how to conduct a hypothesis test for a two-way contingency table based on categorical probabilities. Please refer to page 759 of the text, Example 13.3.

A social scientist wants to determine whether the marital status (divorced or not divorced) of U.S. men is independent of their religious affiliation (or lack thereof). A sample of 500 U.S. men is surveyed and the results are tabulated below.
- a. Test to see whether there is sufficient evidence to indicate that the marital status of men who have been or are currently married is dependent on religious affiliation? Test at $\alpha = 0.01$.
- b. Plot the data and describe the patterns revealed. Is the result of the text supported by the plot?

		Religious Affiliation					
		A	B	C	D	None	Totals
Marital Status	Divorced	39	19	12	28	18	116
	Never Divorced	172	61	44	70	37	384
	Totals	211	80	56	98	35	500

Steps for conducting a two-way Chi-square test:

Step 1: Open a clean workbook.

Step 2: Click on the **Add-Ins** tab.

Step 3: Click on **PHStat**.

Step 4: Drop down to **Multiple-Sample Tests** and click on **Chi-Square Test...**

Step 5: A box for **Chi-Square Test** options will appear.

184

- In the box next to **Level of Significance** enter α. For this example that is 0.01.
- In the box next to **Number of Rows** enter number of rows of data from the table. For this example that is 2.
- In the box next to **Number of Columns** enter number of columns of data from the table. For this example that is 5.
- There is a blank box to enter a title. Use this when computing several tests in order to keep the output organized.
- Click the **OK** button and a new sheet labeled **ChiSquare** will appear. The worksheet that appears is actually a template that the user can manipulate in order to derive several types of chi-square test scenarios.

Step 6: Press the delete key on the computer keyboard to eliminate the PHStat message.

	A	B	C	D	E	F	G	H	I	J	K	L	M
1	Text Example 13.3												
2													
3			Observed Frequencies				PHStat2 User Note:						
4				Column variable			Enter replacement labels for the		ations				
5	Row variable	C1	C2	C3			row and column variables as well as				fo-fe		
6	R1						the observed frequency counts in		ʲ0!	#DIV/0!	#DIV/0!	#DIV/0!	#DIV/0!
7	R2						the table that starts in row 3.		ʲ0!	#DIV/0!	#DIV/0!	#DIV/0!	#DIV/0!
8	Total	0	0	0									
9							Note: The #DIV/0! error messages						
10			Expected Frequencies				will disappear after you enter the						
11				Column variable			observed frequency counts.						
12	Row variable	C1	C2	C3							(fo-fe)^2/fe		
13	R1	#DIV/0!	#DIV/0!	#DIV/0!	#DI		(Before continuing, press the Delete		ʲ0!	#DIV/0!	#DIV/0!	#DIV/0!	#DIV/0!
14	R2	#DIV/0!	#DIV/0!	#DIV/0!	#DI		key to delete this note.)		ʲ0!	#DIV/0!	#DIV/0!	#DIV/0!	#DIV/0!
15	Total	#DIV/0!	#DIV/0!	#DIV/0!	#DI								
16													
17	Data												
18	Level of Significance	0.01											
19	Number of Rows	2											
20	Number of Columns	5											
21	Degrees of Freedom	4											
22													
23	Results												
24	Critical Value	13.2767											
25	Chi-Square Test Statistic	#DIV/0!											
26	p-Value	#DIV/0!											

ChiSquare / Sheet1 / Sheet2 / Sheet3

Step 7: Click on the cells containing **R1** and **R2** in the **Observed Frequencies** table and replace them with the row variable names. For this example that will be **Divorced** and **Never Divorced**.

Step 8: Click on the cells containing **C1, C2, C3, C4** and **C5** in the **Observed Frequencies** table and replace them with the column variable names. For this example that will be **A, B, C, D** and **None**.

Step 9: Next, enter the outcomes from the survey into the body of the **Observed Frequencies** table. The **Expected Frequencies, Calculations,** and **Results** will be modified automatically.

The test statistic is $\chi^2 = 7.135464$ and it does not exceed the critical value of 13.2767. Since the test statistic does not lie within the critical region, the conclusion is to fail to reject the null hypothesis. Therefore, there is not sufficient evidence to indicate that the marital status of men who have been or are currently married is dependent on religious affiliation. Notice that the PHStat output also provides the p-value. In this problem it is calculated to be 0.1289. Since this is greater than the level of significance used in this case ($\alpha = 0.01$), the same conclusion (of course!) is reached: fail to reject the null hypothesis.

13.4 Sample Exercises

The following exercises were selected from the textbook to test the reader's understanding of PHStat and its usage within the chapter. Try to replicate the output in PHStat for the following problems.

Problem 13.13, Top Internet search engines, page 752

Nielsen/NetRatings is a global leader in Internet media and market research. In May 2006, the firm reported on the "search" shares (i.e., the percentage of all Internet searches) for the most popular search engines available on the Web. Google Search accounted for 50% of all searches, Yahoo! Search for 22%, MSN Search for 11%, and all other search engines for 17%. Suppose that, in a random sample of 1,000 recent Internet searches, 487 used Google Search, 245 used Yahoo! Search, 121 used MNS Search, and 147 used another search engine. Do the sample data disagree with the percentages reported by Nielsen/NetRatings? Test, using $\alpha = 0.05$.

Solution for Problem 13.13

Problem 13.33, Politics and religion, page 766.

University of Maryland professor Ted R. Gurr examined the political strategies used by ethnic groups worldwide in their fight for minority rights (*Political Science & Politics*, June 2000). Each in a sample of 275 ethnic groups was classified according to world region and highest level of political action reported. The data are summarized in the accompanying contingency table. Conduct a test at $\alpha = 0.10$ to determine whether political strategy of ethnic groups depends on world region.

World Region	No Political Action	Mobilization	Terrorism, Rebellion
Latin American	24	31	7
Post-Communism	32	23	4
South, SE, and East Asia	11	22	26
Africa/Middle East	39	36	20

Solution for Problem 13.33

	A	B	C	D	E	F	G	H	I	J
1	Text Exercise 13.33									
2										
3		Observed Frequencies								
4			Column variable				Calculations			
5	Row variable	No Act	Mass Act	Terror	Total			fo-fe		
6	Latin America	24	31	7	62		0.101818	5.749091	-5.85091	
7	Post-Communist	32	23	4	59		9.258182	-1.02909	-8.22909	
8	South, SE, East Asia	11	22	26	59		-11.7418	-2.02909	13.77091	
9	Africa/Middle East	39	36	20	95		2.381818	-2.69091	0.309091	
10	Total	106	112	57	275					
11										
12		Expected Frequencies								
13			Column variable							
14	Row variable	No Act	Mass Act	Terror	Total			(fo-fe)^2/fe		
15	Latin America	23.89818	25.25091	12.85091	62		0.000434	1.308945	2.663869	
16	Post-Communist	22.74182	24.02909	12.22909	59		3.769001	0.044073	5.537447	
17	South, SE, East Asia	22.74182	24.02909	12.22909	59		6.062413	0.171343	15.50712	
18	Africa/Middle East	36.61818	38.69091	19.69091	95		0.154925	0.18715	0.004852	
19	Total	106	112	57	275					
20										
21		Data								
22	Level of Significance	0.1								
23	Number of Rows	4								
24	Number of Columns	3								
25	Degrees of Freedom	6								
26										

H ◄ ► H   ChiSquare   Sheet1   Sheet2   Sheet3

	A	B	C	D	E	F	G	H	I	J
13		Column variable						(fo-fe)^2/fe		
14	Row variable	No Act	Mass Act	Terror	Total					
15	Latin America	23.89818	25.25091	12.85091	62		0.000434	1.308945	2.663869	
16	Post-Communist	22.74182	24.02909	12.22909	59		3.769001	0.044073	5.537447	
17	South, SE, East Asia	22.74182	24.02909	12.22909	59		6.062413	0.171343	15.50712	
18	Africa/Middle East	36.61818	38.69091	19.69091	95		0.154925	0.18715	0.004852	
19	Total	106	112	57	275					
20										
21	Data									
22	Level of Significance	0.1								
23	Number of Rows	4								
24	Number of Columns	3								
25	Degrees of Freedom	6								
26										
27	Results									
28	Critical Value	10.64464								
29	Chi-Square Test Statistic	35.41157								
30	p-Value	3.59E-06								
31	Reject the null hypothesis									
32										
33	Expected frequency assumption									
34	is met.									
35										

14.1 Introduction

In Chapter 14 of the text, the authors present techniques for analyzing data in which there are no assumptions about the probability distributions of the populations. The branch of inferential statistics devoted to distribution-free tests is called nonparametrics.

The following examples will be used to illustrate techniques in PHStat and Excel.

> Example 14.2, page 14-14 (Drug Reaction Times)
> Section 14.5, Table 14.7, page 14-29 (Number of Available Beds)
> Example 14.4, page 14-43 (Smoking versus Babies' Weights)
> Sample Exercise #1: Problem 14.109, page 14-57 (Spending on IS Technology)
> Sample Exercise #2: Problem 14.61, page 14-34 (The "name game")

14.2 Comparing Two Populations: The Wilcoxon Rank Sum Test for Independent Samples

The Wilcoxon rank sum test is used to test the hypothesis that the probability distributions associated with two independent populations are equivalent. In certain situations the researcher may not be sure that the population distributions are normal; therefore, the t-test discussed in previous chapters cannot be used. Remember from the text that in order to use the Wilcoxon rank sum test the distributions are assumed to be continuous; however, there is no specification on the shape or type of probability distribution.

The following example from the text will be used to illustrate the use of the Wilcoxon rank sum test. Please refer to page 14-14 of the text, Example 14.2.

Do the data given in Table 14.2 of the text provide sufficient evidence to indicate a shift in the probability distributions for drugs A and B? That is, does the probability distribution corresponding to drug A lies either to the right or left of the probability distribution corresponding to drug B? Test at the 0.05 level of significance.

Steps for completing the Wilcoxon Rank Sum Test:

Step 1: Open a new workbook.

Step 2: Input the data manually. For each observation there will be two pieces of information: the group it belongs to and the numerical outcome. Enter the group designation into the first column and the outcome into the second column. For this example enter "A" or "B" to designate which drug group the subject was assigned and enter the reaction time into the second column. (It is also possible to enter the data for the two groups into two separate columns with a descriptive title in the first row of each column.)

Step 3: Click on the **Add-Ins** tab.

Step 4: Click on **PHStat**.

Step 5: Drop down to **Two-Sample Tests** and click on **Wilcoxon Rank Sum Test....**

Step 6: A box for the **Wilcoxon Rank Sum Test** options will appear.

- In the box next to **Level of Significance**, the default is 0.05. This could be changed if the problem required a different level of significance.
- In the box next to **Population 1 Sample Cell Range** enter the rows and columns where the data for the first group are located. This may be done by typing the range in manually or by clicking and dragging the appropriate range of data cells. In this example the data are located in Sheet1!$A$1:$B$6.
- In the box next to **Population 2 Sample Cell Range** enter the rows and columns where the data for the second group are located. This may be done by typing the range in manually or by clicking and dragging the appropriate range of data cells. In this example the data are located in Sheet1!$A$7:$B$13.
- Note that the box in front of **First cells in both ranges contain label** already has a check in it. In **Step 2** the user was instructed to enter the data in this manner.
- There are four choices under **Test Options:**
  - The first choice, **Two-Tail Test**, corresponds to H_a: D_1 *is shifted to the right or to the left of* D_2. This is the default choice. Since this example asks "does the probability distribution corresponding to drug A lie either to the right or left of the probability distribution corresponding to drug B", this is the correct choice for this problem.
  - The second choice, **Upper-Tail Test**, corresponds to H_a: D_1 *is shifted to the right of* D_2.
  - The third choice, **Lower-Tail Test**, corresponds to H_a: D_1 *is shifted to the left of* D_2.
- There is a blank box to enter a title. For this example enter "Text Example 14.2".
- Click the **OK** button and the results will appear in two new sheets labeled **Calculations** and **Sorted**. The most important results, the p-value and conclusion, are located in the **Calculations** sheet.

	A	B	C
1	Text Example 14.2		
2			
3	Data		
4	Level of Significance	0.05	
5			
6	Population 1 Sample		
7	Sample Size	6	
8	Sum of Ranks	25	
9	Population 2 Sample		
10	Sample Size	7	
11	Sum of Ranks	66	
12			
13	Intermediate Calculations		
14	Total Sample Size n	13	
15	$T1$ Test Statistic	25	
16	$T1$ Mean	42	
17	Standard Error of $T1$	7	
18	Z Test Statistic	-2.42857	
19			
20	Two-Tail Test		
21	Lower Critical Value	-1.95996	
22	Upper Critical Value	1.959964	
23	p-Value	0.015158	
24	Reject the null hypothesis		
25			
26			

Since the p-value = 0.015158 < 0.05, reject the null hypothesis. Therefore, conclude that the probability distributions for drug A and drug B are not identical. Notice that PHStat had done all of the necessary calculations; the user needs only to enter the raw data and choose the appropriate options. The rank for each observation is determined by PHStat. The output for the ranks is contained in the sheet **Sorted.**

TextExample14.2.xlsx

	A	B	C	D
1	Sample	Value	Rank	
2	A	1.62	1	
3	A	1.71	2	
4	A	1.93	3	
5	A	1.96	4	
6	B	2.07	5	
7	B	2.11	6	
8	A	2.24	7	
9	A	2.41	8	
10	B	2.43	9	
11	B	2.5	10	
12	B	2.71	11	
13	B	2.84	12	
14	B	2.88	13	
15				

14.3 The Kruskal-Wallis H-Test for a Completely Randomized Design

According to the text, the Kruskal-Wallis test is used to compare more than two populations where no assumptions are made concerning the population probability distributions.

Please refer to Table 14.7, page 14-29 of the text for an illustrative example of the Kruskal-Wallis H-test.

Suppose a health administrator wants to compare the unoccupied bed space for three hospitals located in the same city. She randomly selects 10 different days from the records of each hospital and lists the number of unoccupied beds for each day in Table 14.7 in the text. Use the Kruskal-Wallis H-test to determine if the distributions for the number of beds available at the three hospitals differ. Use a 0.05 level of significance.

Steps for completing the Kruskal-Wallis H-Test:

Step 1: Open a new workbook.

Step 2: Input the data manually. Enter the data for each group in a separate column with a descriptive heading in the first row.

	A	B	C	D
1	hospital1	hospital2	hospital3	
2	6	34	13	
3	38	28	35	
4	3	42	19	
5	17	13	4	
6	11	40	29	
7	30	31	0	
8	15	9	7	
9	16	32	33	
10	25	39	18	
11	5	27	24	
12				

Step 3: Click on the **Add-Ins** tab.

Step 4: Click on **PHStat**.

Step 5: Drop down to **Multiple-Sample Tests** and click on **Kruskal-Wallis Rank Test....**

Step 6: A box for the **Kruskal-Wallis Rank Test** options will appear.

- In the box next to **Level of Significance**, enter the desired α. For this example the level of significance is 0.05.
- In the box next to **Sample Data Cell Range** enter the rows and columns where the data for the first group are located. This may be done by typing the range in manually or by clicking and

dragging the appropriate range of data cells. In this example the data are located in Sheet1!$A$1:$C$11.

- Note that the box in front of **First cells contain labels** already has a check in it. In **Step 2** the user was instructed to enter the data in this manner.
- There is a blank box to enter a title. For this example enter "Text Section 14.5".
- Click the **OK** button and the results will appear in two new sheet labeled **Kruskal** and **Sorted**. The most important results, the p-value and conclusion, are located in the **Kruskal** sheet.

TextTable14.7.xlsx

	A	B	C	D	E	F	G
1	Text Section 14.5						
2							
3	**Data**						
4	Level of Significance	0.05		Group	Sample Size	Sum of Ranks	Mean Ranks
5				1	10	120	12
6	**Intermediate Calculations**			2	10	210.5	21.05
7	Sum of Squared Ranks/Sample Size	7680.05		3	10	134.5	13.45
8	Sum of Sample Sizes	30					
9	Number of Groups	3					
10							
11	**Test Result**						
12	H Test Statistic	6.097419					
13	Critical Value	5.991465					
14	p-Value	0.04742					
15	Reject the null hypothesis						
16							

Since the p-value $= 0.04742 < 0.05$, we reject the null hypothesis. Therefore, we conclude that at least one of the hospitals tends to have a larger number of unoccupied beds than the others. Notice that PHStat had done all of the necessary calculations; the user needs only to enter the raw data and choose the appropriate options. The rank for each observation is determined by PHStat. The output for the ranks is contained in the sheet **Sorted.**

	A	B	C	D	E	F	G	H
1	Sample	Value	Rank		hospital1	hospital2	hospital3	
2	hospital3	0	1		6	34	13	
3	hospital1	3	2		38	28	35	
4	hospital3	4	3		3	42	19	
5	hospital1	5	4		17	13	4	
6	hospital1	6	5		11	40	29	
7	hospital3	7	6		30	31	0	
8	hospital2	9	7		15	9	7	
9	hospital1	11	8		16	32	33	
10	hospital2	13	9.5		25	39	18	
11	hospital3	13	9.5		5	27	24	
12	hospital1	15	11					
13	hospital1	16	12					
14	hospital1	17	13					
15	hospital3	18	14					
16	hospital3	19	15					
17	hospital3	24	16					
18	hospital1	25	17					
19	hospital2	27	18					
20	hospital2	28	19					
21	hospital3	29	20					
22	hospital1	30	21					
23	hospital2	31	22					

Sheet1 | **Sorted** | Kruskal | Sheet2 | Sheet3

14.4 Rank Correlation Analysis

Spearman's rank correlation coefficient provides a measure of correlation between ranks. This correlation calculation is the nonparametric alternative to the type of analysis done in Chapter 11.

Please refer to Example 14.4, page 14-43 of the text for an example of Spearman's rank correlation.

A study is conducted to investigate the relationship between cigarette smoking during pregnancy and the weights of newborn infants. The 15 women who make up the sample kept accurate records of the number of cigarettes they smoked during their pregnancies, and the weights of their children were recorded at birth. The data are given in Table 14.11.

Steps for computing the Spearman's rank correlation:

Step 1: Open a new workbook.

Step 2: Input the data manually. Enter the ranks for the two variables for each observation in separate columns with a descriptive heading in the first row.

	A	B	C
1	woman	cig rank	lbs rank
2	1	1	5
3	2	2	9
4	3	13	4
5	4	7	10
6	5	5.5	13.5
7	6	3	11.5
8	7	4	15
9	8	15	6
10	9	5.5	11.5
11	10	8.5	1
12	11	14	3
13	12	8.5	7
14	13	12	8
15	14	10	2
16	15	11	13.5

TextExample14.4.xlsx

Step 3: Click on the **Data** tab.

Step 4: Click on **Data Analysis**.

Step 5: A box for **Data Analysis** options will appear. Click on **Correlation** and then click the **OK** button.

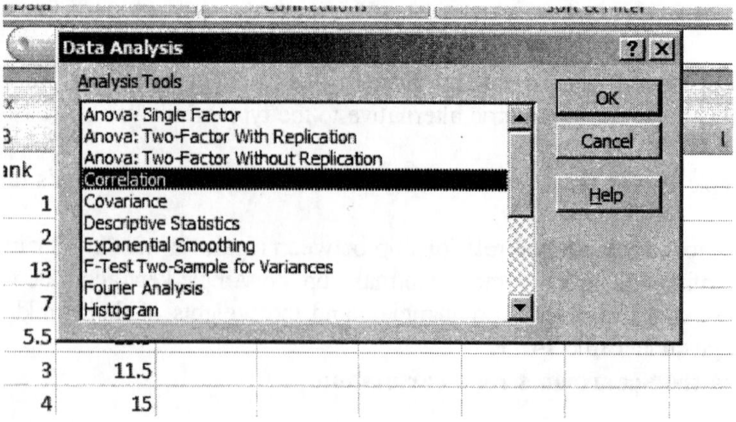

Data Analysis

Analysis Tools

Anova: Single Factor
Anova: Two-Factor With Replication
Anova: Two-Factor Without Replication
Correlation
Covariance
Descriptive Statistics
Exponential Smoothing
F-Test Two-Sample for Variances
Fourier Analysis
Histogram

OK
Cancel
Help

Step 6: A box for the **Correlation** options will appear.

- Click on the entry box by **Input Range**. Enter the rows and column where the data are located. This may be done by typing the range in manually or by clicking and dragging the appropriate range of data cells. In this example the data are located in $B$1:$C$16.
- The default is that data are grouped by columns. In Step 2 the data were entered in this manner.
- Since there is a label in the first row of each column, click on the box in front of **Labels in First Row** so that there is a check mark beside it.
- The default is that the results will be provided in a new sheet.
- Click the **OK** button and the correlation computation will appear in a new sheet.

The sample Spearman rank correlation coefficient is equal to -0.42473. According to Table XIV in Appendix A, if a level of significance of 0.05 is used then the sample statistic does not fall within the rejection region. Therefore, we cannot reject the null hypothesis. This sample does not provide sufficient evidence that a correlation exists between the two variables.

Note that Excel does not have an option specifically for the Spearman rank correlation, but the **Correlation** analysis will produce the correct calculation if the ranks, not the raw data, are entered into the spreadsheet. The ranks will still have to be determined by the user. There is a **RANK** function within Excel, but it does not deal with ties for these types of calculations as instructed by the text.

14.5   Sample Exercises

The following exercises were selected from the textbook to test the reader's understanding of PHStat and its usage within the chapter. Try to replicate the output in PHStat for the following problems.

Problem 14.109, Spending on IS Technology, page 14-57

University of Queensland researchers sampled private sector and public sector organizations in Australia to study the planning undertaken by their information systems departments (*Management Science*, July 1996). As part of the process they asked each sample organization how much it had spend on information systems and technology in the previous fiscal year as a percentage of the organization's total revenues. The results are reported in the text table. Do the two sampled populations have identical probability distributions or is the distribution for the public sector located to the right of Australia's private sector firm? Test using $\alpha = 0.05$.

	A	B
1	Private Sector	Public Sector
2	2.58	5.40
3	5.05	2.55
4	0.05	9.00
5	2.10	10.55
6	4.30	1.02
7	2.25	5.11
8	2.50	24.42
9	1.94	1.67
10	2.33	3.33

Solution for Problem 14.109

	A	B
1	**Wilcoxon Rank Sum Test**	
2		
3	**Data**	
4	**Level of Significance**	0.05
5		
6	Population 1 Sample	
7	Sample Size	9
8	Sum of Ranks	66
9	Population 2 Sample	
10	Sample Size	9
11	Sum of Ranks	105
12		
13	Intermediate Calculations	
14	Total Sample Size n	18
15	$T1$ Test Statistic	66
16	$T1$ Mean	85.5
17	Standard Error of $T1$	11.32475
18	**Z Test Statistic**	-1.72189
19		
20	**Lower-Tail Test**	
21	**Lower Critical Value**	-1.64485
22	**p-Value**	0.042545
23	**Reject the null hypothesis**	
24		
25		
26		

Problem 14.61, The "name game", page 14-34.

Refer to the *Journal of Experimental Psychology – Applied* (June 2000) study of different methods of learning names, presented in Exercise 10.34. Recall that three groups of students used different methods to learn the names of the other students in their group. Group 1 used the "simple name game," Group 2 used the "elaborate name game," and Group 3 used "pairwise introductions." The table at the bottom of p. 14-34 lists the percentage of names recalled (after one year) for each student respondent. Use an alternative nonparametric test to compare the distributions of the percentages of names recalled for the three name-retrieval methods. Take $\alpha = 0.05$.

Solution for Problem 14.61

	A	B	C	D	E	F	G
1	Text Example 14.61						
2							
3	Data						
4	Level of Significance	0.05		Group	Sample Size	Sum of Ranks	Mean Ranks
5				1	50	4156	83.12
6	Intermediate Calculations			2	42	3068	73.047619
7	Sum of Squared Ranks/Sample Size	703174.6		3	47	2506	53.3191489
8	Sum of Sample Sizes	139					
9	Number of Groups	3					
10							
11	Test Result						
12	H Test Statistic	13.61229					
13	Critical Value	5.991465					
14	p-Value	0.001107					
15	Reject the null hypothesis						
16							

202

PHStat v2.7 with Data Files for use with the Technology Manual to accompany Statistics, 11e
James T. Mendenhall & Terry Sincich
ISBN 13: 978-0-13-600181-2
ISBN 10:   0-13-600181-5
CD License Agreement
© 2009 Pearson Education, Inc.
Pearson Prentice Hall
Pearson Education, Inc.
Upper Saddle River, NJ 07458
All rights reserved.
Pearson Prentice Hall™ is a trademark of Pearson Education, Inc.

READ THIS LICENSE CAREFULLY BEFORE OPENING THIS PACKAGE. BY OPENING THIS PACKAGE, YOU ARE AGREEING TO THE TERMS AND CONDITIONS OF THIS LICENSE. IF YOU DO NOT AGREE, DO NOT OPEN THE PACKAGE. PROMPTLY RETURN THE UNOPENED PACKAGE AND ALL ACCOMPANYING ITEMS TO THE PLACE YOU OBTAINED THEM. THESE TERMS APPLY TO ALL LICENSED SOFTWARE ON THE DISK EXCEPT THAT THE TERMS FOR USE OF ANY SHAREWARE OR FREEWARE ON THE DISKETTES ARE AS SET FORTH IN THE ELECTRONIC LICENSE LOCATED ON THE DISK:

****
SYSTEM REQUIREMENTS
****
*Microsoft Windows, Windows 98, Windows NT, Windows 2000, Windows ME, or Windows XP
*In addition to the minimum processor requirements for the operating system your computer is running, this CD requires a Pentium II, 200 MHz or higher processor
*In addition to the RAM required by the operating system your computer is running, this CD requires 64 MB RAM for Windows 98, Windows NT 4.0, Windows 2000, Windows ME, and Windows XP
*Macintosh OS 9.x or 10.x
*In addition to the minimum processor requirements for the operating system your computer is running, this CD requires a PowerPC G3 233 MHz or better
*In addition to the RAM required by the operating system your computer is running, this CD requires 64 MB RAM
*Microsoft Excel 97, 2000, 2002, or 2003 (Excel 97 use must apply the SR-2 or a later free update from Microsoft in order to use PHStat 2.5. Excel 2000 and 2002 must have the macro security level set to Medium) for Windows; MINITAB Version 14 or 12; JMP version 5.1; SPSS versions 13.0, 12.0, 11.0; or other statistics software.
*PHStat will not work on the Macintosh
*Microsoft Excel Data Analysis ToolPak and Analysis ToolPak VBA installed (supplied on the Microsoft Office/Excel program CD)
*CD-ROM or DVD-ROM drive; Mouse and keyboard; Color monitor (256 or more colors and screen resolution settings set to 800 by 600 pixels or 1024 by 748 pixels)
*Internet browser for Windows (Netscape 4.x, 6.x or 7.x, or Internet Explorer 5.x or 6.x) and Internet connection suggested but not required
*Internet browser for Macintosh (Internet Explorer 5.x or Safari 1.x) and Internet connection suggested but not required
*This CD-ROM is intended for stand-alone use only. It is not meant for use on a network.
****
CD-ROM CONTENTS
****
--Data Files : Included within that folder are:
--JMP Data Files (Folder: JMP_Mendenhall_Data)
JMP is required to view and use these files. Information about JMP can be found on the internet at http://www.jmp.com
--SPSS Data Files (Folder: SPSS_Mendenhall_Data)
SPSS is required to view and use these files. Information about SPSS can be found on the internet at http://www.spss.com
--MINITAB Data Files (Folder: minitab_Mendenhall_Data)
MINITAB is required to view and use these files. Information about MINITAB can be found on the internet at http://www.minitab.com/support/index.htm
--Data Files (Folder: ASCII_Mendenhall_Data)
For use with SPSS or other statistics software.
--Excel Files (Folder: Excel_Mendenhall_Data)
Excel is required to view and use these files. Information about Excel can be found on the Internet at http://office.microsoft.com/en-us/default.aspx
--TI-83/84 Files (Folder: TI-8x_Mendenhall_Data)
TI-83/84 calculator is required http://education.ti.com/us/product/main.html.
--PHStat 2.5
Prentice Hall's PHStat statistical add-in system enhances Microsoft Excel to better support learning in an introductory statistics course. http://www.prenhall.com/phstat/
--Readme.txt
****
TECHNICAL SUPPORT
****
If you continue to experience difficulties, call 1 (800) 677-6337, 8 am to 8 pm Monday through Friday and 5 pm to 12 am Sunday (all times Eastern) or visit Prentice Hall's Technical Support Web site at http://247.prenhall.com/mediaform.
Our technical staff will need to know certain things about your system in order to help us solve your problems more quickly and efficiently. If possible, please be at your computer when you call for support. You should have the following information ready:

- Textbook ISBN
- CD-Rom/Diskette ISBN
- Corresponding product and title
- Computer make and model
- Operating System (Windows or Macintosh) and Version
- RAM available
- Hard disk space available
- Sound card? Yes or No
- Printer make and model
- Network connection
- Detailed description of the problem, including the exact wording of any error messages.

NOTE: Pearson does not support and/or assist with the following:
- 3d-party software (i.e. Microsoft including Microsoft Office Suite, Apple, Borland, etc.)
- Homework assistance
- Textbooks and CD-ROMs purchased used are not supported and are non-replaceable.
For assistance with third-party software, please visit:
JMP Support (for JMP): http://www.jmp.com/support/techsup/index.shtml
MINITAB Support: http://www.minitab.com/support/index.htm
SPSS Support: http://www.spss.com/tech/spssdefault.htm
TI-8x Support: http://education.ti.com/us/support/main.html
Excel Support: http://support.microsoft.com/
PHStat Support: http://www.prenhall.com/phstat/phstat2/phstat2(main).htm
----
Windows and Windows NT are registered trademarks of Microsoft Corporation in the United States and/or other countries.